Alessandro Bencini   Dante Gatteschi

Electron Paramagnetic Resonance of Exchange Coupled Systems

With 177 Figures and 38 Tables

Springer-Verlag Berlin Heidelberg New York
London Paris Tokyo Hong Kong

Prof. Dr. Alessandro Bencini
Prof. Dr. Dante Gatteschi

Department of Chemistry, University of Florence
Via Maragliano, 75/77, I-50144 Florence, Italy

ISBN-13: 978-3-642-74601-7   e-ISBN-13: 978-3-642-74599-7
DOI: 10.1007/978-3-642-74599-7

Library of Congress Cataloging-in-Publication Data. Bencini, Alessandro, 1951—Electron paramagnetic resonance of exchange coupled systems Alessandro Bencini, Dante Gatteschi p. cm.

1. Electron paramagnetic resonance spectroscopy. I. Gatteschi, D. (Dante) II. Title. III. Title: Exchange coupled systems.   QC763.B46   1989   538'.364--dc20   89-19732

© Springer-Verlag Berlin Heidelberg 1990
Softcover reprint of the hardcover 1st edition 1990

Typesetting: Macmillan India Ltd., Bangalore-25.
2151/3140(3011)-543210 – Printed on acid-free paper

Preface

This book is intended to collect in one place as much information as possible on the use of EPR spectroscopy in the analysis of systems in which two or more spins are magnetically coupled. This is a field where research is very active and chemists are elbow-to-elbow with physicists and biologists in the forefront. Here, as in many other fields, the contributions coming from different disciplines are very important, but for active researchers it is sometimes difficult to follow the literature, due to differences in languages, and sources which are familiar to, e.g., a physicist, are exotic to a chemist. Therefore, an effort is needed in order to provide a unitary description of the many different phenomena which are collected under the title. In order to define the arguments which are treated, it is useful to state clearly what is not contained here. So we do not treat magnetic phenomena in conductors and we neglect ferro- and antiferromagnetic resonance. The basic foundations of EPR spectroscopy are supposed to be known by the reader, while we introduce the basis of magnetic interactions between spins.

In the first two chapters we review the foundations of exchange interactions, trying to show how the magnetic parameters are bound to the electronic structure of the interacting centers. Chapter 3 is about the spectra of pairs, and Chapter 4 gives a brief introduction to the spectra of systems containing more than two, but less than infinite, centers. Chapter 5 is about relaxation, while Chapter 6 shows how the complicated cases of infinite lattices can be tackled, and how EPR can provide first-hand information on spin dynamics.

The following chapters report a survey of experimental data which hopefully will be of some help as general reference to the field: Chapter 7 is about spectra of pairs, Chapter 8 about systems in which transition metal ions are coupled to stable organic radicals, Chapter 9 reports some examples of magnetically coupled systems found in biological materials, a fascinating and fast expanding area, Chapter 10 surveys low dimensional materials, and Chapter 11 finally reports the use of EPR to characterize excitons and exciton motion. The survey is far from complete, but hopefully it will be a useful introduction to the area.

At the end of a preface it is mandatory to express sincere thanks to all the people who made the authors feel less desperate in their struggle with the literature to follow and the pages to write. So first of all we would like to dedicate the book to Ninetta, Silvia, Alessandra, and Mariella who did not oppose to the frequent retreats from family life.

Many people read in advance some chapters and gave us useful suggestions, which certainly improved the manuscript. The flaws, of course, remain our responsibility. We heartily thank E.I. Solomon, O. Kahn, G.R. Eaton, S.S. Eaton, S. Clement, J.P. Renard, R.D. Willett, and J. Drumheller. All the people in our group in Florence, C. Benelli, A. Dei, C. Zanchini, A. Caneschi, L. Pardi, O. Gouillou, and R. Sessoli must be particularly thanked for acting as guinea pigs reading the manuscript from the first stages.

Finally, it is a tradition to thank the people who typed the manuscript. Unfortunately in this technological era we had to type the manuscript ourselves, so that we can only thank our personal computers which made the burden bearable.

Alessandro Bencini
Dante Gatteschi

Table of Contents

1 Exchange and Superexchange

1.1 The Exchange Interaction

The essential fundament of the exchange (or the superexchange) interaction is the formation of a weak bond. It is well known that spin pairing characterizes bond formation: two isolated hydrogen atoms have a spin $S = 1/2$ each, but when they couple to form a molecule, H_2, the result is a spin singlet state, because the two electrons must pair their spins to obey the Pauli principle. If the bond is strong enough, the possibility of having the two electrons with parallel spins is very low, and the triplet state has a much higher energy than the singlet (ΔE, the singlet-triplet separation, is much larger than kT at room temperature). However, if the bonding interaction is weak, the singlet-triplet energy separation becomes smaller, and eventually of the same order of magnitude as kT. It must be recalled here that although the exchange interaction is a bond interaction, therefore, acting only on the orbital coordinates of the electrons, the spin coordinates are extremely useful for the characterization of the wave functions of the pair. In fact, the Pauli principle imposes that the complete wave function of a system is antisymmetric with respect to the exchange of electrons: in the above example of the hydrogen molecule the symmetric orbital function must be coupled to the antisymmetric spin singlet function, and the antisymmetric orbital function is coupled to the symmetric spin triplet. Therefore, spins act as indicators of the nature of the orbital states.

When the two centers in the pair have individual spins S_i different from $\frac{1}{2}$, as can occur when the number of unpaired electrons is larger than one, the states of the pair are classified by the total spin quantum number S defined by the angular momentum addition rules:

$$|S_1 - S_2| \leq S \leq S_1 + S_2. \tag{1.1}$$

The exchange regime occurs when the interaction between two species, characterized by individual spins S_1 and S_2 before turning on the coupling, yields a number of levels characterized by different total spins, neglecting relativistic effects, which are thermally populated within the normal range of temperatures.

With regards to the intuitively simple example of two identical species with $S_i = 1/2$ (one unpaired electron on each noninteracting species), three cases can occur in the limit of weak interaction. When the interaction is vanishingly small, the two spins are completely uncorrelated and the two centers can be described by their individual spin quantum numbers. A simple way of determining whether this situation holds is through measurements of the magnetic susceptibility which must be the sum of the individual susceptibilities. In principle, EPR as well can be used to this purpose, and one should observe the spectra of the individual spins. However, EPR is a much more sensitive technique than static magnetic susceptibility measurements, and even residual interactions, including magnetic dipolar interactions, as small as a fraction of wave number, can be enough to yield spectra very different from the spectra of the individual spins. In other words, EPR moves the limit for vanishingly small interaction to much lower energy than in the case of magnetic susceptibility. Indeed, for the latter the limit is always of the order of kT, and unless extremely low temperatures are reached, it cannot become much smaller than 1 cm^{-1}. EPR can easily detect interactions of 10^{-2}–10^{-3} cm^{-1} even at room temperature. Even two spins as far apart as 1000 pm can be found to be interacting by the EPR technique.

The second limiting case occurs when the two spins are coupled in such a way that the singlet is the ground state and the triplet is thermally populated. In this case the coupling is said to be antiferromagnetic.

The third limiting case occurs when the triplet is the ground state and the singlet is thermally populated. In this case the coupling is said to be ferromagnetic.

When the two individual spins have $S_i \neq 1/2$, the situation is similar: the antiferromagnetic case is obtained when $S = |S_1 - S_2|$ is the ground state, and the ferromagnetic case is obtained when $S = S_1 + S_2$ has the lowest energy. A simple picture of the three limiting cases for $S_1 = S_2 = 1/2$ is shown in Fig. 1.1.

The exchange regime can be rarely obtained when two paramagnetic atoms are directly bound, but generally this situation is found in more complex molecules. A rather common case is that of two paramagnetic metal ions which are bridged by some intervening, formally diamagnetic, atoms. A relevant example is shown in Fig. 1.2. The two copper (II) ions, which have a ground d^9 configuration, and one unpaired electron each, are bridged by one oxalato ion. It has been found experimentally [1.1] that the Cu(ox)Cu moiety has a ground singlet and an excited triplet at ≈ 385 cm^{-1}. Since the copper-copper distance, > 500 pm, is too long to justify any direct overlap between the two metal ions, it must be concluded that the diamagnetic oxalato ion is effectively transmitting the exchange interaction. This situation, in which the paramagnetic centers are coupled through intervening diamagnetic atoms, or groups of atoms, is referred to as superexchange.

In order to put all the above qualitative conclusions on a more quantitative basis it is necessary to resort to some model for the description of the chemical bond intervening between the two individual species. The first successful attempt in this direction was made by Anderson [1.2], who used a Valence Bond approach and clarified which terms are responsible for the coupling. Much effort has been made [1.3–5] since then to translate Anderson's approach into the Molecular Orbital language, which is much more familiar to chemists, for instance, due to the relative ease with which calculations can be performed in this scheme. In particular, in the last few years, with the advent of fast computers, it

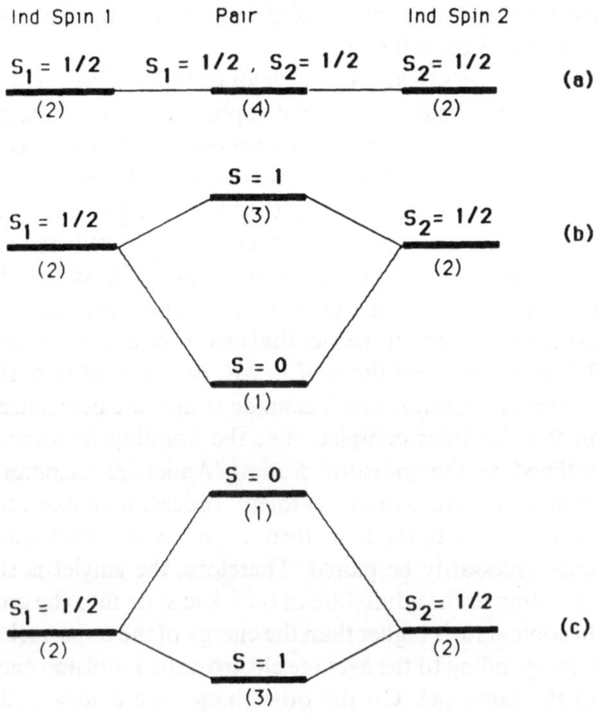

Ind Spin 1
Pair
Ind Spin 2
S1 = 1/2
S1 = 1/2 , S2= 1/2
S2= 1/2
(2)
(4)
(2)
(a)
S = 1
(3)
S1 = 1/2
S2= 1/2
(2)
(2)
(b)
S = 0
(1)
S = 0
(1)
S1 = 1/2
S2= 1/2
(2)
(2)
(c)
S = 1
(3)

Fig. 1.1a–c. Scheme of the energy levels appropriate to a pair of spins $S_1 = S_2 = \frac{1}{2}$. **a** Uncorrelated spins; **b** antiferromagnetic coupling; **c** ferromagnetic coupling

L
O
O
L
Cu
Cu
L
O
O
L

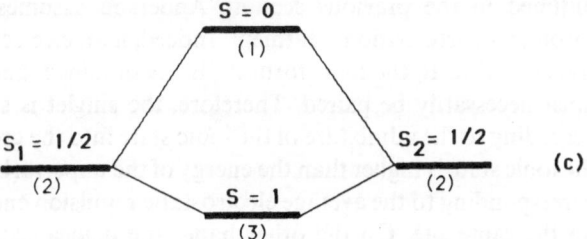

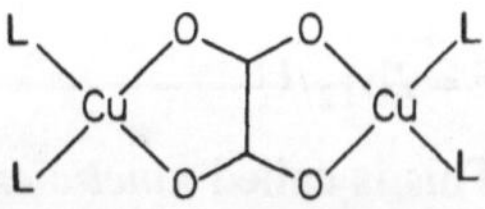

Fig. 1.2. Scheme of two copper ions bridged by one oxalato group

has become possible to pass to the truly quantitative description, actually calculating the extent of the interaction.

In the following sections we will resume the Anderson's theory, then the qualitative MO models, and finally we will give a description of the quantitative MO calculations.

1.2 Anderson's Theory

Anderson's theory [1.2] takes a firm standpoint in the theory of magnetic interactions. Before that many important contributions were provided by Heisenberg [1.6, 7], Dirac [1.8], van Vleck [1.9], and many others [1.10, 11], but the literature was full of confusion regarding the use of orthogonal or non-orthogonal basis functions.

Anderson's theory was developed to describe exchange in insulators, but it is most simply understood when applied to a pair of interacting centers, e.g., to two identical transition metal ions possessing one unpaired electron each. The basis functions are localized on the centers A and B (Fig. 1.3) and correspond essentially to the functions of the two centers in an environment where the other magnetic species is not present. They may as well include orbitals of other atoms, especially the ones of the atoms bridging A and B. These functions are then allowed to interact, and in order to do that it is necessary to correct the functions to take into account the fact that electrons are no longer localized on center A or B, respectively, but those of A, e.g., can now be partially delocalized on B. The central assumption which is made is that the delocalization of the electron of A on B is far from complete, i.e., the bonding interaction is not strong, as was outlined in the previous section. Anderson assumes that this is performed through a perturbation treatment. Indeed, if an electron is removed from A and transferred to B, the ionic form A^+B^- is obtained, and the two electrons on B must necessarily be paired. Therefore, the singlet is stabilized over the triplet according to the admixture of the ionic state into the ground state. The energy of the ionic state is higher than the energy of the unperturbed state by an amount U, corresponding to the average electrostatic repulsion energy for the two electrons on the same site. On the other hand, the delocalization stabilizes the singlet affecting the average kinetic energy of the electrons. This is represented by a transfer integral, b_{12}, between the orbitals 1, localized on A, and the orbital 2, localized on B. In the case of only two interacting orbitals this yields a stabilization of the singlet corresponding to:

$$E = 2b_{12}^2/U. \tag{1.2}$$

This is called *kinetic exchange*, and is intrinsically antiferromagnetic.

The second effect which must be taken into account is the exchange, determined by the self-energy of the charge distribution $\Phi_{A1}^* \Phi_{B2} = \rho_{12}^*$:

$$J_{12} = \int \Phi_{A1}^*(1) \Phi_{B2}^*(2) e^2/r_{12} \Phi_{A1}(2) \Phi_{B2}(1) d\tau_1 d\tau_2. \tag{1.3}$$

This term was called *potential exchange*, and it stabilizes the triplet over the singlet according to Hund's rule. Anderson uses also the term superexchange for the kinetic exchange, but we prefer to avoid it, because it can generate confusion

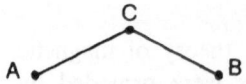

Fig. 1.3. Scheme of two paramagnetic centers, *A* and *B*, bridged by a diamagnetic group *C*

with the current meaning of interaction propagated by intervening atoms. He also uses the term direct for potential exchange, but again we prefer the term used in the text, because generally with direct exchange one understands an interaction not propagated by intervening atoms.

In the case of one unpaired electron on each center, the singlet-triplet separation can be written as:

$$E(\text{triplet}) - E(\text{singlet}) \equiv J = b_{12}^2/U - J_{12},$$ (1.4)

When there is more than one unpaired electron per center, the energies of the various spin multiplets can still be expressed by one parameter:

$$J = \Sigma\,(b_{ij}^2/U_i - J_{ij}),$$ (1.5)

The formalism which allows us to do that will be developed in Chap. 2.

Finally, a third term was taken into consideration. This is due to polarization effects which are determined by the presence of magnetic electrons. In fact, for open shell systems the energy of the spin-up orbitals must be different from the energy of the spin-down ones, because of the exchange terms in the Hartree-Fock equations, which are nonzero for electrons with the same spin. This effect, for instance, is capable of inducing unpaired spin density on a formally diamagnetic ligand, by favoring one spin state over the other. The total effect of this term can be either ferro- or antiferromagnetic, depending on the nature of the interacting orbitals. This is much more difficult to appreciate qualitatively, and it has been generally neglected.

This model developed by Anderson has had the great merit of putting on a firm theoretical basis the exchange interactions, clarifying all the points which had become inextricably entangled in the previous literature. The main drawback of the theory has been that passing to the true quantitative approach has proved to be practically impossible, after some initial attempts [1.12–17]. As Anderson states [1.2]: "it seems wise not to claim more than about 100% accuracy for the theory in view of the uncertainties".

However, this rationalization allowed Goodenough [1.18, 19] and Kanamori [1.20] to express some rules, which for quite some time now have been the bible for experimentalists wishing to understand the magnetic properties of transition metal compounds. They can be expressed as follows:

1. When the two ions have lobes of magnetic orbitals (the orbitals containing the unpaired electrons) pointing toward each other in such a way that the orbitals would have a reasonably large overlap integral, the exchange is antiferromagnetic;
2. When the orbitals are arranged in such a way that they are expected to be in contact but have no overlap integral, the interaction is ferromagnetic;
3. If a magnetic orbital overlaps an empty orbital, the interaction between the two ions is ferromagnetic.

The most thorough and detailed exploitation of these rules has been made in a review article by Ginsberg [1.21], to which the interested reader is referred.

It is perhaps useful at this point to work out an example in order to familiarize the techniques which allow one to understand, in a qualitative way, how exchange operates between magnetic orbitals. Let us focus on one of the possible geometries of pairs, such as that depicted in Fig. 1.4, which corresponds to two octahedra joined thorough one side. If the two metal ions are copper(II), with a d^9 configuration, the magnetic orbitals are of the xy type, in an appropriate reference frame. The bridge atom, which might be an oxygen atom, has both s and p orbitals completely filled. The xy orbital of the left atom overlaps the s orbitals of the oxygens, which in turn overlap the xy orbital of the right copper. According to rule (1) of Goodenough and Kanamori, this corresponds to an antiferromagnetic pathway, which can be symbolically written as: $xy\|s\|xy$. Since the oxygen orbital is spheric, the extent of the coupling does not depend on the copper-oxygen-copper angle, but only on the copper-oxygen distance. Matters are different when we consider the overlap with a p orbital. The overlap with, e.g., the right copper ion can be maximized (Fig. 1.5), but then the overlap with the left xy orbital is determined by the geometry of the bridge. If the O–Cu–O angle, ϕ, is $90°$, the overlap of the latter with the p orbital is zero, but a variation in the value of the angle can restore an overlap different from zero. In the case of $\phi = 90°$ the exchange pathway can be written as: $xy\|p\perp xy$, which according to Goodenough-Kanamori rules, corresponds to a ferromagnetic coupling, while in case of $\phi \neq 90°$ the pathway can be written as: $xy\|p\|xy$, which gives antiferromagnetic coupling. Since the oxygen atoms have both s and p

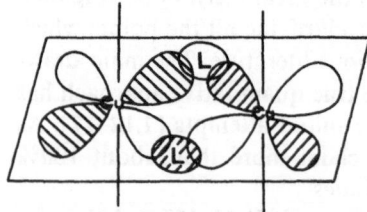

Fig. 1.4. Two octahedrally coordinated copper(II) ions bridged by sharing an edge

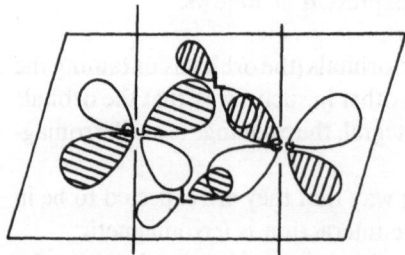

Fig. 1.5. Overlaps of the oxygen p orbitals with the copper xy orbitals

orbitals, both exchange mechanisms must be operative. However, the p orbitals must interact more strongly, because their energies are closer to those of the d orbitals than that of the s orbitals. Therefore, it can be anticipated that for bridges characterized by Cu–O–Cu angles close to 90° the coupling must be ferromagnetic, while when the angle deviates substantially from this value, the coupling becomes antiferromagnetic. This has been experimentally confirmed [1.22, 23].

If one of the copper ions is substituted by another metal ion, such as oxovanadium(IV), an interesting case is obtained. The magnetic orbital of oxovanadium(IV), a d^1 ion, is x^2–y^2. If we consider the s pathway, we immediately recognize that the oxygen orbitals can overlap to the xy orbital of copper, but they are orthogonal to the vanadium x^2–y^2 orbital. Therefore, this provides an $xy \| s \perp x^2$–y^2 pathway, which determines a ferromagnetic coupling. One p orbital of the oxygen atom can overlap to the xy orbital of copper, and at the same time it overlaps to the x^2–y^2 orbitals of vanadium. The same situation holds for the other oxygen as well, so that one might be induced to anticipate an antiferromagnetic coupling between copper and vanadium. However, this prediction is wrong, and both theory and experiment agree on a ferromagnetic coupling [1.24]. In fact, the overlap of the up-oxygen in Fig. 1.6 with the magnetic orbital of vanadium is equal, but of opposite sign, compared to the overlap of the down-oxygen. If one is not convinced of the argument, one can try to overlap the orbitals in different ways, but in any case the final result will be the one expressed above. A safer way to check is by using symmetry arguments. Apparently, considering the symmetry plane passing through copper and vanadium, xy is antisymmetric, while x^2–y^2 is symmetric; therefore, their overlap must be zero. This example is precious because it shows, when more than one bridging atom is present, that it is not enough to look at the possible independent exchange pathways, but the overall system must be taken into consideration.

Another example which is useful to consider, concerns systems with more than one unpaired electron on each center. In general, when there are more than one unpaired electrons on each center, the experimental coupling constant J, is decomposed in a sum of contributions according to:

$$J = (1/n_1 n_2) \Sigma_i \Sigma_j J_{ij}, \tag{1.6}$$

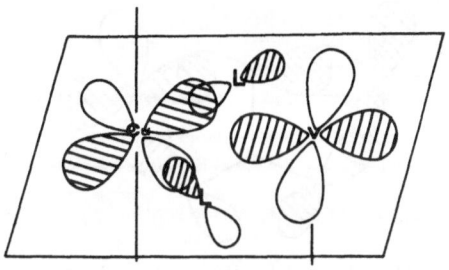

Fig. 1.6. Overlaps of the oxygen p orbitals with the copper xy and oxovanadium $x^2 - y^2$ orbitals

where n_1 and n_2 are the numbers of unpaired electrons on center 1 and 2, respectively, the sums are on all the magnetic orbitals of the two centers. The meaning of the J_{ij} constants is that of the coupling of an electron in the magnetic orbital i with an electron in the magnetic orbital j. If we substitute the oxovanadium(IV) with a manganese(II), we have one unpaired electron on the copper(II) ion, and five on the manganese(II), which has a ground d^5 configuration. Now we have to consider the interactions of the xy magnetic orbital on copper with the five magnetic orbitals on manganese. When the Cu–O–Mn angle is different from 90°, the pathway connecting the xy magnetic orbitals on the two metal ions is antiferromagnetic. All the other possible overlaps of xy with x^2-y^2, xz, yz, and z^2 are zero, therefore, all of these determine ferromagnetic coupling pathways. In general, if there is one antiferromagnetic pathway, it dominates, and indeed the observed coupling is antiferromagnetic in several copper-manganese pairs [1.25].

Finally, we want to discuss the coupling propagated by extended bridges. One interesting example is provided by μ-carbonato bridged copper(II) complexes. In Fig. 1.7 an arrangement, experimentally observed, is shown which corresponds to a very strong antiferromagnetic coupling. In fact, the singlet-triplet separation has been estimated [1.26] to be larger than 1000 cm^{-1}. The origin of this behavior is in the strong overlap of the metal orbitals with the HOMO of the carbonate ion [1.27], as shown in Fig. 1.8. The main interaction is given by the overlap with the p orbital of the oxygen atom which lies approximately on the line connecting the two copper ions.

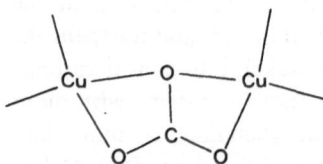

Fig. 1.7. Scheme of the geometry of μ-carbonato copper(II) bridged pairs

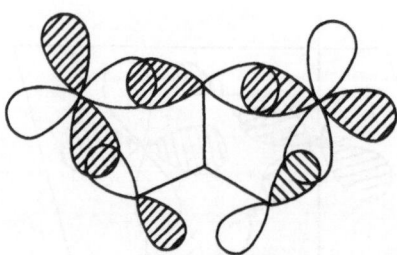

Fig. 1.8. Scheme of the orbitals responsible of the strong coupling in μ-carbonato bridged copper(II) pairs

This last example shows that when the nature of the bridges becomes more complex, simple qualitative considerations such as those we have worked out in this section become very difficult. In fact, beyond the stated impossibility of calculating the exchange interaction, the other drawback of Anderson's theory and the Goodenough-Kanamori rules is that the analysis, even at the qualitative level, becomes cumbersome for low symmetry complexes and for extended polyatomic bridges. It must be stressed that the above theory and rules were produced in the 1950s, and in a physical environment, with the interest directed on ionic lattices. For these compounds the symmetry is relatively high and the ligands bridging transition metal ions are simple anions such as O^{2-}, F^-, S^{2-}, etc.

When chemists became more and more involved in the synthesis of molecular compounds in which paramagnetic transition metal ions are bridged by an incredible variety of different ligands, the simple Goodenough-Kanamori rules became impossible to apply, and new models, based on the MO scheme so familiar to chemists, had to be developed. This happened in the mid-1970s and we shall briefly discuss some of them in the next sections.

1.3 Molecular Orbital Exchange Models

At first glance it might be assumed that the MO treatment of exchange phenomena should not be a very difficult task. Indeed, concentrating for the sake of simplicity on a symmetric bimetallic system with one unpaired electron on each center, and using for the same reason the one-electron approximation, the orbitals containing the unpaired electrons can be described by two linear combinations:

$$\psi_g \approx (\Phi_A + \Phi_B); \tag{1.7}$$

$$\psi_u \approx (\Phi_A - \Phi_B), \tag{1.8}$$

which are depicted in Fig. 1.9. ψ_g and ψ_u will have different energies, and let us assume that ψ_g lies lowest. Φ_A and Φ_B are essentially atomic orbitals of the two metal ions. Actually in ψ_g and ψ_u ligand functions will be present also, although

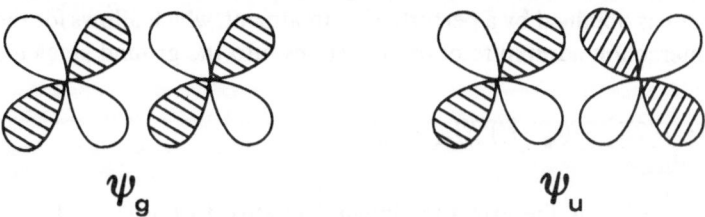

Fig. 1.9. Scheme of the two-center molecular orbitals

we do not write them out explicitly. Now, in the assumption of weak interaction between the two metal centers, which is central to all our treatments, the energy separation between the two MO's, ΔE, is small. As a consequence, the two magnetic electrons should occupy the two orbitals, ψ_g, and ψ_u, yielding a ground triplet state. This is clearly in sharp contrast with the experimental data, which show that both ground triplet and singlet states can be obtained. Further, the latter are much more common than the former! The breakdown of the simple (naive) MO approach outlined here is far from being unexpected, and is due to the neglect of electron correlation which is intrinsic to the MO model at this low level of approximation. This is the same reason why the zero-order VB approximation to the hydrogen molecule is much better than the corresponding MO wave function which introduces "covalent" and "ionic" terms with identical weights. In the present case the MO starting point is even more approximate than in the H_2 molecule due to the small overlap between Φ_A and Φ_B, which makes the probability of having two electrons at the same time on either the left or the right metal ion extremely improbable.

Several corrections to this extremely simple scheme have been suggested, either performing configuration interaction calculations, at various degrees of sophistication, or using a modified model, which makes the MO model closer to the VB.

Among the former, one particularly well suited to a semiquantitative discussion is that put forward by Hay et al. [1.3]. In a short outline of this method we can say that the HOMO's, which, in the treatment of these authors, should be obtained through SCF calculations on the triplet state, are first of all localized on the two metal centers by the transformations:

$$\Phi_a = (\psi_g + \psi_u)/\sqrt{2}; \qquad (1.9)$$

$$\Phi_b = (\psi_g - \psi_u)/\sqrt{2}. \qquad (1.10)$$

It is easy to verify, using (1.7) and (1.8), that Φ_a and Φ_b are indeed largely localized on A and B, respectively, but they include ligand functions as well. Also, it is important to note that Φ_a and Φ_b are orthogonal to each other, while Φ_A and Φ_B were not. In Fig. 1.10 the Φ_a and Φ_A orbitals in a model dinuclear copper(II) complex are shown.

Using these orbitals, the energy separation between the singlet and the triplet state is obtained by a perturbation treatment, which allows for the configuration interaction admixture of excited states into the ground singlet, as:

$$J = -2K_{ab} + (\varepsilon_g - \varepsilon_u)^2/(J_{aa} - J_{ab}), \qquad (1.11)$$

where K_{ab} is the exchange integral relative to the Φ_a and Φ_b orthogonalized molecular orbitals, J_{aa} and J_{ab} are the corresponding Coulomb integrals, and ε_g

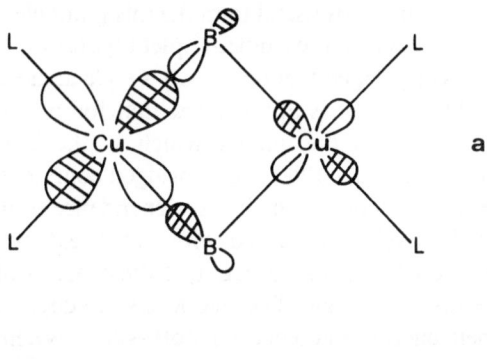

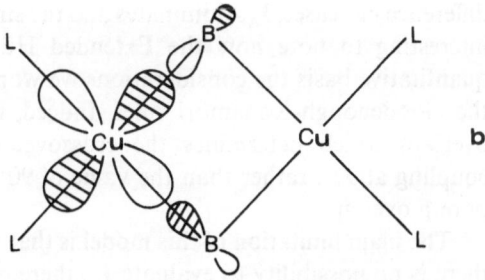

Fig. 1.10 a, b. Scheme of a localized molecular orbital; b non-orthogonal magnetic orbital

and ε_u are the energies of the ψ_g and ψ_u MO's. Equation (1.11) is occasionally written as:

$$J = J_F + J_{AF},\qquad(1.12)$$

where F stands for ferro- and AF for antiferromagnetic contribution, and:

$$J_F = -2K_{ab};\qquad(1.13)$$

$$J_{AF} = (\varepsilon_g - \varepsilon_u)^2/(J_{aa} - J_{ab}).\qquad(1.14)$$

Indeed, with orthogonalized orbitals K_{ab} is always positive and stabilizes the triplet state, while the J_{AF} term is always positive and stabilizes the singlet state.

Central to this model is the appreciation that in a series of complexes which differ essentially for geometrical parameters in the bridge, the most rapidly varying quantity is $(\varepsilon_g - \varepsilon_u)$, therefore, the dependence of the singlet-triplet splitting on the structural parameters of the complexes can be understood by calculating the energies of the MO's as a function of those parameters and observing the variation of $\varepsilon_g - \varepsilon_u$. This can be easily done within an Extended Hückel model.

11

It is perhaps useful to make this point clear with an example. Experimentally it is found that the singlet-triplet separation in di-μ-hydroxo bridged complexes is linearly dependent on the Cu–O–Cu angle, α, of the bridge [1.22, 23], as shown in Fig. 1.11. It is appreciated that for $\alpha < 97°$ the triplet lies lower, while for $\alpha > 97°$ it is the singlet which is the ground state. Beyond di-μ-hydroxo complexes also other μ-oxo bridged complexes are reported here, showing a less regular behavior. The linear dependence of the di-μ-hydroxo bridged species is qualitatively understood with simple extended Hückel calculations performed on model systems. In Fig. 1.12 the energies of the orbitals ε_g and ε_u are plotted versus the α angle. The two levels are degenerate for $\alpha = 97°$ and they increase their energy difference on both sides. When $\varepsilon_g - \varepsilon_u$ is zero (or at least small) $J_{AF} = 0$, and the J_F term dominates, stabilizing the triplet state. When the difference increases, J_{AF} dominates and the singlet becomes the ground state. It is interesting to note how the Extended Hückel calculations put on a semi-quantitative basis the considerations we worked out in Sect. 1.2 for illustrating the Goodenough-Kanamori rules. Indeed, it is the presence of both s and p overlaps which determines the crossover from ferro- to antiferromagnetic coupling at 97°, rather than the value of 90° which would be anticipated for a pure p overlap.

The main limitation of this model is that within the Extended Hückel model there is no possibility to evaluate J_F, therefore, it is not possible to calculate J. However, this model has been used with success also as a basis of an ab initio treatment which estimates J on the basis of a perturbation procedure (see next section). It is also worth noting here that the success of the Extended Hückel model is essentially based on the topological properties of the complexes, therefore, the same results can be qualitatively obtained also within the Angular Overlap Model (AOM) [1.5].

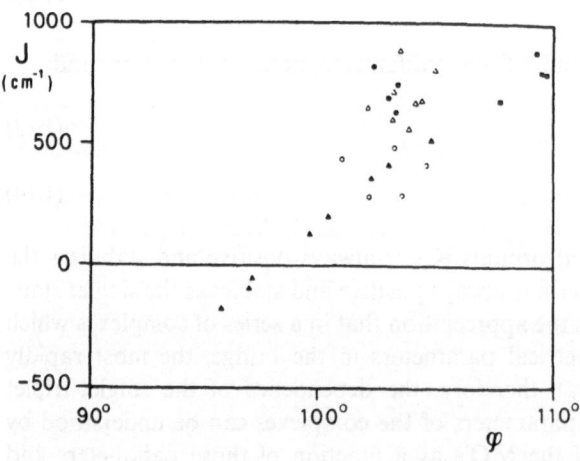

Fig. 1.11. Angular dependence of the J coupling constants in di-μ-oxo bridged copper (II) complexes. ▲ μ-hydroxo; ○ μ-alkoxo; ● μ-phenoxo; ■ μ-pyridine-N-oxo. After [1.28]

12

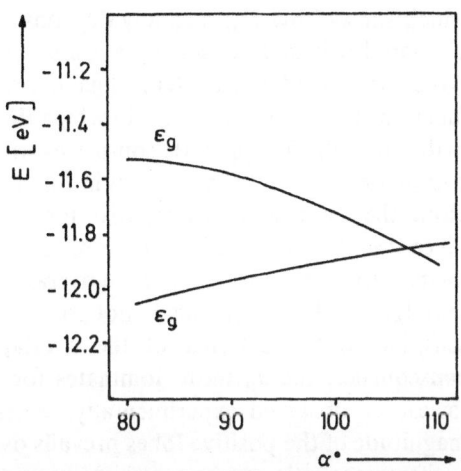

Fig. 1.12. Plot of ε_g and ε_u vs. the Cu–O–Cu angle in di-μ-hydroxo bridged copper (II) complexes. After [1.3]

Another qualitative model was put forward by Kahn et al. [1.4] who use nonorthogonal molecular orbitals of the type shown in Fig. 1.11b. In this frame the singlet-triplet energy separation depends on the overlap between the two orbitals, according to the relation:

$$J_{AF} \propto S\Delta, \tag{1.15}$$

where Δ is the energy difference between the two highest occupied orbitals, and S is the overlap integral between the two nonorthogonal magnetic orbitals. The main advantage of this approach is that the nonorthogonalized magnetic orbitals are much easier to visualize than their orthogonal counterparts, being the MO (or simply the Ligand Field) orbitals of the mononuclear moieties which make up the interacting pair. As such they have been widely used for rationalizing the observed magnetic properties and also for designing new types of interacting pairs with novel properties.

An interesting application of the model has been made to obtain at least an estimation of the ferromagnetic contribution [1.29]. The exchange integral, K_{ab}, can be written as:

$$K_{ab} = \int \Phi_A^*(1)\, \Phi_B^*(2)\, e^2/r_{12}\, \Phi_A(2)\, \Phi_B(1)\, d\tau_1\, d\tau_2. \tag{1.16}$$

By putting:

$$\rho_i = \Phi_A^*(i)\, \Phi_B(i) \tag{1.17}$$

with i = 1, 2, Eq. (1.16) can be rewritten as:

$$K_{ab} = \int \rho_1\, \rho_2\, e^2/r_{12}\, d\tau_1\, d\tau_2, \tag{1.18}$$

13

which shows how K_{ab} strongly depends on the ρ_i quantities, which are called overlap densities. In order to have a large K_{AB} it is necessary that the ρ_i's have large extrema at the bridging ligands, where the overlap of the Φ_A and Φ_B functions is maximum. A case in which this situation is obtained is that of di-μ-hydroxo bridged copper(II) complexes with $\phi = 95°$ and $110°$ whose calculated overlap densities [1.30] are shown in Fig. 1.13. The curves show two positive lobes along the x axis and two negative lobes along the y axis around each bridge. Since the altitude of the positive lobes and the depth of the negative lobes cancel themselves, in the former case the overall overlap between Φ_A and Φ_B is zero, so that $J_{AF} = 0$. However, this is not true for the exchange integral which is only sensitive to the extrema of the overlap density, not to their sign. As a consequence, the J_F term dominates for this compound, and the triplet lies lowest, as observed experimentally. When $\phi = 110°$, on the other hand, the magnitude of the positive lobes prevails over that of the negative lobes, and the antiferromagnetic component is dominant.

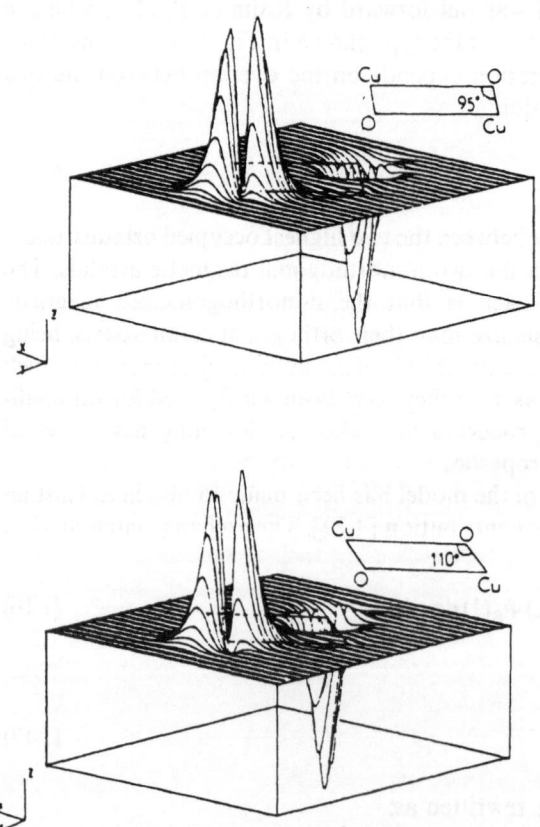

Fig. 1.13. Overlap density for model di-μ-hydroxo bridged copper(II) complexes

In a sense, the advantage of this approach is that of making clear with a picture, obtained for example through simple Extended Hückel calculations, the qualitative statements of rule (2) of Goodenough-Kanamori.

1.4 Quantitative MO Calculations of Singlet-Triplet Splitting

The first ab initio calculation of the singlet-triplet splitting in a symmetric metal dimer was reported in 1981 [1.31]. The energy separation was obtained through a perturbation treatment performed on the restricted Hartree-Fock MO's of the triplet state of copper acetate hydrate, whose structure is shown in Fig. 1.14. In a sense it is extremely pleasant that the first calculation was performed on the first molecular compound which was recognized to be dinuclear through EPR spectroscopy [1.32]. The orbitals ψ_g and ψ_u are first localized, as in Eqs. (1.9–10), and the various terms are calculated as in Anderson's theory. What is important is that the corrections which are needed are not only those due to kinetic exchange, potential exchange, and exchange polarization, but also other terms must be added, in order to reproduce the experimental data. In particular it is necessary to include also excitations from one of the metal orbitals, Φ_a or Φ_b, to a vacant orbital, or from one of the doubly occupied orbitals to one of the metal orbitals. Finally, double excitations different from those corresponding to exchange polarization are also needed. Perhaps for the present book what is now interesting is checking the relative importance of all these terms in a series of complexes. This is done in Table 1.1, where the calculated contributions to the experimental singlet-triplet splitting of di-μ-hydroxo bridged copper(II) complexes are reported [1.33]. If we look at the three canonical terms of Anderson's theory, namely potential exchange, kinetic exchange and exchange polarization, we see that alone they are not able at all to reproduce the experimental data and that the terms that in Table 1.1 are labeled as others are indeed responsible for very large corrections. Also it is in a sense shocking to verify that the potential exchange term, which in most qualitative theories, including Anderson's, is considered as small, is indeed the largest term! All these considerations are a little discouraging, and might induce the reader to complain over the time spent on

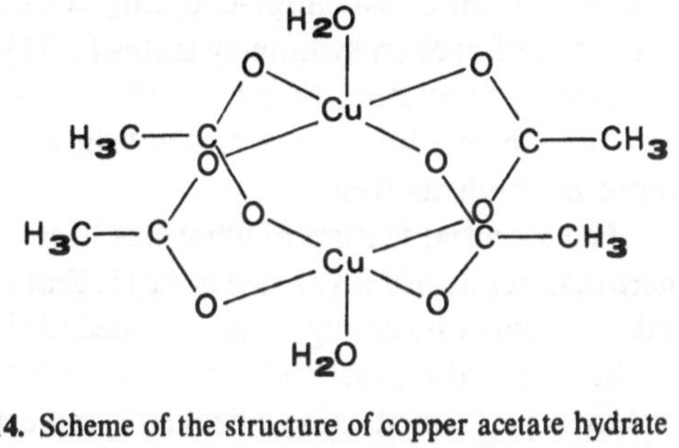

Fig. 1.14. Scheme of the structure of copper acetate hydrate

Table 1.1. Calculated singlet-triplet splittings in di-μ-hydroxo bridged copper(II) complexes[a]

	$\alpha = 95°$	$\alpha = 104°$	Difference
Pot. Exch.	−1274.9	−1380.2	105.3
Kin. Exch.	431.5	1036.3	−605.3
Exch. Polar.	−40.2	−11.3	−29
Others	787.5	747.8	−39.7
Total	−96.1	392.3	−488
Exper.	−161	509	−670

[a] Values in cm^{-1}. After [1.33].

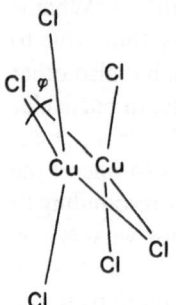

Fig. 1.15. Scheme of $Cu_2Cl_6^{2-}$ pairs. ϕ is the dihedral angle between the planes defined by the terminal and bridging chlorine atoms

the previous pages. However, there is still a consideration to be made which reconciles with the theories of the previous sections: the term which varies more rapidly in the series is the kinetic exchange, whose difference in Table 1.1 is almost six times larger than that of the largest of all the others. Therefore, it seems that focusing on this is a reasonable procedure when looking at the variation of the coupling in a series of compounds.

The only other theoretical approach which has been applied quantitatively to the calculation of exchange-coupling constants of transition-metal systems has been that of broken symmetry states [1.34]. Since this has been used within the Xα density functional theory [1.35–37], we will refer to this in the following, although the method can in principle be applied to unrestricted Hartree-Fock ab initio methods as well.

The essential feature of what has been called [1.34] the VB-Xα method (VB here denotes as usual valence bond) is that of using nonorthogonalized MO's: in this way only one configuration is needed since the excited states already appear in the non-orthogonalized molecular orbitals. The nonorthogonalization procedure is performed using a broken symmetry approach. This is best represented

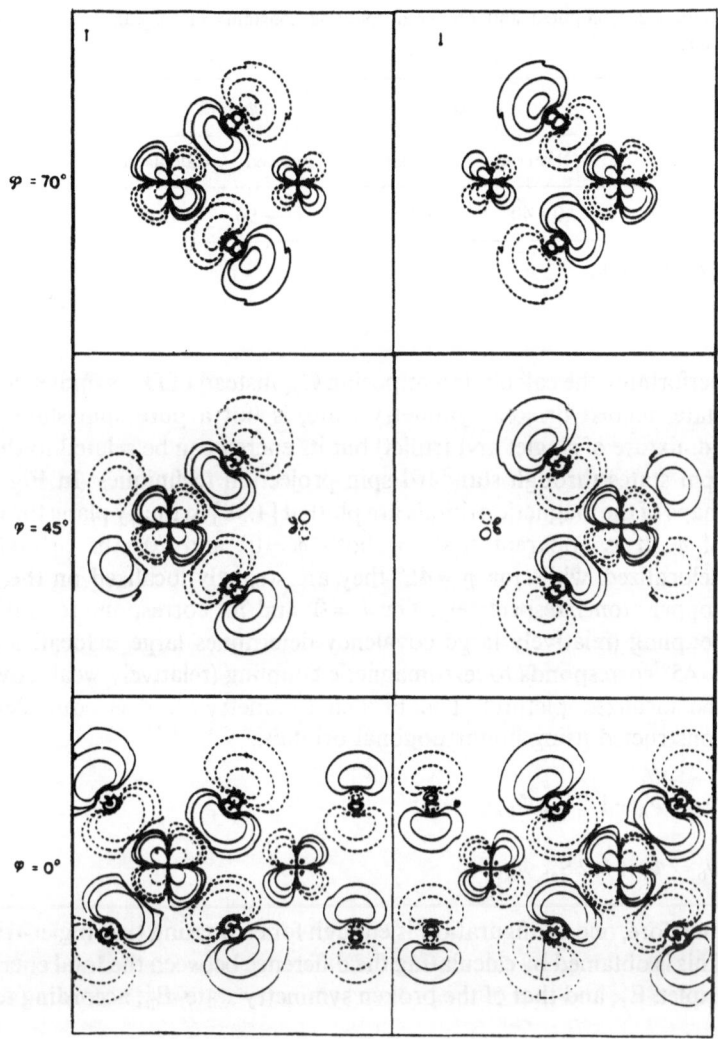

Fig. 1.16. Contour map of the magnetic orbitals of $Cu_2Cl_6^{2-}$ plotted in the xy plane for $\phi = 0°$ (*below*), $\phi = 45°$ (*middle*), and $\phi = 70°$ (*top*). Spin-up levels are plotted on the *left*, spin-down levels on the *right*

by an example. Let us consider a $Cu_2Cl_6^{2-}$ dimer as shown in Fig. 1.15. The molecular symmetry is D_{2d} and the calculation can be performed for the triplet state in this limit. Using a spin-unrestricted approach (different orbitals for different spins), it is possible to define a state in which the magnetic orbital on the left copper atom is occupied by an electron with spin-up and that of the right copper by an electron with spin-down, by removing the symmetry elements which transform the two copper atoms one into the other and by imposing a mirror symmetry to the spin densities. In the $Cu_2Cl_6^{2-}$ case this can be done by

Table 1.2. Computed and observed coupling constants in $Cu_2Cl_6^{2-}$ dimers[a]

| | $\phi = 0°$ | | $\phi = 45°$ | |
	Observed	Calculated	Observed	Calculated
$J\,(cm^{-1})$	$0 \div 40$	197.6	$-10 \div -90$	-210

[a] After [1.38].

performing the calculation imposing C_{2v} instead of D_{2d} symmetry. The resulting state, named broken symmetry state, is not a pure spin state (it will be an admixture of singlet and triplet) but its energy can be related to that of the pure spin states through standard spin projection techniques. In Fig. 1.16 contour maps of the magnetic orbitals are plotted [1.38] in the xy plane for various values of ϕ. It is apparent that for both $\phi = 0°$ and $70°$ the orbitals are largely delocalized, while for $\phi = 45°$ they are strongly localized on the left and right copper atoms, respectively. The $\phi = 0°$ and $70°$ correspond to antiferromagnetic coupling (relatively large covalency determines large delocalization), while $\phi = 45°$ corresponds to ferromagnetic coupling (relatively weak covalency favors the localized picture). The broken symmetry state is equivalent to a state constructed using nonorthogonal orbitals:

$$\Phi_a = \Phi_A + 1/2\,S_{ab}\,\Phi_B; \tag{1.19}$$

$$\Phi_b = \Phi_B + 1/2\,S_{ab}\,\Phi_A, \tag{1.20}$$

therefore, one configuration is enough for estimating the singlet-triplet splitting. This is obtained by calculating the difference between the total energy of the pure triplet, E_T, and that of the broken symmetry state, E_B, according to the relation:

$$E_T - E_B = J/2. \tag{1.21}$$

The values of J have been computed for $Cu_2Cl_6^{2-}$ systems as a function of the dihedral angle, ϕ, between the planes of the terminal and the bridging chlorine atoms. The results shown in Table 1.2 are in fair agreement with the experimental data, and also with ab initio calculations performed on the same systems.

References

1.1 Julve M, Verdaguer M, Kahn O, Gleizes A, Philoche-Levisalles M (1983) Inorg. Chem. 22: 369
1.2 Anderson PW (1963) In: Rado GT, Suhl H (eds) 'Magnetism'. Academic vol 1 p 25 and references therein

1.3 Hay PJ, Thibeault JC, Hoffmann RJ (1975) J. Am. Chem. Soc. 97: 4884

1.4 Kahn O, Briat BJ (1976) J. Chem. Soc. Faraday Trans. II 72: 268 (1976)

1.5 Bencini A, Gatteschi D (1978) Inorg. Chim. Acta 31: 11

1.6 Heisenberg W (1926) Z. Phys. 38: 411

1.7 Heisenberg (1928) Z. Phys. 49: 619

1.8 Dirac PAM (1929) Proc. Roy. Soc. A123: 714

1.9 van Vleck JH (1932) The theory of electric and magnetic susceptibilities, Oxford University Press, London, Chapter XI

1.10 Hulthen L (1936) Proc. Amst. Acad. Sci. 39: 190

1.11 Kramers HA (1934) Physica 1: 182

1.12 Yamashita J (1954) J Phys. Soc. Japan 9: 339

1.13 Yamashita J (1954) Progr. Theoret. Phys. (Kyoto) 12: 808

1.14 Kondo J (1957) Progr. Theoret. Phys. (Kyoto) 18: 541

1.15 Anderson PW (1959) Phys. Rev. 115: 2

1.16 Shulman RG, Sugano S (1963) Phys. Rev. 130: 506

1.17 Knox K, Shulman RG, Sugano S (1963) Phys. Rev. 130: 512

1.18 Goodenough JB (1955) Phys. Rev. 100: 564

1.19 Goodenough JB (1958) Phys. Chem. Solids 6: 287

1.20 Kanamori J (1959) Phys. Chem. Solids 10: 87

1.21 Ginsberg A (1971) Inorg. Chim. Acta Rev. 5: 45

1.22 Hatfield WE (1974) Am. Chem. Soc. Symp. Ser. No 5: 108

1.23 Hodgson DJ (1975) Progr. Inorg. Chem. 19: 173

1.24 Kahn O, Galy J, Journaux Y, Jaud J, Morgenstern-Badarau I (1982) J. Am. Chem. Soc. 104: 2165

1.25 Kahn O (1987) Structure and Bonding (Berlin) 68: 91

1.26 Churchill MR, Davies G, El-Sayed MA, El-Shazly MF, Hutchinson JP, Rupich MW, Watkins KO (1979) Inorg. Chem 18: 2296

1.27 Albonico C, Bencini A (1988) Inorg. Chem. 27: 1934

1.28 Gatteschi D, Bencini A (1985) In Willett RD, Gatteschi D, Kahn O (eds) Magneto-structural correlations in exchange coupled systems. Reidel, Dordrecht, p 241

1.29 Kahn O, Charlot MF (1980) Nouv. J. Chim. 4: 567

1.30 Charlot MF, Journaux Y, Kahn O, Bencini A, Gatteschi D, Zanchini C (1986) Inorg. Chem. 25: 1060

1.31 de Loth P, Cassoux P, Daudey JP, Malrieu JP (1981) J. Am. Chem. Soc. 103: 4007

1.32 Bleaney B, Bowers KD. Proc. R. Soc. A214, 451 (1952)

1.33 Daudey JP, de Loth P, Malrieu JP (1985) In: Willett RD, Gatteschi D, Kahn O (eds) 'Magneto-structural correlations in exchange coupled systems'. Reidel, Dordrecht p 87

1.34 Noodlemann L (1981) J. Chem. Phys. 74: 5737

1.35 Herman F, McLean AD, Nesbet RK (eds) (1979) Computational methods for large molecules and localized states in solids. Plenum, New York

1.36 Johnson KH (1973) Adv. Quantum Chem. 7: 143

1.37 Slater JC (1974) Quantum theory of molecules and solids McGraw Hill, New York

1.38 Bencini A, Gatteschi D (1986) J. Am. Chem. Soc. 108: 5763

2 Spin Hamiltonians

2.1 The Spin Hamiltonian Approach

The replacement of the true hamiltonian of a system with an effective one which operates only on the spin variables is commonplace in all areas of magnetic resonance spectroscopy. This is a parametric approach, which is helpful for the interpretation of sets of experimental data. The parameters which are obtained have no particular meaning per se, but they must be compared with more fundamental theory. When one finds, for example, that the EPR spectra of a copper(II) complex can be interpreted within the spin hamiltonian formalism to yield $g_\parallel = 2.20$, $g_\perp = 2.06$, it is only recurring to ligand field theory that the conclusion can be made that the unpaired electron is located in either a $x^2 - y^2$ or a xy orbital.

It is therefore the great simplicity of the spin hamiltonian approach which makes it so well suited for the analysis of complex systems, allowing at least a first-order rationalization of their properties. For example in Fig. 2.1 the temperature dependence of the magnetic susceptibility of copper acetate hydrate [2.1] is shown whose structure was shown in Fig. 1.14. The experimental points show that the susceptibility increases on decreasing temperature in the range 300–280 K, while it decreases on decreasing further the temperature below 280 K. This behavior is easily interpreted within a simple spin hamiltonian formalism to yield a parameter $J = 280$ cm^{-1}, which is a measure of the energy separation of the singlet and triplet states orginating from the interaction of the two unpaired electrons localized on the two copper ions. The sign of J indicates that the singlet lies lower, i.e., the coupling is antiferromagnetic. Explaining both the sign and the intensity of the interaction can be done only within the theories developed in the previous sections, but some important results have already been achieved with the much simpler spin hamiltonian model.

The main difficulty related to the spin hamiltonian model is the justification of the model itself. Therefore, in order not to complicate the situation too much at this stage, we defer this discussion to Sect. 2.5 and develop the spin hamiltonian formalism first.

Let us assume two centers, A and B, whose individual magnetic properties, before allowing them to interact, can be described by the effective spin operators S_A and S_B, respectively. The spin quantum numbers S_A and S_B may, or may not, correspond to the true spin of the system. In general, we may say that the value of S_A (S_B) is such that $2S_A + 1$ ($2S_B + 1$) equals the multiplicity of the levels which are thermally populated. In the hypothesis that the ground states are orbitally

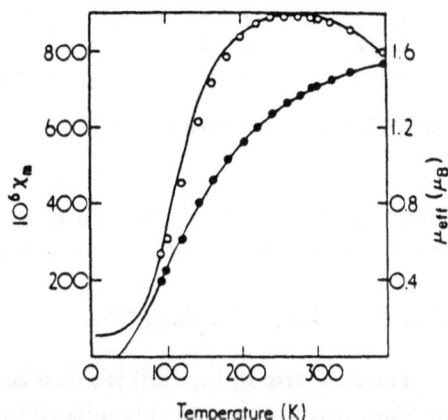

Fig. 2.1. Temperature dependence of the magnetic susceptibility (*open circles*) and moment (*closed circles*) of copper acetate hydrate. After [2.1]

nondegenerate and that the excited levels are much higher in energy, the interaction between the two spins can be described by the hamiltonian:

$$H = S_A \cdot J_{AB} \cdot S_B, \tag{2.1}$$

where J_{AB} is a dyadic, i.e., a general tensor, which contains all the relevant exchange parameters.

Equation (2.1) is a bilinear relation in the S_i operators. We may expect that also higher power terms can be added, and in fact biquadratic terms have sometimes been added (odd powers are not allowed because they do not yield totally symmetric representations). We will discuss these possibilities in Sect. 2.4.

Any second-rank tensor, J, can be decomposed [2.2] into the sum of a symmetric, S, and an antisymmetric, A, tensor, according to:

$$J_{ij} = S_{ij} + A_{ij}, \tag{2.2}$$

where $S_{ij} = \frac{1}{2}(J_{ij} + J_{ji})$ and $A_{ij} = \frac{1}{2}(J_{ij} - J_{ji})$. The antisymmetric tensor is traceless and S can be made traceless by subtracting from the diagonal elements one-third of the trace, so that

$$J = JE + S + A, \tag{2.3}$$

where E is the identity matrix and $J = 1/3 \, \text{Tr}(J)$. Using this decomposition Eq. (2.1) can be rewritten as:

$$H = J_{AB} S_A \cdot S_B + d_{AB} \cdot S_A \times S_B + S_A \cdot D_{AB} \cdot S_B, \tag{2.4}$$

21

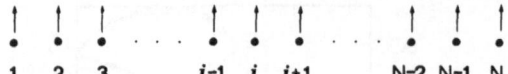

Figure 2.2. A scheme of a magnetic chain

where J_{AB} is a scalar, $\mathbf{d}_{AB}$ is a polar vector, and $\mathbf{D}_{AB}$ is a symmetric traceless tensor. The components of $\mathbf{d}_{AB}$ are given by:

$$d_{AB,x} = A_{yz}; \; d_{AB,y} = A_{zx}; \; d_{AB,z} = A_{xy}. \tag{2.5}$$

The first term in Eq. (2.4) is called isotropic, the second antisymmetric, and the third anisotropic spin-spin interaction. Both Eqs. (2.1) and (2.4) can be easily extended to the case of N interacting spins by summing over all the possible pairs. Quite often only the nearest neighbor interactions are taken into consideration, i.e., only those involving adjacent spins. For instance, in a chain of spins as that shown in Fig. 2.2, only the interactions of the i-th spin with the $(i-1)$-th and $(i+1)$-th will be taken into account on this assumption.

The nature of the interaction between the spins, represented either by J_{AB} or by the set J_{AB}, $\mathbf{d}_{AB}$, $\mathbf{D}_{AB}$, can be twofold, either through space or through bonds. The former is the magnetic interaction between the spins, which at the simplest level can be taken as the interaction between magnetic dipoles centered at the A and B sites, while the latter is the exchange interaction which has been introduced in the previous chapter. The two will be treated separately in the next two sections.

2.2 The Magnetic Spin-Spin Interaction

When the two interacting spins A and B are sufficiently removed one from the other it is conceivable that their magnetic interaction reduces to that of two magnetic dipoles separated by the $A - B$ distance, R. The magnetic dipoles of the two spins are given by:

$$\mathbf{m}_i = -\mu_B \mathbf{g}_i \cdot \mathbf{S}_i, \tag{2.6}$$

where $i = A, B$, μ_B is the Bohr magneton, and $\mathbf{g}_i$ is the $\mathbf{g}$ tensor of the individual i center. The classic form of interaction between the two spins can be written as:

$$H = \mathbf{S}_A \cdot \mathbf{J}_{AB}^{dip} \cdot \mathbf{S}_B, \tag{2.7}$$

where

$$\mathbf{J}_{AB} = \mu_B^2 / R^3 (\mathbf{g}_A \cdot \mathbf{g}_B - 3(\mathbf{g}_A \cdot \mathbf{R})(\mathbf{R} \cdot \mathbf{g}_B)), \tag{2.8}$$

$\mathbf{R}$ is a unit vector parallel to the $A - B$ direction.

An example may help to clarify the meaning of Eq. (2.8). Let us consider a hypothetic dinuclear species as shown in Fig. 2.3. A and B can be, for instance, copper(II) and manganese(II). A possible reference frame for the pair is that shown in Fig. 2.3. The **g** tensor for the manganese ion can be reasonably assumed to be isotropic and equal to $g_e = 2.00$:

$$g_{Mn} = g_e\, E, \tag{2.9}$$

where E is the identity matrix. the g tensor for the copper ion can be assumed to be axial: $g_{zz} = 2.00$, $g_{xx} = g_{yy} = 2.20$. The vector R has components $(0, \sin\alpha, \cos\alpha)$. Therefore, the matrix J_{AB}^{dip} in this case takes the form:

$$\mu_B{}^2/R^3 \begin{vmatrix} g_{xx}\,g_e & 0 & 0 \\ 0 & g_{yy}g_e(1-3\sin^2\alpha) & -3\sin\alpha\cos\alpha\ g_{yy}g_e \\ 0 & -3\sin\alpha\ \cos\alpha\ g_{zz}g_e & g_{zz}g_e(1-3\cos^2\alpha) \end{vmatrix} \tag{2.10}$$

The matrix can be decomposed according to Eqs. (2.2) and (2.3) into an isotropic, an anisotropic, and an antisymmetric part. The isotropic part of Eq. (2.10) is:

$$J_{AB}^{dip} = 1/3\,Tr J_{AB}^{dip} = \mu_B{}^2/(3R^3)\,g_e[g_{xx}+(1-3\sin^2\alpha)g_{yy}+(1-3\cos^2\alpha)g_{zz}]. \tag{2.11}$$

If R is expressed in Ångstrøm and J in cm^{-1}, the numerical value of $\mu_B{}^2$ is 0.433, which means that for a metal-metal distance of 3.50 Å and $\alpha = 45°$ the isotropic part of the dipolar interaction is only $7 \times 10^{-4}\ cm^{-1}$, as shown in Table 2.1.

The antisymmetric and the anisotropic parts of the J_{AB}^{dip} matrix are:

$$D_{AB}^{dip}(xx)=[2/3\ g_{xx}-(1-3\sin^2\alpha)\,g_{yy}/3-(1-3\cos^2\alpha)\,g_{zz}/3]\,g_e;$$
$$D_{AB}^{dip}(xy)=D_{AB}^{dip}(xz) = 0;$$
$$D_{AB}^{dip}(yy)=[-g_{xx}/3+2/3(1-3\sin^2\alpha)g_{yy}-(1-\cos^2\alpha)\,g_{zz}]g_e; \tag{2.12}$$
$$D_{AB}^{dip}(yz)= -3/2\ g_e\ \sin\alpha\ \cos\alpha\,(g_{zz}+g_{yy});$$
$$D_{AB}^{dip}(zz)=[-g_{xx}/3-(1-3\sin^2\alpha)g_{yy}/3+2/3(1-3\cos^2\alpha)\,g_{zz}]g_e;$$

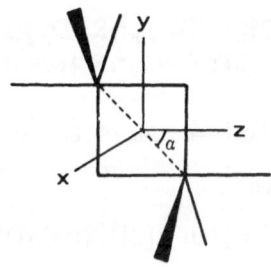

Fig. 2.3. Scheme of a dinuclear species formed by two trigonal bipyramidal complexes

Table 2.1. Calculated values of the dipolar matrix for a pair of interacting copper (II)-manganese (II) ions[a]

	R = 250	R = 350	R = 450
J^{dip}	18	7	3
D_{xx}^{dip}	1201	438	206
D_{yy}^{dip}	−628	−229	−108
D_{zz}^{dip}	−573	−209	−98
D_{yz}^{dip}	−1746	−636	−299
D	−3519	−1283	−603
E	28	11	5
d_{x}^{dip}	−83	−30	−14

[a] The distances in pm, the spin hamiltonian parameters in $10^{-4}\,cm^{-1}$.
The values were calculated setting $g_{cu,x} = g_{cu,y} = 2.2$; $g_{cu,z} = 2.0$; $g_{Mn,x}$
$= g_{Mn,y} = g_{Mn,z} = 2.0$, $\alpha = 45°$.

and the matrix, which is expressed in units μ_B^2/R^3, is symmetric by construction.

$$A_{AB}^{dip}(yz) = -A_{AB}^{dip}(zy) = -3/2\,\sin\alpha\,\cos\alpha g_e(g_{yy} - g_{zz}),\tag{2.13}$$

all the other elements being zero.

In the above example the only component of $\mathbf{d}_{AB}$ which is different from zero is that parallel to x. This is in accord with the symmetry rules for antisymmetric spin-spin interactions. These have been given by Moryia [2.3] and state that $\mathbf{d}_{AB}$ is zero if a center of symmetry relates the two interacting spins. Further, when a mirror plane is present, including A and B, $\mathbf{d}_{AB}$ is orthogonal to the mirror plane (this is the case for our example); when a mirror plane perpendicular to AB bisects it, then $\mathbf{d}_{AB}$ is parallel to the mirror plane; when a twofold rotation axis perpendicular to A−B passes through the midpoint of A−B, then $\mathbf{d}_{AB}$ is orthogonal to the twofold axis; when there is an n-fold axis (n > 2) along A−B, then $\mathbf{d}_{AB}$ is parallel to A−B.

The anisotropic component of the dipolar interaction in the above example is not diagonal. However, it can be reduced to a diagonal form by a standard procedure yielding the principal values:

$$D_{AB}^{dip}(XX) = \mu_B{}^2/R^3\,D_{AB}^{dip}(xx);$$
$$D_{AB}^{dip}(YY) = \mu_B{}^2/R^3[\cos^2(\Theta)D_{AB}^{dip}(yy) + \sin^2(\Theta)D_{AB}^{dip}(zz) + \sin(2\Theta)D_{AB}^{dip}(yz)];\tag{2.14}$$

$$D_{AB}^{dip}(ZZ) = \mu_B{}^2/R^3[\sin^2(\Theta)D_{AB}^{dip}(yy) + \cos^2(\Theta)D_{AB}^{dip}(zz) - \sin(2\Theta)D_{AB}^{dip}(yz)],$$

where

$$\Theta = \tfrac{1}{2}\tan^{-1}[2D_{AB}^{dip}(yz)/(D_{AB}^{dip}(yy) - D_{AB}^{dip}(zz))].\tag{2.15}$$

The X axis is parallel to x, while Z is practically parallel to the A−B direction.

The largest component of D_{AB}^{dip}, in absolute value, is along the Z direction, and the sign of $D_{AB}^{dip}(ZZ)$ is negative. This is a general result, which we must always expect whenever the point dipolar approximation is valid: the largest anisotropic component has a negative sign and is directed approximately along the $A-B$ direction, when the anisotropy of the **g** tensors is not large, such as is the case for orbitally non-degenerate ground states. The other two directions are determined by the relative orientations of the $\mathbf{g_A}$ and $\mathbf{g_B}$ tensors and by their anisotropies. In the same approximation as above the two principal values along the X and Y axes are not expected to be too different from each other, yielding a substantially axial $\mathbf{D_{AB}^{dip}}$ tensor.

The values of all the above quantities, calculated for the copper-manganese example, are given in Table 2.1. They are calculated for three different distances R and, due to the R^{-3} dependence, they rapidly decrease on increasing the $A-B$ distance.

The point dipolar approximation fails when the distance between the two spins is not large compared to the average distance of the unpaired electrons from their nuclei. This can be the case for some organic biradicals, such as dinitroxides, e.g., in Fig. 2.4. Also, we want to mention a transition metal complex in which the metal ion is bound to an organic radical such as the one shown in Fig. 2.5. In this compound the central copper(II) ion is octahedrally coordinated by four oxygen atoms of two hexafluoroacetylacetonato ligands, by an OH group, and by an NO group of two different TEMPOL ligands [2.4]. TEMPOL is the nitroxide 4-hydroxy-2,2,6,6-tetramethyl piperidinyl-N-oxy. Each TEMPOL ligand binds to two different copper(II) ions so that a linear chain is formed. The magnetic data have shown [2.5] that there is a weak ferromagnetic interaction between the copper ion and the organic radical (the singlet-triplet separation has been estimated to be $15\,cm^{-1}$), and the EPR spectra are typical of a triplet with a zero field splitting characterized by D $=0.1710\,cm^{-1}$, and E/D$=1/3$. Now the distance between the copper ion and the oxygen atom of the NO group, which formally carries a good portion of the unpaired spin density of the radical, is fairly short, 245 pm, so that the point dipolar assumption in this case appears to be questionable.

An alternative procedure to the point dipolar approximation requires an MO approach. The molecule must be treated within a suitable MO formalism,

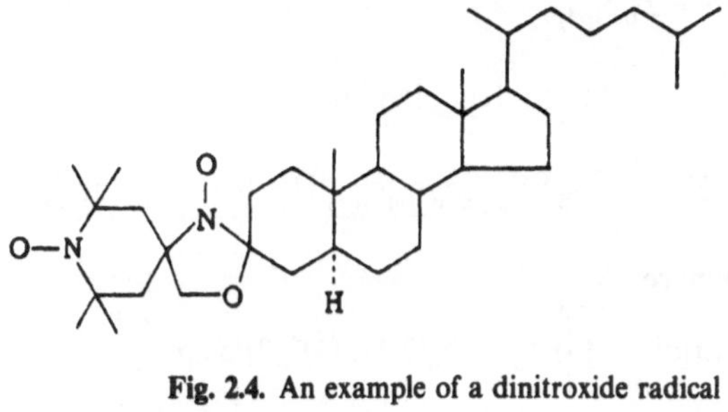

Fig. 2.4. An example of a dinitroxide radical

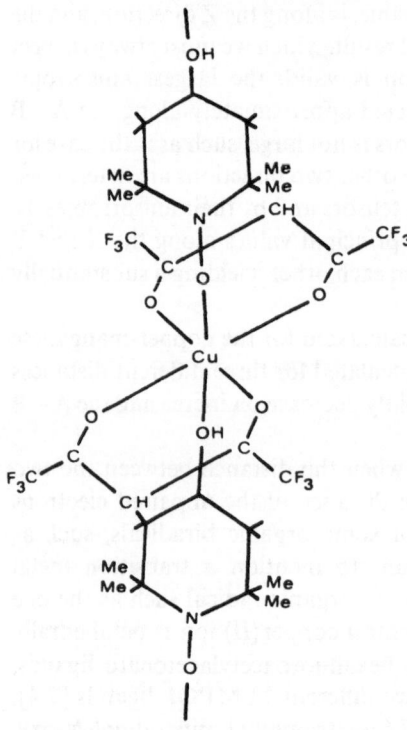

Fig. 2.5. Scheme of the structure of Cu(hfac)$_2$ (TEMPOL)

obtaining the eigenvectors of the magnetic orbitals, i.e., of the orbitals carrying the unpaired electrons, Φ_i. The correct matrix of the spin-spin interaction must then be calculated through the elements:

$$\langle \Phi_i(1)\,\Phi_j(2)|H|\Phi_i(1)\,\Phi_j(2)\rangle - \langle \Phi_i(1)\,\Phi_j(2)|H|\Phi_i(2)\,\Phi_j(1)\rangle, \tag{2.16}$$

where H is defined through Eqs. (2.7) and (2.8). The orbitals Φ_i and Φ_j are expressed as linear combinations of atomic orbitals:

$$\Phi_i = \Sigma_p\, c_{ip}\, f_p, \tag{2.17}$$

so that the matrix elements of Eq. (2.16) become:

$$\langle H\rangle = g^2\mu_B^2\,\Sigma_{pqrs}\,c_{ip}c_{jq}c_{ir}c_{js}[(pr|qs)-(ps|qr)], \tag{2.18}$$

where

$$(pr|qs) = \iint f^*_p(1)f^*_q(2)Hf_r(1)f_s(2)\,dv_1dv_2. \tag{2.19}$$

In the literature several methods have been reported to calculate these integrals [2.6 – 8] and the methods applied to several aromatic hydrocarbons [2.9, 10], and other systems [2.11, 12]. However, for systems characterized by relatively weak exchange, like the ones in which we are interested, the main difficulty is that of obtaining reliable molecular orbitals, as discussed in Chap. 1. Therefore, the method has been relatively seldom used, and empirical methods have been employed.

Attempts have been made of taking into account the fact that the dipoles cannot be regarded as point dipoles by introducing two [2.13] (or four in the case of transition metal ions [2.14]) negative charges delocalized along the lobes of the $p(d)$ orbitals which formally carry the unpaired electrons. The charge localized along the lobes of the orbital is $1/2(1/4)$ the overall charge assumed to be present in the orbital. The distance of the negative charge from the nucleus has been assumed to be 35 pm for p and 50 pm for d orbitals, considering that for purely atomic orbitals the $p(d)$ electron density maximizes at that distance.

Finally, another complication must be mentioned here with regards to the use of empirical methods. So far it has been assumed that the magnetic dipoles are essentially localized on two centers. Now let us take into consideration two metal ions bridged by some intervening ligand: it is apparent that the unpaired electrons, although mainly localized on the metal ions, will have a finite probability also on the bridging and the remaining ligands. Although the fraction of electrons transferred to the ligand may be small, the distance of the ligand from the metal is much smaller than that of the other metal so that a relevant contribution can also result. It is worth mentioning here that this is a general problem in magnetic resonance spectroscopy which has been discussed at length also in the paramagnetic NMR literature [2.15]. However, this problem is, alas, difficult to solve. In fact, in order to calculate the ligand contribution to the magnetic spin-spin interaction fairly accurate functions are needed and, as outlined in the previous chapter, the era of MO calculations on actual dinuclear species has just started. Recently Extended Hückel calculations have been applied to fluoro-bridged copper(II) dimers, but the values did not deviate significantly from those expected for the point-dipolar approximation [2.16].

2.3 The Exchange Contribution

The isotropic part of the exchange-determined component of $\mathbf{J_{AB}}$, J_{AB}^{ex} is mainly determined by the weak bonding interaction described in the previous chapter, and we will not expand on it further. The anisotropic and the antisymmetric parts of $\mathbf{J_{AB}^{ex}}$, on the other hand, are determined by relativistic effects, i.e., by the admixture of excited states into the ground state by spin-orbit coupling. The rigorous inclusion of spin-orbit coupling effects in exchange-coupled systems is very difficult, and no completely rigorous attempt to do this has been made. In the following we will report the essential lines of a treatment suggested by

Kanamori [2.17], which although simplified, has been used with some success in the interpretation of the EPR spectra of transition metal compounds.

From the elementary theory of EPR it is well known that excited states can be admixed into the orbitally non-degenerate ground state by spin-orbit coupling, yielding g values different from the free electron value g_e. Spin-orbit coupling is more important for transition metal ions than for organic radicals, as shown by the g values which for the latter are generally quasi-isotropic and close to g_e. The situation is much more complicated in the case of lanthanides and actinides and will not be considered here.

For transition metal ions and for organic radicals spin-orbit coupling can be treated as a perturbation. Therefore, in the lowest $|g_i\rangle$ state (i denotes the spin center) the excited states will be admixed through the spin-orbit hamiltonian, H_{so}, for which a convenient form is:

$$H_{so} = \Sigma_i \xi(r_i) \, l_i . s_i, \tag{2.20}$$

where the sum is over all the unpaired electrons of the configuration, l_i and s_i are the orbital and spin angular momentum operators, respectively, and $\xi(r_i)$ is a radial function. The perturbed functions can be written as:

$$|f_i\rangle = |g_i\rangle + \Sigma_l \langle e_{il}|H_{so}|g_i\rangle |e_{il}\rangle / \Delta_{ie}, \tag{2.21}$$

where the sum is over all the excited states, and Δ_{ie} is the energy difference between the excited and the ground state.

If the exchange interaction is introduced as a perturbation, leaving the relative hamiltonian H_{ex} unexplicit, the first terms which are relevant to the anisotropic contribution to J_{AB}^{ex} appear in third order and are given by:

$$\frac{\langle g_A g_B|H_{A,so}|e_{Al} g_B\rangle \langle e_{Al} g_B|H_{ex}|e_{Aj} g_B\rangle \langle e_{Aj} g_B|H_{A,so}|g_A g_B\rangle}{\Delta e_{Al} \Delta e_{Aj}}, \tag{2.22}$$

where the suffix of H_{so} indicates that it operates on the coordinates of the electron centered on A. Of course, analogous terms for B will also be found, and a sum must be performed on all the excited states.

The final result is that it is possible to rewrite the sum of terms similar to Eq. (2.22) by separating the spin and orbital variables according to the effective spin hamiltonian:

$$H_{ex} = S_A . J_{AB} . S_B, \tag{2.23}$$

where

$$J_{AB}^{ex}(kl) = \Sigma_\alpha \Sigma_i \Sigma_j \lambda_\alpha^2 \frac{\langle g_\alpha|L_{\alpha,k}|e_{\alpha i}\rangle \langle e_{\alpha i}|L_{\alpha,l}|g_\alpha\rangle \, J(e_{\alpha i} g_\beta e_{\alpha j} g_\beta)}{\Delta e_{\alpha i} \Delta e_{\alpha j}}, \tag{2.24}$$

28

where the α sum is over the two centers ($\alpha = A, B$; β indicates the center different from α); the i and j sums are over all the excited states; k and l are cartesian components. This matrix is not traceless nor symmetric, when the A and B centers are different, therefore, it must be reduced according to the procedure outlined in the previous section. Equation (2.24) has been obtained in the simplifying assumption that the excited states mixed into the ground state by spin-orbit coupling belong to the same ground spectral term, so that the relative hamiltonian can be written as:

$$H_{\alpha, so} = \lambda_\alpha L_\alpha \cdot S_\alpha. \tag{2.25}$$

$J(e_{\alpha i}g_\beta e_{\alpha j}g_\beta)$ is identical to $\langle e_{\alpha i}g_\beta|H_{ex}|e_{\alpha j}g_\beta\rangle$. It is *not* a simple exchange integral, but rather the exchange interaction with an excited state. We will come back to this below.

Equation (2.24) can take a more appealing form when the symmetry of the system is at least orthorhombic ($J_{AB}^{ex}(kl) = \delta_{kl}$) and only one excited state is admixed into the ground state by each $L_{\alpha, k}$ component. In this case (2.24) reduces to:

$$J_{AB}^{ex}(kk) = \Sigma_\alpha \lambda_\alpha^2 \frac{|\langle g_\alpha|L_{\alpha, k}|e_\alpha\rangle|^2 J(e_\alpha g_\beta e_\alpha g_\beta)}{(\Delta e_\alpha)^2}. \tag{2.26}$$

Comparing the expression which gives the **g** tensors for the individual centers α:

$$\Delta g_\alpha(kk) = g_\alpha(kk) - g_e = \lambda_\alpha \frac{|\langle g_\alpha|L_{\alpha, k}|e_\alpha\rangle|^2}{(\Delta e_\alpha)}, \tag{2.27}$$

we see that finally Eq. (2.26) can be written as:

$$J_{AB}^{ex}(kk) = \Sigma_\alpha \frac{[\Delta g_\alpha(kk)]^2 \; J(e_\alpha g_\beta e_\alpha g_\beta)}{|\langle g_\alpha|L_{\alpha, k}|e_\alpha\rangle|^2}, \tag{2.28}$$

which shows that the elements of the $\mathbf{J_{AB}^{ex}}$ matrix are of the order of $(\Delta g)^2$, which means that for orbitally nondegenerate ground states the matrix elements are very small. For example for organic radicals, manganese(II), or gadolinium(III) ions, for which $\Delta g = 0$, the elements of the $\mathbf{J_{AB}^{ex}}$ matrix are practically zero. Equation (2.28) also shows that the principal axes of the exchange contribution to $\mathbf{J_{AB}}$ are parallel to the principal axes of **g**, provided that the two **g** tensors of the A and B centers are parallel to each other.

In the decomposition of $\mathbf{J_{AB}^{ex}}$ the scalar component adds to the scalar component originating from the exchange interaction between the ground, $|g_\alpha\rangle$, states. When the latter is of the order of at least 10° cm^{-1}, then the additional component brought about by spin-orbit coupling can be safely neglected. However, this cannot be the case when the ground state exchange interaction is smaller.

A caveat must be clearly stressed at this point, and it pertains to the $J(e_\alpha g_\beta e_\alpha g_\beta)$ parameter. According to its definition it describes the exchange interaction between the ground, $|g_\beta\rangle$, orbital on center β, with the excited, $|e_\alpha\rangle$, orbital on center α. This interaction can be completely different from the interaction between the ground states g_α and g_β, i.e., it can have different sign and different intensity. This point can be made clear with one example. Let us take into consideration a dinuclear copper(II) complex like the one described in Sect. 1.2 and in Fig. 1.4. We saw that the ground orbital can be described to a satisfactory approximation as xy and that the exchange interaction between the two magnetic orbitals can be either ferro- or antiferro-magnetic, depending on the Cu-L-Cu angle. The excited state $x^2 - y^2$ is mixed into the ground state through spin-orbit coupling by the z component of L and $|\langle xy|L_z|x^2 - y^2\rangle|^2 = 4$. The exchange interaction between these two orbitals is dominated by the fact that they are orthogonal: according to the Goodenough-Kanamori rules, this means that the coupling between the two must be ferromagnetic. Indeed, we already considered this case in Sect. 1.2, for a copper(II)-oxovanadium(IV) pair. Therefore, when the Cu-L-Cu angle is large enough, $J(xy, xy, xy, xy)$ is antiferromagnetic and $J(x^2 - y^2, xy, x^2 - y^2, xy)$ is ferromagnetic. It is apparent that any attempt to estimate $J_{CuCu}^{ex}(zz)$ by replacing the unknown value of $J(x^2 - y^2, xy, x^2 - y^2, xy)$ in (2.28) by the value of $J(xy, xy, xy, xy)$, determined for instance through magnetic susceptibility measurements, is destined to fail.

Since for orbitally nondegenerate ground states the main contribution of $\mathbf{J_{AB}^{ex}}$ is to the zero field splitting tensor, it has been customary to have an order of magnitude estimate of the $\mathbf{D}$ tensor, using for instance the D parameter, according to:

$$D \propto (\Delta g/g)^2 J, \tag{2.29}$$

where J is $J(g_A g_B g_A g_B)$. From the considerations above it is apparent that this estimation can be completely wrong and must be used, *faut de mieux*, with extreme circumspection.

Indeed, the EPR spectra of a few copper(II) complexes possessing the geometry of Fig. 1.4 have been studied [2.18, 19]. In all cases the largest zero field splitting component has been observed to be orthogonal to the coordination plane, in agreement with a dominant exchange contribution to the zero field splitting, and not along the copper-copper direction, as would be required by dominant dipolar interaction. The results are summarized in Fig. 2.6, where the experimental D_{zz} values are plotted vs the copper-copper distance. The dotted area corresponds to the calculated dipolar interaction: it is apparent that all the experimental points are well above that, even at the largest distances [2.20]. The experimental exchange contributions can be fitted with an exponential regression [2.18], but it is safer to use this result as indicative that D_{zz} decreases on increasing the metal-metal distance.

A corollary to the use of (2.29) has been that when J is small, the exchange contribution to D can be neglected, and the experimental value can be safely

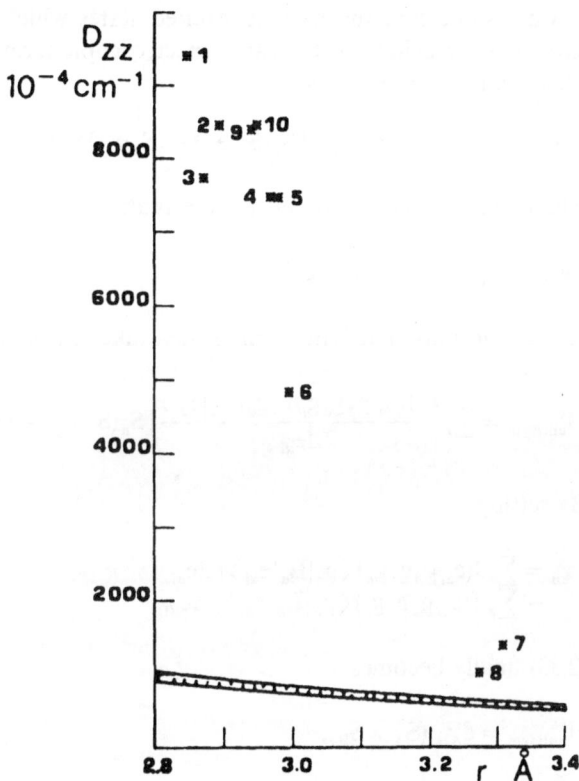

Fig. 2.6. Experimental D_{zz} values for a series of dinuclear copper (II) complexes. After [2.20]

assumed to be due to the dipolar component, thus allowing one to determine the metal-metal distance through the R^{-3} dependence of D. For instance, it has been common practice to neglect the exchange contribution to the zero field splitting when $J < 30$ cm^{-1} [2.21]. Although it may work sometimes, there are several examples in the literature which show how dangerous it can be to rely on it in order to obtain structural information.

Finally, it may be interesting to note that the derivation outlined above to the anisotropic exchange is not the only one which is possible. Keijzers showed [2.22] that specializing to pairs, the contribution of excited states appears in second order, and confirmed that the $J(e_\alpha g_\beta e_\alpha g_\beta)$ parameters refer to the exchange interaction rather than to the exchange integral.

In the original Kanamori treatment beyond the third-order contribution outlined above, a second-order exchange contribution may also be operative, but this, as we will show below, is relevant only to the antisymmetric part of the J_{AB} matrix. In fact, the second-order terms are of the type:

$$[\langle g_\alpha g_\beta | H_{\alpha, so} | e_{\alpha i} g_\beta \rangle \langle e_{\alpha i} g_\beta | H_{ex} | g_\alpha g_\beta \rangle +$$
$$\langle g_\alpha g_\beta | H_{ex} | e_{\alpha i} g_\beta \rangle \langle e_{\alpha i} g_\beta | H_{\alpha, so} | g_\alpha g_\beta \rangle] / \Delta e_{\alpha i}. \tag{2.30}$$

If we assume that the relevant excited states which can be admixed into the ground state belong to the same spectroscopic term as the ground state, then (2.30) can be rewritten as:

$$\lambda_\alpha J(e_{\alpha i} g_\beta g_\alpha g_\beta) \langle g_\alpha | L_\alpha | e_{\alpha i} \rangle [S_\alpha (S_\alpha \cdot S_\beta) - (S_\alpha \cdot S_\beta) S_\alpha] / \Delta e_{\alpha i}, \qquad (2.31)$$

where use has been made of the fact that:

$$\langle g_\alpha | L_\alpha | e_{\alpha i} \rangle = - \langle e_{\alpha i} | L_\alpha | g_\alpha \rangle \qquad (2.32)$$

for real orbitals. The hamiltonian then takes the form:

$$H^{ex}_{antisym} = \sum_\alpha \frac{\lambda_\alpha J(e_{\alpha i} g_\beta g_\alpha g_\beta) \langle g_\alpha | L_\alpha | e_{\alpha i} \rangle}{\Delta e_{\alpha i}} [S_\alpha (S_\alpha \cdot S_\beta) - (S_\alpha \cdot S_\beta) S_\alpha]. \qquad (2.33)$$

By setting:

$$\begin{aligned} d^{ex}_{AB} = &\sum_\alpha J(e_{\alpha i} g_\beta g_\alpha g_\beta) \langle g_\alpha | L_\alpha | e_{\alpha i} \rangle / \Delta e_{\alpha i} \\ &- \sum_\beta J(e_{\beta i} g_\alpha g_\beta g_\alpha) \langle g_\beta | L_\beta | e_{\beta i} \rangle / \Delta e_{\beta i}, \end{aligned} \qquad (2.34)$$

(2.33) finally becomes:

$$H^{ex}_{antisym} = d^{ex}_{AB} \cdot (S_A \times S_B), \qquad (2.35)$$

which yields the second-order contribution of the exchange interactions to the antisymmetric spin-spin interactions. From (2.35) we learn that d^{ex}_{AB} is identical to zero when the two ions are related by an inversion center, and more generally that the symmetry rules for d^{ex}_{AB} are the same as outlined above for the magnetic contribution.

These rules have been expressed in a more general way by Bencini and Gatteschi [2.23]. Two cases can be distinguished: one in which the two paramagnetic centers are related by a symmetry element and the other where they are not. In the former the symmetry of the pair is higher than the symmetry of the individual centers, while in the latter the symmetry of the pair is identical to that of the single centers. In this case the orientation of d is determined recurring to the character table of the symmetry point group of the pair. In fact, d may be different from zero only if some of the individual d_i's are different from zero. In order to have this it is necessary that $|e_i >$ and $|g_i >$ span the same irreducible representation of the symmetry group of the molecule, since the exchange integral must be different from zero, and that a component of L_α spans the totally symmetric representation of the group in order to have $\langle e_\alpha | L_{\alpha,k} | g_\alpha \rangle \neq 0$. Considering, for example, a pair possessing C_{2v} symmetry, as shown in Fig. 2.7, it is easy to show that d must be zero, because there is no totally symmetric component of L in C_{2v} symmetry. This result is also clearly stated in Moriya's rules [2.3], but with the present approach it is easier to recognize when

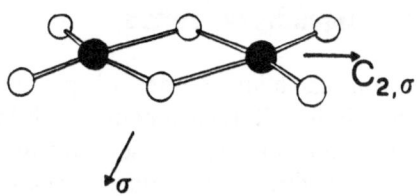

Fig. 2.7. Schematic view of a dinuclear complex of C_{2v} symmetry. After [2.14]

d is zero, not by symmetry arguments, but only because of the actual nature of the ground states. If, for instance, we consider a pair of metal ions with a z^2 ground state in C_n symmetry, the rules of Moriya demand that **d** is parallel to the symmetry axis, but since $L_z|z^2> = 0$, **d** will actually be zero in this case.

In the case in which the two centers are related by a symmetry element, it is the latter which determines the conditions under which **d** is different from zero. Since d_i is an axial vector, a **d** component different from zero can be found only in the directions in which the scalar components of the d_i's are not transformed one into the other by the symmetry elements of the pair.

The second important information contained in (2.34) is that the antisymmetric exchange contribution is proportional to Δg, rather than to $(\Delta g)^2$ as the anisotropic part. Since for orbitally nondegenerate cases, $\Delta g < 1$, the antisymmetric contribution can be fairly large. Since, however, in (2.34) a difference is present, the relative signs of $\langle g_\alpha|L_\alpha|e_{\alpha i}\rangle$ and $J(e_{\alpha i}g_\beta g_\alpha g_\beta)$ are also extremely important in determining the size of $\mathbf{d}_{AB}^{ex}$. The physical meaning of $J(e_{\alpha i}g_\beta g_\alpha g_\beta)$ is less well established as compared to $J(g_\alpha g_\beta g_\alpha g_\beta)$ and $J(e_{\alpha i}g_\beta e_{\alpha i}g_\beta)$, discussed in the previous sections, and at the moment no attempt has been made to relate the former to exchange pathways, nor are simple rules available to anticipate the extent or even the sign of this parameter.

Before closing this section it must be recalled that also other perturbations may be relevant to the general J_{AB} matrix. One is the electric quadrupole-electric quadrupole interaction, and the other is determined by vibronic effects [2.24].

The electronic quadrupole interaction is bound to the electrostatic interaction resulting from the charge distribution on one ion of the pair contributing to the electric field gradient at the other. It has an R^{-5} dependence and it increases with the increase of the orbital contribution to the ground state. Therefore, it proved to be of some importance in the analysis of the EPR spectra of lanthanide ions.

The vibronic-determined interaction has its origin in the modulation of the crystal field at one spin center A induced by phonons. In a pair the modulation at the two centers is correlated in such a way that a phonon emitted, e.g., by A, is immediately absorbed by center B. This yields a component depending on R^{-3}, which in the case of nickel (II) Tutton salt has been calculated to be of the same order as the magnetic dipolar interaction.

2.4 Biquadratic Terms

Beyond the bilinear terms it is possible to introduce also higher-order terms in the spin hamiltonian, among which biquadratic terms are the most important. Also in this case there are several different possible origins, but the most relevant are the higher order intrinsic exchange and the exchange striction effects.

The former enters naturally Anderson's theory [2.25] when it is extended to the fourth order in perturbation and physically represents the admixture into the ground state of excited states corresponding to a double excitation in the superexchange process. This process can be represented by a spin hamiltonian:

$$H = -j(S_A \cdot S_B)^2. \tag{2.36}$$

Several attempts to estimate j for different cases have been made and, although there are large discrepancies in the calculated values, there seems to be a fairly general agreement that the j/J ratio is of the order of 10^{-2} at best [2.26]. Although it is a small effect, it can be observed in the analysis of the EPR spectra of systems with large S values since the inclusion of (2.36) in the total spin hamiltonian induces variations in the S manifold splitting pattern.

The other important physical phenomenon which can give rise to a term like (2.36) is exchange striction [2.27], i.e., the change in the R distance between the two spin centers due to the exchange stabilization. In general, |J| increases on decreasing the R distance: therefore, exchange tends to bring the two spins closer but the process is not indefinite because the restoring forces oppose that. Assuming a simple Hooke's law for the restoring force yields a biquadratic form of the effective hamiltonian identical to (2.36).

2.5 Justification of the Spin Hamiltonian Formalism

The most elegant justification of the spin hamiltonian formalism has been provided by Stevens using a second quantization perturbational approach [2.28, 29]. We will try to provide here a concise illustration of the method, using the mathematical formalism as little as possible. We provide in Appendix A a short resumé of the foundations of the second quantization formalism in order to provide the readers who are not familiar with it the possibility of following the line of reasoning, although at the expenses of some rigour.

Central to Stevens treatment is a reformulation of the perturbation problem for the case of two interacting ions. Throughout the treatment the orbitals on the two spin centers are assumed to be orthogonal. The true hamiltonian appropriate to the system is denoted H, and is not further specified, except to say that it is as complete as possible. In order to perform a perturbation treatment a suitable unperturbed hamiltonian H_0 is defined, such that

$$H = H_0 + H'. \tag{2.37}$$

H_0 is not, as often assumed, simply the sum of the hamiltonians appropriate to the A and B species separately, because, if this intuitively simple procedure is followed, the unusual result is obtained that the perturbed hamiltonian has higher symmetry than the unperturbed one! One can be easily convinced that this would be the case considering that the $H_A + H_B$ hamiltonian is not invariant to the exchange of electrons between A and B, while the hamiltonian including the perturbation must necessarily be invariant to electron exchange. Therefore, H_0 is chosen according to different criteria: it must be invariant to electron exchange and it must be suitable for a perturbation treatment. These two conditions are met by the hamiltonian:

$$H_0 = \Sigma_n \Omega_n |n> <n|, \tag{2.38}$$

where $|n>$ is an eigenstate of H and the sum extends over all the states. The quantity Ω_n is defined as:

$$\Omega_n = <n|H|n>. \tag{2.39}$$

and it is the n-th eigenvalue of H or the mean of the eigenvalues taken over a group of *quasi*-degenerate levels. The energy of the unperturbed state does not need to be known: the only relevant information is that all the $|n>$ functions have the same energy with the hamiltonian H_0. It should be noted that this procedure applies to symmetric as well as to nonsymmetric A–B species. The perturbation hamiltonian is:

$$H' = H - H_o. \tag{2.40}$$

Defining the projection operators:

$$P_o = \sum_n |n_o> <n_o|; \tag{2.41}$$

$$P_i = \sum_n |n_i> <n_i|; \tag{2.42}$$

where the sums are over the ground and the excited states specified by the 0 and i indexes, respectively, it is possible to express the correction to energy up to second order by an effective hamiltonian defined as:

$$H_{eff} = P_o H' P_o + \sum_i P_o H' P_i H' P_o / \Delta_i, \tag{2.43}$$

where Δ_i is the energy difference between the ground and the excited manifold:

$$\Delta_i = E_o - E_i, \cdot \tag{2.44}$$

Using (2.43) it is possible to arrive at the required spin hamiltonian. It is at this point that second quantized operators are needed. For the reader who is not familiar with them it can be stated that second quantized operators provide a formalism for handling Slater determinants, which, as is well known, become a rather cumbersome tool for providing antisymmetrization when numerous orbitals are involved.

Let us suppose to have defined the Slater determinant appropriate to a system, for instance of three electrons, and then we wish to pass to a four-electron system: formally this can be easily done by defining an operator $a^*_{\alpha\sigma}$, called creation operator, which simply performs what we wish. This means that operating with $a^*_{\alpha\sigma}$ on the system of three electrons a new Slater determinant, 4 × 4, is obtained which differs from the previous one due to the addition of one electron in the $\alpha\sigma$ (α orbital σ spin part) orbital. Conversely, if we wish to go back to the system of three electrons we have just to define an annihilation operator, $a_{\alpha\sigma}$, which performs exactly what we want. If we define a vacuum state (a state with no electrons) we may construct from that all the Slater determinants we want by simply using all the electron creation operators we need. So, for instance, applying the operator:

$$H = a^*_{z2+}\, a^*_{yz+}\, a^*_{xz+}\, a^*_{xy+}\, a^*_{x2-y2+} \tag{2.45}$$

on the vacuum state we obtain a 5 × 5 Slater determinant in which the five d orbitals are singly occupied with spin-up.

Using the appropriate theory, one- and two-electron operators can be expressed in second quantization as:

$$H_1 = \langle r_1\sigma_1|H_1|r_2\sigma_2\rangle a^*_{r1\sigma1}\, a_{r2\sigma2}; \tag{2.46}$$

$$H_{12} = \langle r_1\sigma_1(1)r_2\sigma_2(2)|H_{12}|r_3\sigma_3(1)r_4\sigma_4(2)\rangle$$
$$a^*_{r1\sigma1}\, a^*_{r2\sigma2}\, a_{r4\sigma4}\, a_{r3\sigma3}; \tag{2.47}$$

where H_1 is a one-electron and H_{12} is a two-electron operator.

A general hamiltonian for a pair contains both one- and two-electron operators: some of them involve orbitals localized on the same center, while some involve orbitals belonging to both centers.

Considering the perturbation hamiltonian (2.43) the first-order contribution, given by $P_o H' P_o$, contains many terms, which, however, keep the number of electrons on center A and B respectively fixed. This means that if an electron is annihilated by, e.g., a_i, another one must be created at the same site by, e.g. a^*_j. Pairs of operators $a^*_j a_i$ do not necessarily commute. Among two-electron operators sets of the type $a^*_j a_i b^*_k b_l$, where k and l are localized on B while j and i are located on A, are present. It is easy to prove that pairs of operators on different sites necessarily commute. It is this property which can be used for the comparison of the spin hamiltonian formalism. Indeed, in the latter case the hamiltonian is expressed as a sum of terms, some of which include couples of spin operators on the same center, which do not necessarily commute, and couples of spin operators on different centers, which do necessarily commute. These commutation rules are analogous to those of angular momentum operators, and indeed a general expression is available which relates angular momentum and second quantized operators. This is given in Appendix A. Through this approach it is possible to substitute to the true hamiltonian, expressed by the second quantized operators, a spin hamiltonian, expressed through angular momentum operators.

The first-order perturbation hamiltonian is:

$$P_o H' P_o = \sum |n_o> <n_o|H - H_o|n_o> <n_o|. \tag{2.48}$$

This can be split into two components, one collecting the parts of the hamiltonian which operate on the electrons of the same center, either A and B, and the other which contains operators which mix the functions of the two centers. In the second quantized form these components are given by:

$$
\begin{aligned}
H_L = & \sum_{\alpha\alpha'\sigma} A_{\alpha\alpha'} a^*_{\alpha\sigma} a_{\alpha'\sigma} + \\
& \tfrac{1}{2} \sum_{\alpha\alpha'\alpha''\alpha'''\sigma\sigma'} B_{\alpha\alpha'\alpha''\alpha'''} a^*_{\alpha\sigma} a^*_{\alpha'\sigma'} a_{\alpha''\sigma'} a_{\alpha'''\sigma} \\
& + \sum_{\beta\beta'\sigma} A_{\beta\beta'} a^*_{\beta\sigma} a_{\beta'\sigma} \\
& + \tfrac{1}{2} \sum_{\beta\beta'\beta''\beta'''\sigma\sigma'} B_{\beta\beta'\beta''\beta'''} a^*_{\beta\sigma} a^*_{\beta'\sigma'} a_{\beta''\sigma'} a_{\beta'''\sigma}
\end{aligned}
\tag{2.49}
$$

$$H_p = \sum_{\alpha\alpha'\beta\beta'\sigma\sigma'} B_{\alpha\beta'\beta\alpha'} a^*_{\alpha\sigma} a_{\alpha'\sigma'} a^*_{\beta'\sigma'} a_{\beta\sigma}, \tag{2.50}$$

where H_L contains all of the ligand field interaction and single ion electron repulsion which is contained in H_o. $A_{\delta\delta'}$ and $B_{\delta\delta'\delta''\delta'''}$ are defined as matrix elements of one- and two-electron operators, respectively, according to (2.46) and (2.47), respectively. The indices σ refer to the spin coordinates and α and β to the orbitals centered on A and B, respectively.

The second-order term in (2.43) can be rewritten as:

$$H_K = \sum_{\alpha\beta\alpha'\beta'\sigma\sigma'} A_{\alpha'\beta'\sigma\sigma'} A^*_{\alpha\beta} a^*_{\alpha\sigma} a_{\alpha'\sigma'} a^*_{\beta'\sigma'} a_{\beta\sigma}/\Delta_i. \tag{2.51}$$

Retaining only the terms in which one electron is transferred from A to B or vice-versa, the spin hamiltonian becomes:

$$H_{eff} = \sum_{\alpha\alpha'\beta\beta'\sigma\sigma'} [(2/U) A_{\alpha\beta} A^*_{\alpha\beta} - B_{\alpha\beta'\beta\alpha'}] a^*_{\alpha\sigma} a_{\alpha'\sigma'} a^*_{\beta'\sigma'} a_{\beta\sigma}, \tag{2.52}$$

where U has exactly the same meaning as in Anderson's theory.

Using the relation between angular momentum and second quantized operators given in Appendix A, H_{eff} reduces to:

$$H_{eff} = \sum_{\alpha\beta} \{2/U \, A^2_{\alpha\beta} - B_{\alpha\beta\beta\alpha}\} \{1/2 + 2S_\alpha \cdot S_\beta\}. \tag{2.53}$$

Neglecting the scalar terms, and passing from one-electron to all-electron spin operators, one finally finds:

$$H = J\, S_A \cdot S_B \tag{2.54}$$

with

$$J = 2f \sum_{\alpha\beta} [(2/U)A_{\alpha\beta}^2 - B_{\alpha\beta\beta\alpha})], \tag{2.55}$$

where f is a positive proportionality factor dependent on the electron configuration. It can be simply expressed as:

$$f = 1/(4S_A S_B) \tag{2.56}$$

and the term in parenthesis in (2.55) can be indicated as $J_{\alpha\beta\beta\alpha}$, i.e., as the exchange interaction involving the α orbital on A and the β orbital on B. $(2/U)A_{\alpha\beta}^2$ corresponds to Anderson's kinetic exchange and $B_{\alpha\beta\beta\alpha}$ corresponds to potential exchange. Therefore, (2.54) corresponds exactly to the Heisenberg exchange hamiltonian. No anisotropy is present in (2.53) because in the true hamiltonian we did not include either spin-orbit or magnetic terms.

2.6 Exchange Involving Degenerate States

Up to now we have always assumed that the interacting states are orbitally non degenerate and have shown that the true hamiltonian can be replaced by an effective spin hamiltonian. The question at hand now is: how can the case of orbital degeneracy be addressed. The formalism developed in the previous section now offers us the possibility to answer this question.

Let us consider first the case when only one of the centers has orbital degeneracy. Starting from the hamiltonian (2.52), we have to replace the second quantized operators by angular momentum operators. Let us assume that the orbital degeneracy of one of the two centers, e.g. A, is associated with a cubic T term, either T_1 or T_2. It is a well-known property of these states that they behave as a spherical angular momentum state characterized by $L = 1$, multiplied by a proportionality constant [2.30]. Therefore, the α orbitals can be labeled as $L = 1$ orbitals, with components x, y, and z. When we substitute second quantized operators, using the the same formalism which we used for the orbitally nondegenerate case, beyond spin operators we will need also orbital angular momentum operators. After some passages which are worked out in Appendix A, we find that (2.52) becomes:

$$
\begin{aligned}
H_{eff} = -2\,[&(1 - L_z^2)\,J_{zz} + (1/2\,L_z^2 - 1/2\,L_x^2 - 1/2\,L_y^2)\,J_{xx} + \\
& + (1/2\,L_z^2 + 1/2\,L_x^2 - 1/2\,L_y^2)\,J_{yy} - (L_x L_z + L_z L_x)\,J_{zx} - \\
& - (L_y L_z + L_z L_y)\,J_{yz} - (L_x L_y + L_y L_x)\,J_{xy}\,]\,(S_A \cdot S_B),
\end{aligned} \tag{2.57}
$$

where $J_{ij} = -2/U\, A_i A_j^* - 2B_{ji}$.

This hamiltonian operates on a given basis providing in an effective way the exchange interaction between A and B. The presence of orbital angular momentum operators in (2.57) makes the hamiltonian anisotropic, even without including dipolar terms. In other words, the presence of orbital degeneracy in one of the ground states causes the anisotropic exchange interaction to appear in the effective hamiltonian in first order, rather than in higher perturbation order as in orbitally nondegenerate cases.

In order to fully express the energy levels of the pair it is necessary to add also the low symmetry ligand field components to (2.57), which have been omitted in (2.52), and the spin-orbit coupling operator. If we assume that the energies of the Russel-Saunders terms of the two centers are well separated from each other, then a suitable form of the latter can be:

$$H_{so} = \lambda_A \, L_A \cdot S_A + \lambda_B \, L_B \cdot S_B, \tag{2.58}$$

where λ_A and λ_B are scalars which depend on the spin center and on the nature of the ground term. The overall hamiltonian is then applied to the chosen basis and the matrix can be diagonalized, yielding the required eigenvalues and eigenvectors.

It is instructive to work out a simple case, in order to appreciate the operation of the effective hamiltonian. Let us consider a pair of coupled low spin iron (III) and copper(II) [2.31]. This example may be relevant, for instance, to the description of the electronic structure of the cyanide form of oxidized cytochrome oxidase, in which an iron(III) porphyrin is coupled to a copper(II) ion in a so far ill-characterized coordination environment [2.32]. Iron (III) is a d^5 ion, which in an octahedral low spin configuration has a $^2T_{2g}$ ground state. In the actual case the symmetry is lower than cubic so that the concurrent effect of the ligand field and spin-orbit coupling yields three Kramers doublets, whose overall splitting is in the range of $1000-2000$ cm^{-1}. The lowest energy doublet can be expressed [2.33] as:

$$|\Phi^+ > = \alpha|yz^+ > - i\,\beta|xz^+ > - \delta|xy^- > \tag{2.59}$$

$$|\Phi^- > = -\alpha|yz^- > - i\,\beta|xz^- > - \delta|xy^+ > \tag{2.60}$$

where α, β, and δ are coefficients.

A convenient set of basis functions for the Fe-Cu pair can be obtained by the direct product of $\{\Phi^+, \Phi^-\}$ with the two spin orbitals of copper $\{\phi^+, \phi^-\}$ which have the same orbital composition. Operating on this 4×4 basis with the (2.57–58) hamiltonian yields a matrix which is given in Table 2.2. If we compare the matrix elements of Table 2.2 with those for the usual spin hamiltonian for two coupled $S = \frac{1}{2}$ spins we find a relation between the various parameters (Table 2.3). It is apparent that the spin hamiltonian approach is still possible in this case, but that the assumption that J is larger than all the other terms is untenable, because all of them have comparable values in Table 2.3. In the case of

Table 2.2. Hamiltonian matrix for coupled low-spin iron (III)-copper (II) pairs

$H_{11} = -\alpha^2/2\,J_{xx} - \beta^2/2\,J_{yy} + \delta^2/2\,J_{zz}$
$H_{12} = \alpha\delta J_{xz} + i\beta\delta J_{yz}$
$H_{13} = \alpha\delta J_{xz} - i\beta\delta J_{yz}$
$H_{14} = -\delta^2 J_{zz}$
$H_{22} = \alpha^2/2\,J_{xx} + \beta^2/2\,J_{yy} - \delta^2/2\,J_{zz}$
$H_{23} = \alpha^2 J_{xx} - \beta^2 J_{yy} - 2i\alpha\beta\,J_{xy}$
$H_{24} = -\alpha\delta J_{xz} + i\beta\delta J_{yz}$
$H_{33} = H_{22}$
$H_{34} = -\alpha\delta J_{xz} - i\beta\delta J_{yz}$
$H_{44} = H_{11}$

Table 2.3. Relations between the bilinear spin hamiltonian and the orbital parameters in low spin iron (III)-copper (II) pairs

$J = 2/3\,\alpha^2 J_{xx} - 2\beta^2 J_{yy} + 2/3\delta^2\,J_{zz}$
$D_{xx} = 4/3\,\delta^2 J_{zz} - 8/3\,\alpha^2 J_{xx}$
$D_{yy} = 4/3\,\delta^2 J_{zz} + 4/3\,\alpha^2 J_{xx}$
$D_{zz} = -8/3\,\delta^2 J_{zz} + 4/3\,\alpha^2 J_{xx}$
$D_{xy} = D_{yz} = 0;\ D_{xz} = 4\alpha\delta J_{xz}$
$d_x = 4\beta\delta J_{yz};\ d_y = 0;\ d_z = -4\alpha\beta J_{xy}$

orthorhombic symmetry all the J_{ij} parameters vanish (i, j being cartesian components), and the effective spin hamiltonian can be rewritten in the simple form [2.34]:

$$H = \Sigma_i\,J_{ii}\,S_A \cdot S_B. \tag{2.61}$$

The case of two orbitally degenerate ions has been considered, for instance, for the characterization of the magnetic properties of $Ti_2 X_9^{3-}$ (X = Cl, Br) [2.35]. The two ions are octahedrally coordinated by six halogen ions. Three of the latter are bridging in such a way that the two octahedra share a face (Fig. 2.8.) The two titanium(III) ions have a ground $^2T_{2g}$ state in octahedral symmetry, which is split by the actual trigonal symmetry of the dimer. In this case the ligand field terms in (2.49) and (2.51) must be included, adding to the total hamiltonian:

$$H_{trig} = D/3\,(2 - 3L_{A,z^2}), \tag{2.62}$$

where D is defined as the difference:

$$D = A_{oo} - A_{11}. \tag{2.63}$$

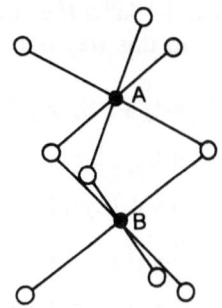

Fig. 2.8. Scheme of the structure of $Ti_2Cl_9^{3-}$ pairs. After [2.35]

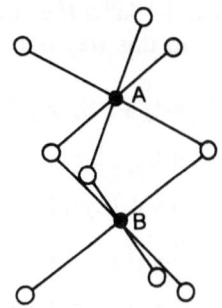

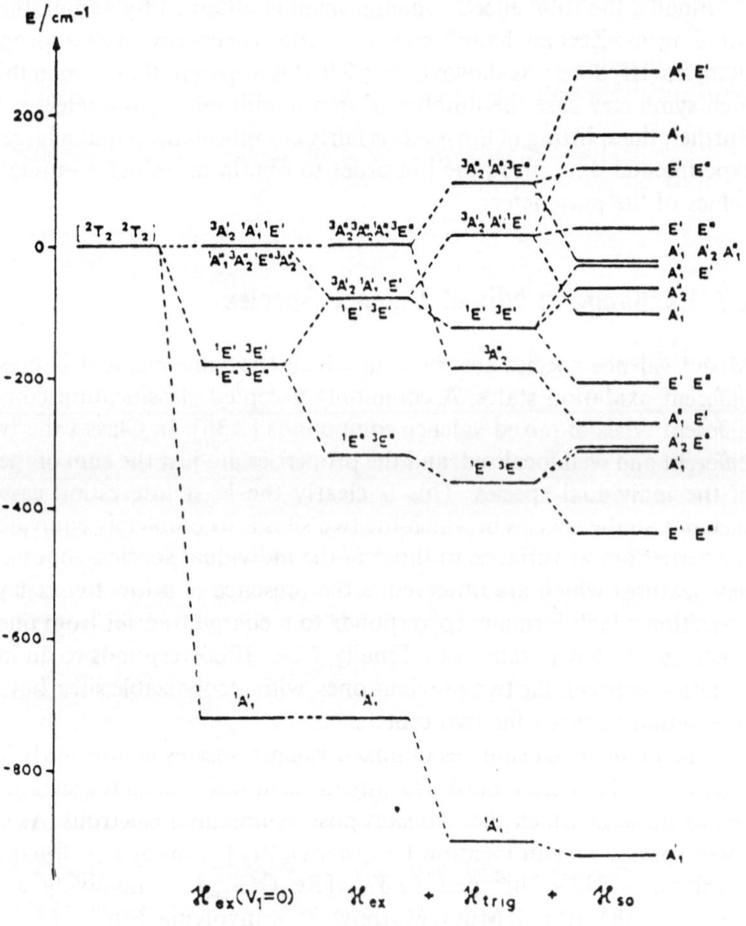

Fig. 2.9. Calculated energy splitting pattern for $Ti_2X_9^{3-}$. After [2.35]

0 and 1 label the components of the t_{2g} orbitals.

In this way the effective hamiltonian becomes:

$$
\begin{aligned}
H_{ex} = (\tfrac{1}{2} + 2\,\mathbf{S_A \cdot S_B})\,\{V_1(L_{A,z}L_{B,z} + L_{A,z}^2 L_{B,z}^2 + \tfrac{1}{2}(L_{A,+}^2 L_{B,-}^2 \\
+ L_{A,-}^2 L_{B,+}^2)) + 2V_0(1 - L_{A,z}^2 - L_{B,z}^2 + L_{A,z}^2 L_{B,z}^2) \\
+ \tfrac{1}{2}V_2(L_{A,+}L_{B,-} + L_{A,-}L_{B,+} + (L_{A,z}L_{A,+} + L_{A,+}L_{A,z}) \\
(L_{B,z}L_{B,-} + L_{B,-}L_{B,z}) + (L_{A,z}L_{A,-} + L_{A,-}L_{A,z}) \\
(L_{B,z}L_{B,+} + L_{B,+}L_{B,z}))\}
\end{aligned}
\tag{2.64}
$$

where

$$
V_0 = A_{oo}^2/U; \quad V_1 = A_{11}^2/U; \quad V_2 = -A_{oo}A_{11}/U.
\tag{2.65}
$$

Finally, the total effective hamiltonian is obtained by adding the spin-orbit coupling and Zeeman hamiltonians. A pattern of energy levels appropriate to the titanium(III) dimers is shown in Fig. 2.9. It is apparent that even in this relatively high symmetry case the number of spin hamiltonian parameters is fairly high. Further, the splitting of the levels is fairly complicated, so that a large number of experimental data is required in order to obtain meaningful estimations of the values of the parameters.

2.7 Exchange in Mixed Valence Species

Mixed valence species are those in which the same element is present in two different oxidation states. A commonly accepted classification considers three different types of mixed valence compounds [2.36]. In Class I the two sites are different and well localized, and the properties are just the sum of the properties of the individual species. This is clearly the least interesting case. Class III includes all the species in which the two sites are completely equivalent, and the properties are at variance to those of the individual species. Among the typical new features which are observed is the presence of a low frequency electronic transition which formally corresponds to a charge transfer from one site to the other (intervalence transition). Finally, Class II corresponds to an intermediate situation between the two previous ones, with recognizable sites, but with strong interaction between the two centers.

The number of examples of mixed valence species is now fairly large, but of interest for the present book are only those in which both the sites are magnetic, i.e., to those in which the two sites possess unpaired electrons. As examples of these species we can mention $[Ni_2(napy)_4 Br_2]^+$, (napy = 1, 8-naphthyridine), involving a $Ni^I - Ni^{II}$ pair [2.37], $[Re_2OCl_{10}]^{3-}$, involving a $Re^{IV} - Re^V$ species [2.38]; $[(bpy)_2 MnO_2 Mn(bpy)_2]^{3+}$, involving $Mn^{III} - Mn^{IV}$ [2.39]; reduced 2Fe–2S ferredoxins [2.40], etc.

The theoretical basis for the description of the electronic properties of mixed valence binuclear species considers the interplay of vibrational and electronic degrees of freedom of the systems [2.41]. The reason why vibrational motion must be taken into account is that, in the limit of equivalent sites, we are in the presence of two electronically degenerate states, ϕ_a on site A, ϕ_b on site B. As is always the case when electronic degeneracy is present the Born-Oppenheimer approximation breaks down, because it is only valid on the assumption that the energy difference of the electronic levels is much larger than the vibrational splitting.

The potential energy surfaces for a dinuclear species which would be symmetric in the absence of vibronic interaction are depicted in Fig. 2.10 vs a vibrational coordinate q, which describes the deviation of the two coupled systems from the symmetric situation. The physical meaning of q should appear clear from Fig. 2.11. When q = 0, the two sites are identical; when q > 0, site A has longer distances than site B; while when q < 0, the reverse is true.

The upper curve in Fig. 2.10 denotes that two equivalent minima are present in the absence of electron coupling, one at q = −f and the other at q = +f. Therefore, f is a measure of the difference, on the two centers, of the equilibrium displacement along the totally symmetric coordinate. If electron interaction is included, the two parabolas split into the curves depicted in Fig. 2.10. Again two minima are present, separated by the activation energy. Indicating with β the electron interaction between the two centers, the activation energy is given by:

$$E_a = \tfrac{1}{2}f^2 + \tfrac{1}{2}\beta^2/f^2 - |\beta| \qquad (2.66)$$

for $|\beta|/f^2 \leq 1$. If E_a is large, a situation which obtains when f^2 is large, and

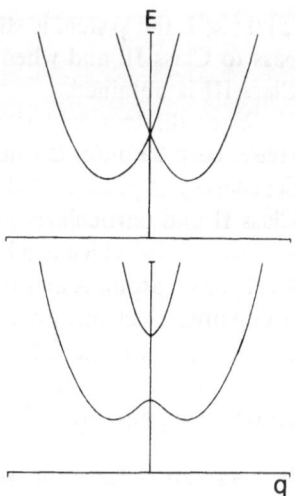

Fig. 2.10. Potential energy surface for a mixed valence species: *upper* no electron interaction; *lower* with electron interaction

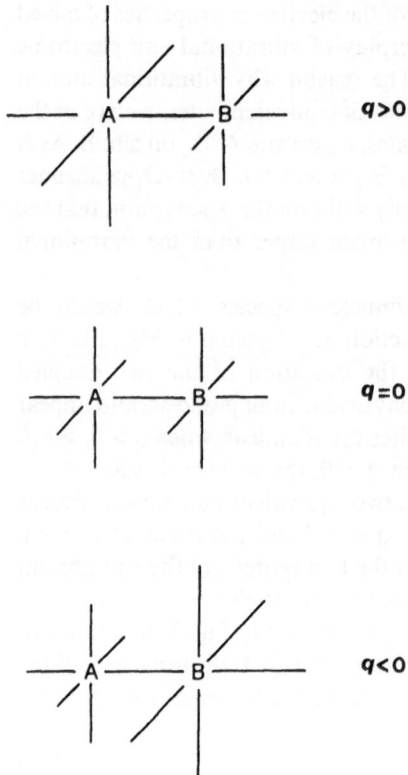

$q > 0$

$q = 0$

$q < 0$

Fig. 2.11. Scheme of the geometry of a dinuclear species under the effect of a vibrational coordinate q

$|\beta|/f^2 \ll 1$, the system is strongly localized (Class I). When $|\beta|/f^2$ increases, we pass to Class II, and when $|\beta|/f^2 \geq 1$ the activation energy goes to zero, and Class III is obtained.

When the two oxidation states are magnetic, it can be anticipated that the Heisenberg hamiltonian must be substantially valid for Class I, unless orbital degeneracy is present in the two sites, but complications must be foreseen for Class II and particularly for Class III systems. Let us consider [2.42] a mixed valence species which can be described by a set of orthogonal molecular orbitals localized on atoms A and B, respectively. The ground configuration may contain n unpaired electrons on site A and $n-1$ on site B. In order to define a spin hamiltonian, we can follow a procedure similar to that of Sect. 2.5–2.6, using second quantized operators. A hamiltonian appropriate for the system is the Hubbard hamiltonian [2.43]:

$$H = \Sigma_{\alpha\sigma} a^*_{\alpha\sigma} a_{\beta\sigma} A_{\alpha\beta} + \Sigma_{\beta\sigma} a^*_{\beta\sigma} a_{\alpha\sigma} A_{\beta\alpha} +$$
$$\Sigma_{\alpha\sigma\sigma'} B_{\alpha\alpha\alpha\alpha} a^*_{\alpha\sigma} a_{\alpha\sigma} a^*_{\alpha\sigma'} a_{\alpha\sigma'} + \tag{2.67}$$

$$\Sigma_{\beta\sigma\sigma'} B_{\beta\beta\beta\beta} a^*_{\beta\sigma} a_{\beta\sigma} a^*_{\beta\sigma'} a_{\beta\sigma'},$$

44

where the A's are the one electron and the B's the repulsion integrals between two electrons in the same localized orbital. The orbitals α and β are differentiated according to the center on which they are localized. The A integrals are of fundamental importance in the treatment. The first $n-1$ ($\alpha, \beta = 1 \ldots n-1$) participates in the exchange phenomenon, while the n-th governs the electron transfer between orbitals α_n and β_n. Let us take into consideration a simple example to understand this point. Suppose we have simply two orbitals on each site (Fig. 2.12). The left-hand side, taken alone, might describe the ground configuration of a heterodinuclear pair, such as for instance $Ni^{II} - Cu^{II}$. In this case the operators which allow the passage of one electron from α_n to β_n or vice versa couple the ground configuration with excited ones. Therefore, the corres-ponding terms will appear in H_{eff} defined in (2.52) only in the second-order term because in the first-order term they are annihilated by the projection operator. In the mixed valence case, on the other hand, the ground configuration comprises both the left and the right cases, therefore, the transfer of one electron from α to β, or vice versa, is allowed in first order. The effective hamiltonian, with a small change in notation, takes the form:

$$H_{eff} = \Sigma_\sigma A_{nn}(a^*_{\alpha n\sigma} a_{\beta n\sigma} + a^*_{\beta n\sigma} a_{\alpha n\sigma}) + J[S_A \cdot S_B],\tag{2.68}$$

where
$$S = n/2 + (n-1)/2 \tag{2.69}$$

with n the maximum number of unpaired electrons on one center. The second term in (2.69) is the usual effective exchange hamiltonian, but now also a one-electron term is present. The effect of the one-electron term will be that of favoring the high spin configurations. This can be understood considering that the passage of an electron from A to B on going from the configuration of Fig. 2.12 (left) to the configuration of Fig. 2.12 (right) will be the most favorable energetically, because the passage will occur without any spin flip. Therefore, the possibility of easy passage of one electron from α to β and vice versa will stabilize the high spin state on both sites. This phenomenon, peculiar to the mixed valence systems, has been termed double exchange.

The eigenvalues of the first term in (2.68) are given by:

$$E_\pm(S) = \pm(2S+1)/2n \, A_{nn}.\tag{2.70}$$

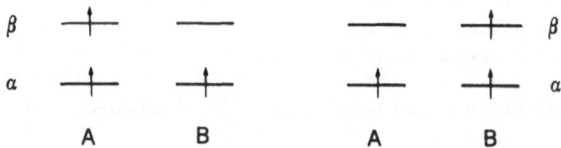

Fig. 2.12. Ground configuration of a dinuclear species with two half-filled orbitals

45

This gives a spectrum of levels with the one of maximum multiplicity lying lowest.

When all the terms in (2.68) are taken into account, and also a vibronic hamiltonian is added, the expression for the energies becomes:

$$E_\pm(S) = \tfrac{1}{2}f^2 q^2 \pm [(2S+1)^2/(4n^2)\,A^2 nn + q^2]^{1/2} + \tfrac{1}{2}J[S(S+1)$$
$$- S_A(S_A+1) - S_B(S_B+1)] \tag{2.71}$$

and the activation energy transforms to:

$$E_a(S) = \tfrac{1}{2}f^2 + \tfrac{1}{2}[\beta^2(2S+1)^2]/(4f^2 n^2) - \tfrac{1}{2}|\beta|\,(2S+1) \tag{2.72}$$

where we have transformed A_{nn} to β to compare directly (2.72) with (2.66). From this comparison we learn that in the case of antiferromagnetic exchange there is a competition between double exchange, which favors the high spin state and exchange, and the activation energy is increased compared to the case of no exchange. In other terms the antiferromagnetic interaction slows down the transfer rate.

The energy of the minima can be expressed as:

$$E_{min}(S) = -\tfrac{1}{2}f^2 - \tfrac{1}{2}\beta^2/f^2 - \tfrac{1}{2}(J - \beta^2/(n^2 f^2))\,[(n^2 - (2S+1)^2/4] \tag{2.73}$$

which shows that vibronic coupling adds a negative contribution to the exchange constant.

References

2.1 Martin R L (1968) In: Ebsworth EAV, Maddock AG, Sharpe AG (eds) New pathways in inorganic chemistry., Cambridge University Press, London, p 175

2.2 Silver BL (1976) Irreducible tensor methods Academic, New York

2.3 Moriya (1963) In: Rado GT, Suhl H (eds) Magnetism, Academic New York, Vol 1 p 85

2.4 Anderson OP, Kuechler T S (1980) Inorg. Chem. 19: 1417

2.5 Bencini A, Benelli C, Gatteschi D, Zanchini C (1984) J. Am. Chem. Soc. 106: 5813

2.6 Geller M (1963) J. Chem. Phys. 39: 84

2.7 Matcha RL, Kern CW, Schrader DM (1969) J. Chem. Phys. 51: 2152

2.8 Matcha RL, Kern CW (1970) Phys. Rev. Lett. 25: 981

2.9 Van der Waals JH, ter Maten G (1964) Mol. Phys 8: 301

2.10 Boorstein SA, Goutermann M (1965) J. Chem. Phys. 42: 3070

2.11 McWeeny R (1961) J. Chem. Phys. 34: 399

2.12 de Jager G, de Jong J, Mac Lean C, Ros P (1977) Theoret. Chim. Acta 20: 57

2.13 Brustolon M, Pasimeni L, Corvaja C (1973) Chem. Phys. Letters 21: 194

2.14 Belford RL, Chasteen ND, So H, Tapscott RE (1969) J. Am. Chem. Soc. 91: 4675

2.15 Bertini I, Luchinat C (1986) NMR of paramagnetic molecules in biological systems Benjamin/Cummings: Menlo Park, CA

2.16 Gribnau M (1988) Ph.D. Thesis, Nijmegen

2.17 Kanamori J (1963) In: Rado TG, Suhl H (eds) Magnetism. Academic, New York, vol 1 p 161

2.18 Bencini A, Gatteschi D, Zanchini C, Haase W (1985) Inorg. Chem. 24: 3485

2.19 Charlot MF, Journaux Y, Kahn O, Bencini A, Gatteschi D, Zanchini C (1985) Inorg. Chem. 25: 1060

2.20 Bencini A, Gatteschi D, Zanchini C (1985) Inorg. Chem. 24: 700

2.21 Smith TD, Pilbrow JR (1974) Coord. Chem. Rev. 13: 173

2.22 Keijzers CP (1986) In: Electron spin resonance The Royal Society of Chemistry, London, vol 10B p 1

2.23 Bencini A, Gatteschi D (1982) Mol. Phys. 47: 161

2.24 Owen J, Harris EA (1972) In: Geschwind S (ed.) Electron paramagnetic resonance. Plenum, New York, p 427

2.25 Anderson PW (1963) In: Seitz F, Turnbull (eds) Solid state physics. New York, vol 14 p 99

2.26 Huang NL, Orbach R (1964) Phys. Rev. Letters 12: 275

2.27 Kanamori J (1957) Progr. Theor. Phys. (Japan) 17: 197

2.28 Stevens KWH (1976) Phys. Rep. 24c, 1

2.29 Stevens KWH (1985) In: Willett RD, Gatteschi D, Kahn O (eds.), Magneto-structural correlations in exchange coupled systems. Reidel, Dordrecht p 105

2.30 Griffith JS (1961) The theory of transition metal ions Cambridge University Press

2.31 Bencini A, Gatteschi D, Zanchini C (1985) Mol. Phys. 56: 97

2.32 Malmstrom BG (1980) In: (1980) In: Spiro TG (ed) Metal ion activation of dioxygen., Ed; Wiley, New York p 181

2.33 Palmer G (1979) In: Dolphin D (ed) The Porphyrins; Academic press, New York, p 313

2.34 Kahn O, Toller P, Coudanne H. (1979) Chem. Phys. 42: 355

2.35 Leuenberger B, Güdel HU (1984) Mol. Phys. 51: 1

2.36 Robin MB, Day P (1967) Adv. Inorg. Chem. Radiochem. 10: 248

2.37 Sacconi L, Mealli C, Gatteschi D. (1974) Inorg. Chem. 13: 185

2.38 Campbell RJ, Clark R J M (1978) Mol. Phys. 36: 1133

2.39 Cooper SR, Calvin M (1977) J. Am. Chem. Soc. 99: 6623

2.40 Mascharak PK, Papaefthymiou GC, Frankel RB, Holm RH (1981) J. Am. Chem. Soc. 103: 6110

2.41 Wong KY, Schatz PN (1981) Progr. Inorg. Chem. 28: 369

2.42 Girerd J-J (1983) J. Chem. Phys. 79: 1766

2.43 Hubbard J (1963) Proc. R. Soc. London Ser. A 276: 238

3 Spectra of Pairs

3.1 The Spin Hamiltonian for Interacting Pairs

The spin hamiltonian appropriate to describe the EPR spectra of interacting pairs can be written as the sum of the spin hamiltonians appropriate for the interpretation of the EPR spectra of the individual spin centers and the spin hamiltonian describing the exchange interaction outlined in the previous chapter. A convenient form of this hamiltonian is:

$$H = \mu_B \mathbf{B} \cdot \mathbf{g}_A \cdot \mathbf{S}_A + \Sigma_k \mathbf{I}^k \cdot \mathbf{A}_A^k \cdot \mathbf{S}_A + \mathbf{S}_A \cdot \mathbf{D}_A \cdot \mathbf{S}_A + \tag{3.1}$$

$$+ \mu_B \mathbf{B} \cdot \mathbf{g}_B \cdot \mathbf{S}_B + \Sigma_k \mathbf{I}^k \cdot \mathbf{A}_B^k \cdot \mathbf{S}_B + \mathbf{S}_B \cdot \mathbf{D}_B \cdot \mathbf{S}_B + \tag{3.2}$$

$$+ J_{AB} \mathbf{S}_A \cdot \mathbf{S}_B + \mathbf{S}_A \cdot \mathbf{D}_{AB} \cdot \mathbf{S}_B + \mathbf{d}_{AB} \cdot \mathbf{S}_A \times \mathbf{S}_B, \tag{3.3}$$

where $\mathbf{g}_I$, $\mathbf{A}_I^k$ and $\mathbf{D}_I$ are the Zeeman, hyperfine (superhyperfine), and zero field splitting tensors of spin center I (the other symbols have been defined in the previous chapters). The sums over k extend to all the nuclei: a given nucleus k will therefore appear both in (3.1) and (3.2), although in general the hyperfine interaction will be described by different tensors.

Other terms describing smaller interactions such as nuclear Zeeman, nuclear quadrupole, higher order spin interactions have been neglected in (3.1–3).

This form of the spin hamiltonian is derived from the physical consideration that for the weak exchange interactions described by (3.3), the eigenstates of the dinuclear system can be approximately described by product kets $|S_A \Pi_k I_A^k S_B \Pi_k I_B^k\rangle = |S_A \Pi_k I_A^k\rangle |S_B \Pi_k I_B^k\rangle$ which form a good basis for the description of the ground states of A and B, respectively or, perhaps better, by product kets $|ab\rangle = |a\rangle |b\rangle$ where $|a\rangle$ and $|b\rangle$ are the eigenvectors of (3.1) and (3.2), respectively. Usually the $|a\rangle$ and $|b\rangle$ kets are expressed as linear combinations of $|S_A \Pi_k I_A^k\rangle$ and $|S_B \Pi_k I_B^k\rangle$, respectively.

In order to analyze the EPR spectra of interacting pairs using (3.1–3) we must determine the transition fields to be compared to the experimental ones as a function of all the spin hamiltonian parameters (g's, J, D's, . . .). The direct procedure requires the calculation of the representation matrix of H in a product basis $\{S_A, \Pi_k I_A^k\} \times \{S_B, \Pi_k I_B^k\}$. By direct diagonalization of this matrix we obtain the eigenvalues and eigenvectors corresponding to a given set of spin hamiltonian parameters and for a fixed value of the magnetic field from which it is possible to compute the transition frequencies and intensities. The whole procedure must then be repeated by varying the spin hamiltonian parameters until a reasonable agreement is obtained between the computed and observed transition frequencies. It is apparent that even neglecting hyperfine coupling this procedure requires long computer time.

Another possible approach is the so-called eigenfield method [3.1-2], which gives the resonance fields for each setting of the spin hamiltonian parameters as solutions of a generalized eigenvalue equation. The solution of this equation requires the diagonalization of an N^2 matrix. Since the computer time needed for matrix diagonalization increases roughly as $(N^2)^3$, the eigenfield method becomes unworkable as N increases. It seems therefore appropriate to look for approximate solutions which are valid in some limit cases and to reserve the accurate calculations only for desperate situations in which the approximate methods cannot be applied. Furthermore, the approximate methods often allow one to obtain analytical solutions which greatly help experimentalists in rationalizing the EPR spectra.

The most common approximation is to consider the isotropic exchange as the leading term in (3.1-3). J_{AB} can actually take any value usually ranging from a few wave numbers to hundreds of wave numbers both in transition metal complexes or in radical ion pairs. For normal operating frequencies the Zeeman energy is in the range $0.3-1.2$ cm^{-1}, hyperfine interactions are normally much smaller than this, and anisotropic exchange for orbitally nondegenerate states hardly exceeds 1 cm^{-1}, usually being much smaller. The single ion zero field splitting can be fairly large in couples involving transition metal ions even in the case of orbitally nondegenerate ground states. For instance, the zero field splitting measured in octahedral high spin nickel(II) or tetrahedral high spin cobalt(II) complexes is in the range 10^0-10^1 cm^{-1} [3.3].

In the following we will solve first the hamiltonian (3.1-3) assuming that J_{AB} is larger than all the other terms in the spin hamiltonian (strong exchange limit). Later we will show how to extend the formalism in the case of small J_{AB} values (weak exchange limit) and in the last section of this chapter we will present solutions in some limiting cases in which anisotropic interactions, including single ion zero field splitting, are considerably larger than the other terms in the spin hamiltonian.

3.2 Spin Levels in the Strong Exchange Limit

In the strong exchange limit we consider the effect of the other operators in (3.1-3) as a perturbation on the eigenvalues of $J_{AB}S_A \cdot S_B$. Since the z component, S_Z, of the total spin operator $S = S_A + S_B$ and $S^2 = S \cdot S$ commute with $J_{AB}S_A \cdot S_B$, the eigenstates of S^2, with eigenvalues $S(S+1)$, and of S_Z, with eigenvalues $-S \leq M \leq S$, are also eigenstates of $J_{AB}S_A \cdot S_B$. The total spin quantum numbers, S, which follow the vector addition relation

$$|S_A - S_B| \leq S \leq S_A + S_B \tag{3.4}$$

and the M values can be used to label the energy levels of $J_{AB}S_A \cdot S_B$. The isotropic exchange interaction removes the degeneracy of the eigenvalues of S^2 and originates a number of states whose energies are given by

$$E(S) = \tfrac{1}{2}J_{AB}[S(S+1) - S_A(S_A+1) - S_B(S_B+1)], \tag{3.5}$$

each state being $(2S+1)$-fold degenerate. The energy differences between adjacent states are:

$$E(S) - E(S-1) = SJ_{AB}, \tag{3.6}$$

The energy levels arising from the exchange interaction between two different pairs, one with $S_A = S_B = 1/2$ and the other with $S_A = S_B = 5/2$, are shown in Fig. 3.1.

S_z and S^2 do not form a complete set of commuting observables and S_A^2 and S_B^2 must be added to complete the set. Any eigenvalue of $J_{AB}S_A \cdot S_B$ should thus be written as $|S_A S_B SM\rangle$. In the following we will generally omit the S_A and S_B labels when there is no ambiguity on the spin states S_A and S_B. The $|SM\rangle$ states can be easily built in the product space according to

$$|SM\rangle = \Sigma_{m_1, m_2}\langle S_1 S_2 SM | S_1 S_2 m_1 m_2\rangle | S_1 S_2 m_1 m_2\rangle, \tag{3.7}$$

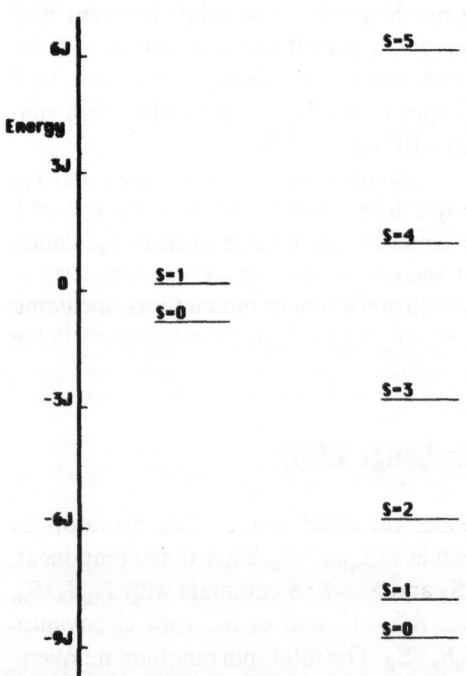

Fig. 3.1. Energy levels arising from the exchange interaction between two spins 1/2 (*left*) and 5/2 (*right*). The levels are labeled according to the total spin quantum number S

50

where $\langle S_1 S_2 SM | S_1 S_2 m_1 m_2 \rangle$ are the vector coupling or Clebsch–Gordan coefficients relative to the coupling of the S_1 and S_2 vector operators. They are defined in Appendix B.

In the strong isotropic exchange limit the energy separation between adjacent levels given by (3.6) will be assumed to be much larger than $h\nu$ and no direct transition between states of different S will be observable. The EPR spectra are just the superposition of the spectra observed for the different total spin states which are thermally populated at the temperature of the experiment. The spectra can thus be phenomenologically described by one or more S spin hamiltonians of the form

$$H_S = \mu_B B \cdot g_s \cdot S + S \cdot D_s \cdot S + \Sigma_k I^k \cdot A_s^k \cdot S, \tag{3.8}$$

instead of (3.1–3). It is now our purpose to show how g_s, D_s, and A_s^k can be expressed as linear combinations of the individual spin centers and exchange spin hamiltonian parameters appearing in (3.1–3). The reader who is not interested in the derivation can skip the next pages and pass directly to the results which are given in (3.20–23).

It is convenient to transform the spin hamiltonian (3.1–3) by substituting

$$S = S_A + S_B; \tag{3.9}$$

$$V = S_A - S_B; \tag{3.10}$$

for S_A and S_B in (3.1–3). This procedure [3.4] gives:

$$H = \mu_B(B \cdot g_+ \cdot S + B \cdot g_- \cdot V)/2 + (S \cdot D_+ \cdot S + V \cdot D_- \cdot V)/4 + (S \cdot D_- \cdot V)/2 + \tag{3.11}$$

$$+ \Sigma_k(I^k \cdot A_A^k \cdot S + I^k \cdot A_A^k \cdot V)/2 + \Sigma_k(I^k \cdot A_B^k \cdot S - I^k \cdot A_B^k \cdot V)/2 + \tag{3.12}$$

$$+ J_{AB}/2(S \cdot S - S_A(S_A + 1) - S_B(S_B + 1)) + (S \cdot D_{AB} \cdot S - V \cdot D_{AB} \cdot V) + \tag{3.13}$$

$$- d_{AB} \cdot S \times V/2 + i/2\, d_{AB} \cdot V, \tag{3.14}$$

where (3.11) and (3.12) are the Zeeman, individual spin center zero field splitting and hyperfine terms, (3.13) contains the isotropic and anisotropic terms, and (3.14) contains the antisymmetric terms. In (3.11) $g_\pm = g_A \pm g_B$ and $D_\pm = D_A \pm D_B$. In the following we will not explicitly consider the antisymmetric term whose effect on the energy levels will be discussed at the end of the section.

In order to set up the matrix representation of (3.11–14) on the $\{|SM\rangle\}$ basis we note that all the operators containing S have nonzero matrix elements only within an S manifold, or $\langle S'M'|f(S)|SM\rangle = \delta_{S'S}\langle S'M'|f(S)|SM\rangle$, where $f(S)$ is any operator containing S, while matrix elements of operators containing V obey the selection rules $\Delta S = 0, \pm 1, \pm 2$. Instead of using the cartesian components, $T_{ij}(i, j$

$= x, y, z)$, of the various tensors and tensor operators appearing in (3.11–14), we will adopt in the following their irreducible tensor components, $T_{kq}(k = 0, 1, 2; -k \leq q \leq k)$, i.e., those components which span the irreducible representations of the real orthogonal rotation group SO(3) [3.5]. Tensors (tensor operators) expressed in this form are called irreducible tensors (tensor operators) and their components are easily expressed as linear combinations of cartesian components as shown in Table 3.1 for zero-, first-, and second-rank tensors (tensor operators). Since irreducible tensor operators can be defined on different operator variables, it is necessary to specify the type of variable, and this is done by including it in parentheses.

The irreducible tensor operators have the following properties:

1. Commutation relationships

$$[S_z, T_{kq}(O)] = q T_{kq}(O) \tag{3.15}$$

$$[S_\pm, T_{kq}(O)] = [k(k+1) - q(q \pm 1)]^{1/2} T_{kq \pm 1}(O) \tag{3.16}$$

2. Wigner–Eckart theorem [3.6]

$$\langle SM|T_{kq}(O)|S'M' \rangle = (-1)^{S-M} \langle S\|T_k(O)\|S' \rangle \begin{pmatrix} S & k & S' \\ -M & q & M' \end{pmatrix}, \tag{3.17}$$

where $T_{kq}(O)$ is the q-th component of the k-th rank tensor operator built up from the operator variable O. The integral $\langle S\|T_k(O)\|S' \rangle$ appearing in (3.17) is called the reduced matrix element and it is independent of M, M', and q. The last term in (3.17) is called 3j symbol and it is defined in Appendix B. Equation (3.17)

Table 3.1. Cartesian representation of irreducible tensor operators up to second rank[a]

Zero-rank tensor operators (scalar operators)
$\quad T_{000}(O) = O$

First-rank tensor operators (vector operators)
$\quad T_{11}(O) = -1/\sqrt{2}(O_x + iO_y) = -1/\sqrt{2} \, O_+$
$\quad T_{10}(O) = O_z$
$\quad T_{1-1}(O) = 1/\sqrt{2}(O_x - iO_y) = 1/\sqrt{2} \, O_-$

Second-rank tensor operators (tensor product of two vector operators)
$\quad T_{22}(O' \cdot O'') = T_{11}(O')T_{11}(O'') = 1/2(O'_+ O''_+)$
$\quad T_{21}(O' \cdot O'') = 1/\sqrt{2}[T_{11}(O')T_{10}(O'') + T_{10}(O')T_{11}(O'')] =$
$\qquad = -1/2(O'_+ O''_z - O'_z O''_+)$
$\quad T_{20}(O' \cdot O'') = \sqrt{(2/3)}[T_{10}(O')T_{10}(O'')] + 1/\sqrt{6}[T_{11}(O')T_{1-1}(O'') +$
$\qquad + T_{1-1}(O')T_{11}(O'')] = 1/\sqrt{6} \,(3O'_z O''_z - O' \cdot O'')$
$\quad T_{2-1}(O' \cdot O'') = 1/\sqrt{2}[T_{1-1}(O')T_{10}(O'') + T_{10}(O')T_{1-1}(O'')] =$
$\qquad = 1/2(O'_- O''_z + O'_z O''_-)$
$\quad T_{2-2}(O' \cdot O'') = T_{1-1}(O')T_{1-1}(O'') = 1/2(O'_- O''_-)$

[a]O is a scalar operator; $O\,(O', O'')$ are vector operators with cartesian components O_x, O_y, O_z.

greatly reduces the time needed for evaluating matrix elements of irreducible tensors once the reduced matrix elements are known for any value of S and S'. Reduced matrix elements relevant to the calculation of the representation matrix of (3.11–13) are given in Appendix B.

By applying (3.17) to calculate the matrix elements $\langle SM|T_{kq}(O)|S'M'\rangle$ and $\langle SM|T_{kq}(S)|S'M'\rangle$ and taking the ratio of the two expressions we obtain the fundamental result:

$$\frac{\langle SM|T_{kq}(O)|S'M'\rangle}{\langle SM|T_{kq}(S)|S'M'\rangle} = \frac{\langle S\|T_k(O)\|S'\rangle}{\langle S\|T_k(S)\|S'\rangle}\delta_{SS'}, \tag{3.18}$$

where the δ symbol appears since the denominator vanishes for $S \neq S'$. Equation (3.18) allows the evaluation of the matrix elements of irreducible tensor operators built up from the operator variable O from the knowledge of the matrix elements of the corresponding operators built up with S, which are easy to calculate, and of some reduced matrix element. Applying (3.18) to evaluate the matrix elements of the irreducible tensor operators containing V as operator variable in (3.11–13) we can substitute S to V obtaining the following spin hamiltonian valid within each S spin manifold:

$$H_{S=S'} = \mu_B[\mathbf{B}\cdot(\mathbf{g}_+ + c\mathbf{g}_-)\cdot\mathbf{S}]/2 + \{\Sigma_k\mathbf{I}^k\cdot[(1+c)\mathbf{A}_B^k + (1-c)\mathbf{A}_B^k]\cdot\mathbf{S}\}/2 + \tag{3.19}$$
$$+ J_{AB}[\mathbf{S}\cdot\mathbf{S} - S_A(S_A+1) - S_B(S_B+1)]/2 + \{\mathbf{S}\cdot[(1-c_+)\mathbf{D}_{AB} + c_+\mathbf{D}_+ +$$
$$+ c_-\mathbf{D}_-]\cdot\mathbf{S}\}/2,$$

where the c, c_+, and c_- coefficients depend only on S, S_A, and S_B and are defined in Table 3.2.

Comparing (3.19) with (3.8) we obtain the following relationships between spin hamiltonian parameters:

$$\mathbf{g}_s = c_1\mathbf{g}_A + c_2\mathbf{g}_B; \tag{3.20}$$

$$\mathbf{D}_s = d_1\mathbf{D}_A + d_2\mathbf{D}_B + d_{12}\mathbf{D}_{AB}; \tag{3.21}$$

$$\mathbf{A}_s^k = c_1\mathbf{A}_A^k + c_2\mathbf{A}_B^k; \tag{3.22}$$

$$c_1 = (1+c)/2; \quad c_2 = (1-c)/2;$$
$$d_1 = (c_+ + c_-)/2; \quad d_2 = (c_+ - c_-)/2; \tag{3.23}$$
$$d_{12} = (1-c_+)/2.$$

when referring to the metal hyperfine coupling k = A, B, only one term is often retained in (3.22) on the assumption that the coupling of nucleus A to the electrons of B is small and vice versa. When these terms, $\mathbf{A}_A^B$, $\mathbf{A}_B^A$, are included they are called supertransferred hyperfine coupling.

Table 3.2. Reduced matrix elements relevant to the calculation of the spin hamiltonian matrix within an S manifold

---

Coefficients in Eq. (3.19)[a]

$c = \langle S \| T_1(V) \| S \rangle / \langle S \| T_1(S) \| S \rangle = [S_A(S_A+1) - S_B(S_B+1)]/S(S+1)$

$c_+ = 1/2\{[\langle S \| T_2(V \cdot V) \| S \rangle / \langle S \| T_2(S \cdot S) \| S \rangle] + 1\} = \{3[S_A(S_A+1)$
$\quad - S_B(S_B+1)]^2 + S(S+1)[3S(S+1) - 3 - 2S_A(S_A+1) - 2S_B(S_B+1)]\}/$
$\quad [(2S+3)(2S-1)S(S+1)]$

$c_- = \langle S \| T_2(S \cdot V) \| S \rangle / \langle S \| T_2(S \cdot S) \| S \rangle = \{\{4S(S+1)[S_A(S_A+1)$
$\quad - S_B(S_B+1)]\} - 3[S_A(S_A+1) - S_B(S_B+1)]\}/[(2S+3)(2S-1)S(S+1)]$

Reduced matrix elements of zero-, first- and second-rank irreducible tensor operators containing **S**

$\langle S \| 1 \| S' \rangle = \delta_{SS'} \sqrt{(2S+1)}$

$\langle S \| T_1(S) \| S' \rangle = \delta_{SS'} \sqrt{[(2S+1)S(S+1)]}$

$\langle S \| T_2(S \cdot S) \| S' \rangle = \delta_{SS'} \sqrt{[(2S+1)(2S+3)(2S-1)S(S+1)/6]}$

---

[a]The coefficients are zero for $S = 0$. When the denominator is zero, the coefficients must be taken as zero.

The coefficients appearing in (3.20–23) are easily evaluated for any two-spin system using Table 3.2. As an example we report in Table 3.3 the coefficients computed for two equivalent spins $S_A = S_B$ ranging from 1/2 to 5/2, and for two inequivalent spins $S_A = 1/2$ and S_B ranging from 1/2 to 5/2. The effect of the zero field splitting $\mathbf{D_S}$ on the energy levels of a triplet state arising from the exchange interaction between two 1/2 spins is shown in Fig. 3.2.

Equations (3.20–23) are relationships between tensors and are always valid even when the tensors are neither diagonal nor collinear, but care must be taken to refer all the tensors to the same reference system. The symmetry of the system can impose some relationships between the tensors appearing in (3.20–23). The presence of an inversion center in the molecule, for example, causes the second rank tensors on A to be equal to the second-rank tensors on B and to have parallel principal axes. This means that also the principal axes system of the coupled tensors is collinear to that of A and B. In this case the $\mathbf{T_-}$ tensors in (3.11–13) are zero and the $\mathbf{T_+}$ tensors are twice the tensors of the individual spins.

The c and d coefficients in (3.20–23) are not linearly independent but $c_1 + c_2 = 1$ and $d_1 + d_2 = c_+$. The first of these relations shows that when the spin centers are equal, the **g** tensor of the pair will be identical to the tensors of the individual spins, as already noted from symmetry arguments. However, if the two individual g tensors are different, the $\mathbf{g_s}$ tensor may be remarkably different from each of them. It must be recalled here that "different" means also two otherwise identical tensors, but differently oriented. For the highest multiplicity spin state $S = S_A + S_B$ we have also $c_1/c_2 = S_A/S_B$ and, since in this spin state S_A, S_B, and S are simply proportional to the number of electrons on A and B, n_A and n_B, respectively, we obtain the simple relationship:

$$c_1 = n_A/(n_A + n_B); \quad c_2 = n_B/(n_A + n_B), \tag{3.24}$$

Table 3.3. Numerical values of the coefficients in Eqs. (3.20-23) for selected dinuclear systems[a]

S_A	S_B	S	C_1	C_2	d_1	d_2	d_{12}
1/2	1/2	1	1/2	1/2	0	0	1/2
1	1	1	1/2	1/2	−1/2	−1/2	1
		2	1/2	1/2	1/6	1/6	1/3
3/2	3/2	1	1/2	1/2	−6/5	−6/5	17/10
		2	1/2	1/2	0	0	1/2
		3	1/2	1/2	1/5	1/5	3/10
2	2	1	1/2	1/2	−21/10	−21/10	13/5
		2	1/2	1/2	−3/14	−3/14	5/7
		3	1/2	1/2	1/10	1/10	2/5
		4	1/2	1/2	3/14	3/14	2/7
5/2	5/2	1	1/2	1/2	−16/5	−16/5	37/10
		2	1/2	1/2	−10/21	−10/21	41/42
		3	1/2	1/2	−1/45	−1/45	47/90
		4	1/2	1/2	1/7	1/7	5/14
		5	1/2	1/2	2/9	2/9	5/18
1/2	1	1/2	−1/3	4/3	0	0	0
		3/2	1/3	2/3	0	1/3	1/3
1/2	3/2	1	−1/4	5/4	0	3/2	−1/4
		2	1/4	3/4	0	1/2	1/4
1/2	2	3/2	−1/5	6/5	0	7/5	−1/5
		5/2	1/5	4/5	0	3/5	1/5
1/2	5/2	2	−1/6	7/6	0	8/6	−1/6
		3	1/6	5/6	0	4/6	1/6

[a] Symmetric dinuclear couples with $S_A = S_B$ from 1/2 to 5/2 and heterodinuclear couples with $S_A = 1/2$ and S_B from 1 to 5/2.

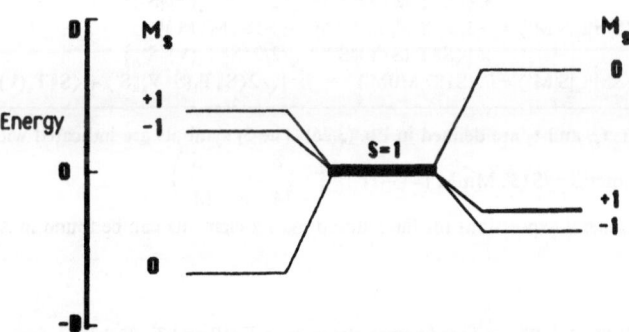

Fig. 3.2. The effect of the zero field splitting on a triplet state. The levels are labeled according to their M value. The $M = \pm 1$ levels are not pure levels since they are admixed by rhombic components. The energies have been computed for E/D = 0.07. *Left*: D > 0; *right*: D < 0

which shows that c_1 and c_2 for the highest spin multiplicity state of the couple are always smaller than 1.

Since c_1 and c_2 apply also to the hyperfine tensors, it is apparent that in the highest multiplicity state of the couple the hyperfine splitting will be scaled from

55

the value observed in a corresponding mononuclear species, or, perhaps better, in a dinuclear species in which one of the two paramagnetic centers has been substituted by a diamagnetic one.

In the preceding discussion we have neglected the antisymmetric terms in (3.14). This term can be written as:

$$H_a = d_x t_x + d_y t_y + d_z t_z, \tag{3.25}$$

where

$$
\begin{aligned}
t_x &= -i\{T_{11}(S \cdot V) - T_{1-1}(S \cdot V) + T_{11}(V)/\sqrt{2} - T_{1-1}(V)/\sqrt{2}\}/2; \\
t_y &= -\{T_{1-1}(S \cdot V) + T_{11}(S \cdot V) - T_{11}(V)/\sqrt{2} + T_{1-1}(V)/\sqrt{2}\}/2; \\
t_z &= i\{T_{10}(S \cdot V)/\sqrt{2} + T_{10}(V)/2\}.
\end{aligned}
\tag{3.26}
$$

The matrix elements of (3.26) are reported in Table 3.4 as a function of reduced matrix elements of $T_{1q}(S \cdot V)$ and $T_{1q}(V)$. In Table 3.5 we show the values of the reduced matrix elements for $S_A = 1/2$ and S_B ranging from 1/2 to 5/2.

In the strong exchange limit the effect of d_{AB} on the energy levels of an S spin manifold can be understood by considering that the general bilinear exchange interaction $S_A \cdot J_{AB} \cdot S_B$ is represented by a positive definite dyadic J_{AB} which can be decomposed by polar decomposition into a product of a real symmetric, J_S

Table 3.4 Matrix elements of (3.25)[a]

$\langle SM\|t_x\|S'M'\rangle = i/2\{(S1S'; M-1M') - (S1S'; M1M')\}$
$\qquad \times \{\langle S\|T_1(S \cdot V)\|S'\rangle + \sqrt{2}/2\langle S\|T_1(V)\|S'\rangle\}$
$\langle SM\|t_y\|S'M'\rangle = -1/2\{(S1S'; M-1M') - (S1S'; M1M')\}$
$\qquad \times \{\langle S\|T_1(S \cdot V)\|S'\rangle + \sqrt{2}/2\langle S\|T_1(V)\|S'\rangle\}$
$\langle SM\|t_z\|S'M'\rangle = i/2(S1S'; M0M') \qquad \times \{\sqrt{2}\langle S\|T_1(S \cdot V)\|S'\rangle + \langle S\|T_1(V)\|S'\rangle\}$

[a] t_x, t_y, and t_z are defined in Eq. (3.26). The 3j symbols are indicated with the shorthand

notation: $(S1S'; MmM') = (-1)^{S-M} \begin{pmatrix} S & 1 & S' \\ -M & m & M' \end{pmatrix}$.

General expressions for the reduced matrix elements can be found in Appendix B.

Table 3.5. The reduced matrix elements of $T_1(V)$ and $T_1(S \cdot V)$ between the S states obtained from $S_A = 1/2$ and S_B from 1/2 to 5/2

S_A	S_B	S	S'	$\langle S\|T_1(V)\|S'\rangle$	$\langle S\|T_1(S \cdot V)\|S'\rangle$
1/2	1/2	1	0	$\sqrt{3}$	$-\sqrt{6}$
1/2	1	3/2	1/2	$4\sqrt{3/3}$	$-5\sqrt{6/3}$
1/2	3/2	2	1	$\sqrt{30}/2$	$-3\sqrt{15/2}$
1/2	2	5/2	3/2	$4\sqrt{15/5}$	$-7\sqrt{30/5}$
1/2	5/2	3	2	$\sqrt{105/3}$	$-2\sqrt{210/3}$

$= (\mathbf{J} \cdot \tilde{\mathbf{J}})^{\frac{1}{2}}$, and a real orthogonal $\mathbf{R}$ matrix and the interaction term becomes [3.7]:

$$H_i = S_A \cdot J_S \cdot S_B', \tag{3.27}$$

where $S_B' = \mathbf{R} \cdot S_B$. Equation (3.27) shows that in the strong exchange limit the effect of the antisymmetric term on the energy levels of an S spin manifold can be represented by a symmetric tensor which, upon making it traceless, gives an additive contribution to the zero field splitting. A closed formula has been worked out for the case of two spins $S_A = 1/2$, S_B any value, and parallel g_A and g_B tensors. In these systems it was found that the antisymmetric coupling determines in the S manifolds an additional axial zero field splitting with the unique axis parallel to d_{AB}. The D value was found to be:

$$D = \{[(S_A + S_B + S + 2)(S_A - S_B + S + 1)(S_B - S_A + S + 1)(S_A + S_B - S) + \\ - (S_A + S_B + S + 1)(S_A - S_B + S)(S_B - S_A)(S_A + S_B - S + 1)]/ \\ [8(2S + 1)(2S + 3)(S + 1)]\} |d_{AB}|^2/J_{AB}. \tag{3.28}$$

It must be remembered now that this is the effect on the energy levels only. In fact, since (3.25) admixes states with $\Delta S = \pm 1$ it affects the transition intensities of the EPR spectra as well as the angular variation of the line widths. In general, it will be difficult to recognize these effects on the EPR spectra and then it will be possible to measure the antisymmetric d_{AB} vector from (3.28) only if we can safely estimate all the other contributions to the zero field splitting both in value and in space orientation. The symmetry of the molecule imposes severe conditions on the antisymmetric exchange term as already elucidated in Chap. 2 and can be of help in solving this problem.

3.3 Spectra of Pairs in the Strong Exchange Limit

Spectra of pairs in the strong exchange limit are generally formed by one or more spectra arising by the spin multiplets which are thermally populated at the temperature of the experiment [3.8]. The theory which is therefore needed to interpret the spectra of pairs in the strong exchange limit is essentially the theory used to interpret the EPR spectra of $S \geq 1/2$ spin systems [3.3]. It must be mentioned that often spin states as rare as $S = 3$ or higher can arise from the exchange interaction, which are not commonly observed in simple systems.

Some further complications must, however, be expected since: (1) the spectra at a given temperature can be a superposition of spectra arising from different multiplets; (2) the g and D tensors will in general be diagonal in different reference frames. The first point generally requires that the EPR spectra are recorded at different temperatures, in order to individuate which signal belongs to which spin state. The total intensity of the signals arising from a given multiplet is temperature-dependent according to the Boltzmann law:

$$I_s \propto e^{-E(S)/kT}/Z, \tag{3.29}$$

where Z is the partition function

$$Z = \Sigma_S (2S+1) e^{-E(S)/kT}, \tag{3.30}$$

and k is the Boltzmann constant. The temperature dependence of I_S for the states arising from the coupling between two $S_A = S_B = 5/2$ spins is shown in Fig. 3.3. By measuring the dependence of the signal intensity of the EPR spectra it is in principle possible to determine the relative energies of the thermally populated levels. This procedure has been used several times in the literature, but the results are generally less accurate than those obtained from the measurement of the temperature variation of the bulk magnetic susceptibility, mainly due to inaccuracies in the determination of the intensity of the signals. In fact, this must be obtained by a double numerical integration of the usual derivative output of the spectrometer, extended to the entire range of the signal, especially on the wings. When several different multiplets are populated it may easily occur that more signals overlap, thus making the measurement inaccurate.

Point (2) above is also responsible for several complications when analyzing the EPR spectra of pairs. In fact, when **g** and **D** are not collinear, polycrystalline powder or glassy matrix spectra do not generally allow the determination of the principal values and single crystal measurements are needed. The following example should clarify the possible origin of the misalignment between **g** and **D**.

The copper(II) ion is often four-coordinated in a square planar environment (additional ligands may also be present, but at longer distances than the equatorial ones). Three fairly common ways of coupling two square planar copper ions in a pair are shown in Fig. 3.4. Figure 3.4a shows the geometry of the copper acetate hydrate dimer.

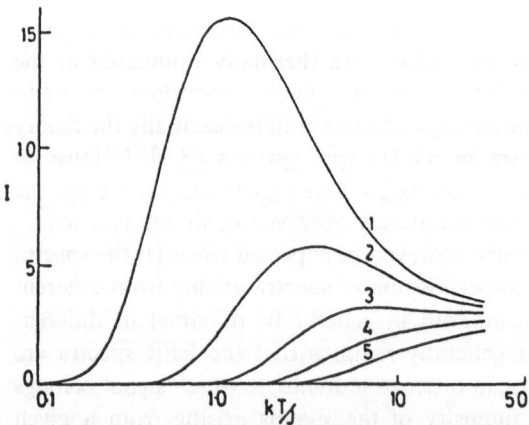

Fig. 3.3. The relative intensities of the EPR transitions for a pair of $S_i = S_j = 5/2$ spins computed with Eq. (3.29) plotted against the reduced temperature kT/J_{AB}. Transitions within different total spin manifolds $S = 1, 2, 3, 4, 5$ are labeled [3.8]

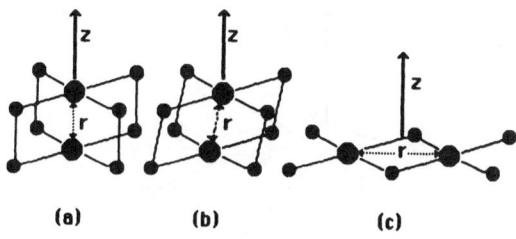

Fig. 3.4. Three common geometries found in copper(II) dimers formed by square planar moieties

The isotropic exchange interaction in these systems couples the two $S = 1/2$ spins to give a singlet and a triplet state. The threefold degeneracy of the triplet is lifted by zero field splitting (Fig. 3.2). The zero field splitting tensor, $\mathbf{D}_1$, according to (3.21) is determined by the $\mathbf{D}_{CuCu}$ tensor, which in Chap. 2 was shown to be due to anisotropic exchange and dipolar contributions. In centrosymmetric systems the $\mathbf{g}_1$ tensor is parallel to the $\mathbf{g}$ tensors of the individual moieties, which in the present case have their z axes orthogonal to the coordination planes with the x and y axes roughly parallel to the bond directions. The z direction is shown in Fig. 3.4 for the three geometries.

The largest component of the dipolar zero field splitting tensor is expected to be parallel to the copper-copper direction, r, while the z principal direction of the anisotropic exchange tensor is generally found parallel to the z axis of the $\mathbf{g}$ tensor (see Chap. 2). It is apparent that only in geometries like in (a) the $\mathbf{g}$ and $\mathbf{D}$ tensors can be roughly parallel to each other, while in (b) and (c), if dipolar and exchange contributions are comparable in magnitude, they must have different principal axes since r and z are misaligned. In particular, in geometries like (c) the principal axes of $\mathbf{g}_1$ and $\mathbf{D}_1$ can be oriented as shown in Fig. 3.5. The D_{1z} axis is determined by the anisotropic exchange, and the D_{1x} axis is determined by the dipolar interaction. In Fig. 3.5 we show also the principal axes system of $\mathbf{g}_1$. It is apparent that the two tensors have the z axis in common, but they have x and y axes rotated by approximately 45°.

A detailed analysis of the spectra which can occur for the various spin states is not in order here, since it would require a whole text of its own. However, the interested reader can refer to general references [3.3].

It is now relevant to provide some examples of applications of Eqs. (3.20–23) showing their experimental validity and their limits. In order to apply these relationships one should know with some accuracy the single ion spin hamiltonian parameters since when the spectra of a pair are observed it is not possible to decompose experimentally the observed spin hamiltonian parameters into a sum of different contributions unless some of them are independently known. In a discrete dinuclear unit one can often substitute one of the two paramagnetic centers with a diamagnetic one, obtaining an isomorphous complex in which only one paramagnetic species is present. This procedure gives the possibility of measuring the spin hamiltonian parameters of at least one of the ions forming the

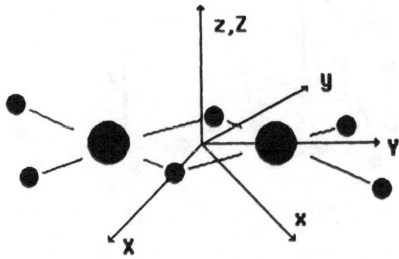

Fig. 3.5. Orientation of the principal axes of the $\mathbf{D_1}$ (X,Y,Z) and $\mathbf{g_1}$ (x,y,z) tensors in square planar copper(II) dimers

dinuclear species. These can be directly compared to the experimental spin hamiltonian parameters in homodinuclear centrosymmetric couples according to (3.20–23). In these systems it is also possible to measure the $\mathbf{D_{AB}}$ tensor. In heterodinuclear species, on the other hand, in order to perform a direct check of (3.20–23) it must be possible to substitute both the paramagnetic centers independently and to evaluate both the single-ion spin hamiltonian parameters independently. If this is not possible, but, for instance, only one of the paramagnetic centers can be easily substituted, then (3.20–23) can be used to calculate the parameters for the paramagnetic center. Comparing these values with those observed in similar mononuclear species, it is possible to check the limit of validity of the above equations.

A beautiful example where Eqs. (3.20–23) have been experimentally verified is provided by the pair spectra observed in the host lattice of di-μ-pyridine-N-oxide bis[dichloro aquo copper(II)], $[Cu(pyO)Cl_2(H_2O)]_2$ [3.9]. The structure of $[Cu(pyO)Cl_2(H_2O)]_2$ is illustrated in Fig. 3.6.

The two copper(II) ions are strongly antiferromagnetically coupled: the singlet is the ground state with the triplet level lying $\simeq 500 \text{ cm}^{-1}$ above. The EPR spectra of the triplet level can be observed at room temperature, and disappear rapidly on cooling according to (3.29) [3.9–10]. They have been interpreted using $g_x = 2.063$, $g_y = 2.076$, $g_z = 2.32$, $|D| = 0.154 \text{ cm}^{-1}$, and E/D = -0.19. The g tensor has quasi-axial symmetry with the z axis orthogonal to the Cl_2O_2 molecular plane and the x and y axes roughly parallel to the in-plane copper-pyridine-N-oxide bonds. If the sample is doped with a little zinc(II) (d^{10} configuration, diamagnetic) in the lattice will be present Cu–Cu, Cu–Zn, and Zn–Zn pairs. At low temperature all the Cu–Cu pairs must be in the ground singlet state, therefore, an EPR spectrum is observed at a temperature below 77 K, arising from a doublet state due to copper(II)-zinc(II) couples. The measured g tensor is identical within experimental error with the tensor observed in the copper(II)-copper(II) couple in agreement with (3.20). Unfortunately, no hyperfine coupling is resolved for the latter, so it is not possible to test in this way the validity of (3.21).

Copper and manganese hyperfine is, however, resolved in the EPR spectra of manganese(II) doped $[Cu(pyO)Cl_2(H_2O)]_2$ [3.9a, 3.10c]. Manganese(II) is a d^5

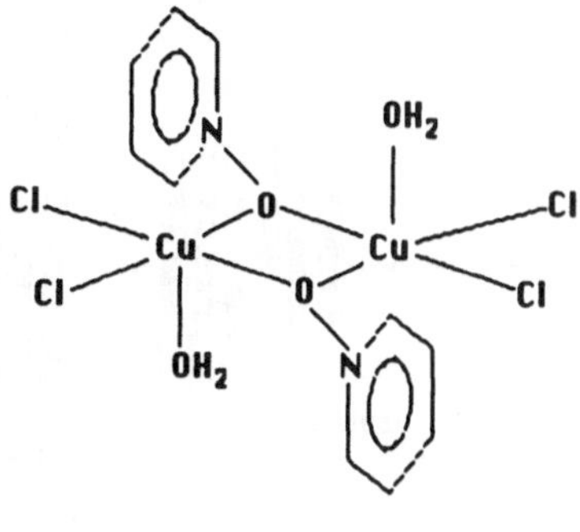

Fig. 3.6. The structure of $[Cu(pyO)Cl_2(H_2O)]_2$

ion, with a ground $S_{Mn} = 5/2$ state, which can be coupled to $S_{Cu} = 1/2$ to give S $= 2,3$. The spectra have been observed at 77 K and have been interpreted using a $S = 2$ spin hamiltonian. A typical single crystal spectrum is shown in Fig. 3.7. Since no signal attributable to the $S = 3$ spin state was observed in the whole temperature range, it was estimated from (3.29) that $J \geq 250$ cm^{-1}. The four main lines in Fig. 3.7 correspond to the $-2 \rightarrow -1$, $-1 \rightarrow 0$, $0 \rightarrow 1$, and $1 \rightarrow 2$ allowed $\Delta M_S = \pm 1$ transitions within the ground $S = 2$ manifold. Each fine structure line is actually split into 24 lines, with the structure corresponding to a sextet of quartets. This structure was attributed to the coupling of the unpaired electrons with the $^{55}Mn(I_{Mn} = 5/2)$ and $^{63, 65}Cu(I_{Cu} = 3/2)$ nuclei. The observed **g** tensor is parallel to that observed in the Cu–Zn couple with principal values: $g_x = 1.991$, $g_y = 1.986$, $g_z = 1.952$. Since manganese (II) is a ^6S ion, the $\mathbf{g}_{Mn}$ tensor can be safely assumed to be isotropic and equal to 2.0. Using (3.20), with $c_1 = 7/6$ and $c_2 = -1/6$, and taking for $\mathbf{g}_{Cu}$ the tensor obtained from the Cu–Zn couple we can calculate $g_{2x} = g_{2y} = 1.99$ and $g_{2z} = 1.95$ in excellent agreement with the experimental values. It must be remarked that although neither ion has g values smaller than 2, the pair does. This is simply the result of the antiferromagnetic alignment of the two spins in the ground $S = 2$ state. The observed copper hyperfine splitting is 23×10^{-4} cm^{-1}, measured along the z axis, which corresponds perfectly to 1/6 the hyperfine of the individual copper ion, $A^{Cu}_{Cu,z} = 139 \times 10^{-4}$ cm^{-1} observed in the Cu–Zn pair spectra. The manganese (II) hyperfine coupling cannot be directly tested since no Mn–Zn couple was reported. However, the experimental value $A^{Mn}_{2Mn,z} = 90 \times 10^{-4}$ cm^{-1} requires by (3.22) $A^{Mn}_{2Mn,z} = 76 \times 10^{-4}$ cm^{-1} in nice agreement with the values observed in MnO_2Cl_2 complexes [3.3].

The previous examples show how well relations (3.20–23) work in interpreting the EPR spectra of couples in the strong exchange limit. It must be recalled here that strong exchange means that the isotropic exchange interaction is much larger than all the other interactions in (3.1–3). We will show in Sect. 3.6 that in the presence of single ion anisotropy this approximation can lose its validity even for large J_{AB} values depending essentially on the J_{AB}/D_A ratio. The validity of

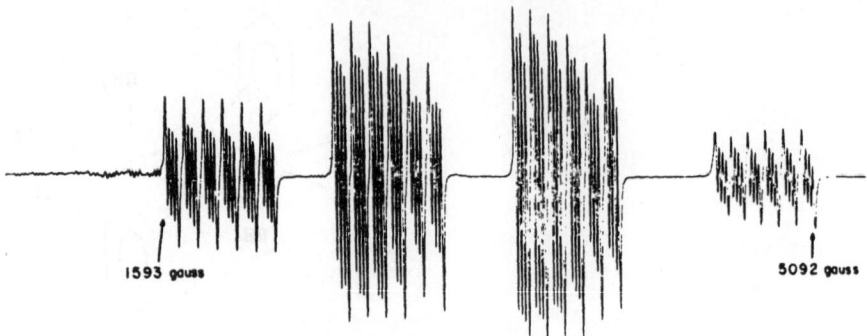

Fig. 3.7. Single crystal EPR spectrum of manganese(II) doped [Cu(pyO)Cl$_2$(H$_2$O)]$_2$ at X-band frequency and 77 K. The static magnetic field is parallel to the x axis of the zero field splitting tensor. After [3.9a]

Eqs. (3.20–23) must always be checked experimentally in each case before using them.

Equation (3.21), relating the measured zero field splitting tensor of the couple to single ion and anisotropic exchange contributions, is certainly the most difficult relationship to verify experimentally since a large number of physical effects, for example, antisymmetric exchange and magnetic dipolar interactions, are represented by a bilinear spin form and contribute to the zero field splitting observed by EPR spectroscopy.

One of the best examples in which Eq. (3.21) seems to hold is the chromium(III) dimer tris(μ-hydroxo)bis(1,4,7-trimethyl-1,4,7-triazacyclononane chromium(III)) whose structure [3.12] is represented in Fig. 3.8. The chromium(III) centers, $S_{Cr} = 3/2$, are antiferromagnetically coupled with $J_{CrCr} = 64$ cm^{-1} leaving $S = 0$ as the ground spin state [3.13]. The zero field splitting of the excited $S = 1$, 2, and 3 spin states has been measured from the EPR spectra to be $|D_1| = 2.28$ cm^{-1}, $|D_2| = 0.08$ cm^{-1}, and $|D_3| = 0.23$ cm^{-1}, respectively [3.14]. Application of (3.21) yields:

$$D_1 = 17/10\, D_{CrCr} - 12/5\, D_{Cr}; \tag{3.31}$$

$$D_2 = 1/2\, D_{CrCr}; \tag{3.32}$$

$$D_3 = 3/10\, D_{CrCr} + 2/5\, D_{Cr}. \tag{3.33}$$

Equation (3.32) has a particularly fortunate form because it depends on one parameter only, D_{CrCr}. Therefore, the measured D_2 splitting can be used to estimate the exchange contribution directly yielding $|D_{CrCr}| = 0.16$ cm^{-1}. One of the main limitations in using (3.21) to evaluate the contributions to the zero field splitting of pairs is now apparent: we do not know experimentally the sign of the zero field splitting parameter. In the present case by assuming that D_1 and D_2 are

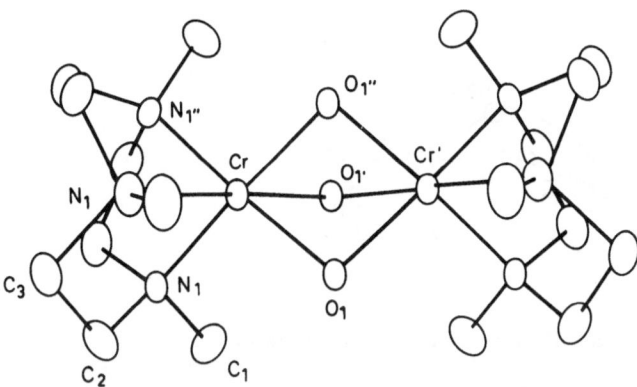

Fig. 3.8. ORTEP view of tris(μ-hydroxo)bis(1,4,7-trimethyl-1,4,7-triazacyclononane chromium (III) [3.12]

positive and D_3 negative and using (3.31) and (3.32) together we obtain $D_{CrCr} = 0.16 \text{ cm}^{-1}$ and $D_{Cr} = -0.84 \text{ cm}^{-1}$ yielding $D_3 = -0.29 \text{ cm}^{-1}$, in nice agreement with the experimental data. The calculated value of D_{Cr} is comparable to the zero field splitting observed in other trigonally distorted complexes such as Cr(acac)$_3$ ($|D| = 0.59 \text{ cm}^{-1}$) [3.15] showing that (3.31–33) can be used to interpret the EPR spectra of this complex.

A situation in which the knowledge of the sign of the zero field splitting tensor can be irrelevant to the analysis of the zero field splitting can be experimentally found in the copper (II)-manganese (II) couples when EPR spectra from both the $S = 2$ and 3 spin states are observed. These complexes can thus in principle provide good systems for the experimental verification of (3.21). In fact, by applying (3.21) to these systems, we get:

$$\mathbf{D_2} = 4/3\,\mathbf{D_{Mn}} - 1/6\,\mathbf{D_{CuMn}} \tag{3.34}$$

$$\mathbf{D_3} = 2/3\,\mathbf{D_{Mn}} + 1/6\,\mathbf{D_{CuMn}} \tag{3.35}$$

and adding together (3.34) and (3.35) gives

$$\mathbf{D_2} + \mathbf{D_3} = 2\mathbf{D_{Mn}}, \tag{3.36}$$

irrespective of the relative signs of the components of $\mathbf{D_2}$ and $\mathbf{D_3}$. Equation (3.36) should be easy to check by directly measuring $\mathbf{D_{Mn}}$ in a Zn–Mn couple or by comparing the measured value with literature reports. Equation (3.36) was used to interpret the measured zero field splitting in Cu(prp)$_2$enMn(hfa)$_2$, where (prp)$_2$en is the Schiff base formed by 2-hydroxypropiophenone and ethylenediamine [3.16]. The structure of Cu(prp)$_2$enMn(hfa)$_2$ is illustrated in Fig. 3.9. The coupling between Cu (II) and Mn (II) is antiferromagnetic with $J_{CuMn} = 26 \text{ cm}^{-1}$.

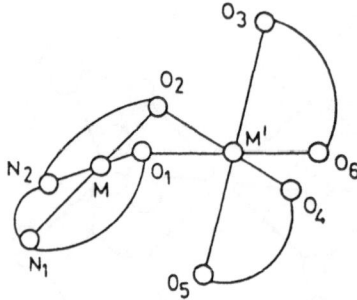

Fig. 3.9. The structure of $Cu(prp)_2enMn(hfa)_2$

The measured zero field splitting parameters are $|D_2| = 0.034(4)$ cm^{-1}, E_2/D_2 $= 0.28(3)$, and $|D_3| = 0.047(5)$ cm^{-1}, $E_3/D_3 = -0.23(2)$ for the $S = 2$ and $S = 3$ states, respectively. Equation (3.36) gives $|D_{Mn}| = 0.040$ cm^{-1} which is of the correct order of magnitude, as we learn from the comparison with the values reported in the literature. An independent check was also tried by measuring the EPR spectra of $Ni(prp)_2enMn(hfa)_2$ which is isomorphous with $Cu(prp)_2enMn(hfa)_2$. The nickel(II) (d^8 electronic configuration) is in a square planar environment and is diamagnetic. Therefore, the zero field splitting measured in the EPR spectra is due to the $[Mn(hfa)_2]^{2+}$ moiety. The measured zero field splitting parameter was $|D_{Mn}| = 0.049(5)$ cm^{-1} which is $\simeq 20\%$ larger than the value computed through (3.36). An obvious explanation for this might be that the zero field splitting in Ni–Mn is not the same as that in Cu–Mn due to some small structural differences. It is therefore apparent that an exact experimental verification of (3.21) can be extremely difficult, but on the other hand the difference between experimental and computed results is only $\simeq 20\%$ which can after all be considered as a reasonable agreement between theoretical predictions and experimental data.

A further complication in the application of (3.20–23) can be brought about by exchange striction effects defined in Sect. 2.4. A discussion of this point will be given in Sect. 9.1.

The antisymmetric term (3.25) has been suggested [3.7] to contribute to the EPR spectra of manganese(II) doped $[Cu(pyO)Cl_2(H_2O)]_2$ which have already been described above. The measured zero field splitting of the $S = 2$ is $|D_2|$ $= 0.051$ cm^{-1}, $E_2/D_2 = -0.26$ with the direction of maximum zero field splitting, D_{2z} orthogonal to the g_{2z} axis, which is parallel to the Cu-O(axial) bond (Fig. 3.10). In the pure copper complex the direction of maximum zero field splitting D_{1z} makes $\simeq 26°$ with the g_{1z} direction and D_{1y} is near the Cu–Cu direction. Since the Cu–Cu complex is centrosymmetric the only contributions to $\mathbf{D_1}$ come from anisotropic exchange and dipolar interactions and the observed misalignment between $\mathbf{g_1}$ and $\mathbf{D_1}$ can be rationalized by a nondiagonal $\mathbf{D_{CuCu}}$ tensor due to the low symmetry of the complex. It is apparent that the same explanation does not hold for the Cu–Mn couple. In fact, purely anisotropic

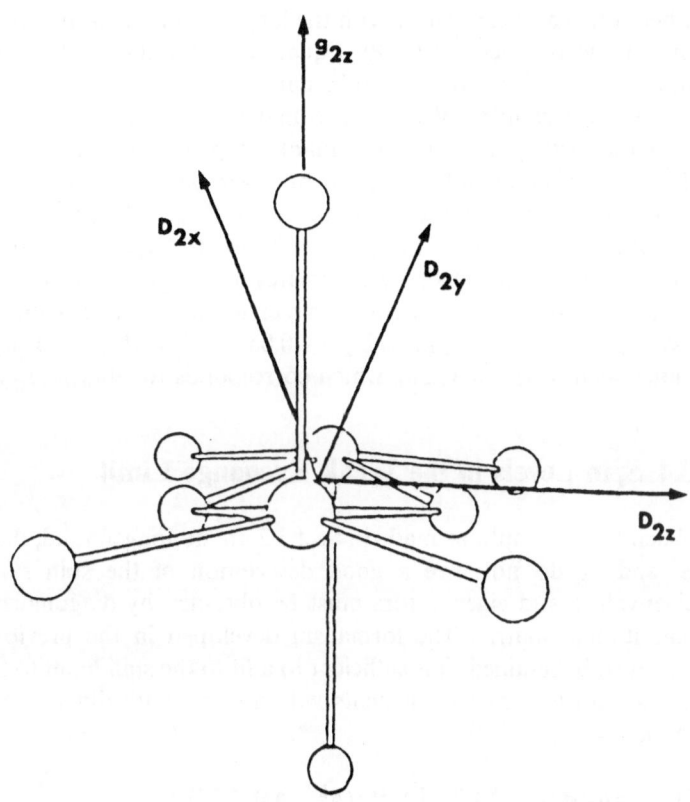

Fig. 3.10. Relative orientation of the $\mathbf{D}_2$ and $\mathbf{g}_2$ tensors in manganese(II) doped [Cu(pyO) Cl$_2$(H$_2$O)]$_2$

exchange interactions would require D_{2z} parallel to g_{2z}, or slightly misaligned from it like in the Cu–Cu couple, while purely dipolar interactions would put D_{2z} parallel to the Cu–Mn direction. Both of these interactions cause the D_{2z} to be nearly orthogonal to the observed one. Furthermore, since the exchange contributions to $\mathbf{D_{AB}}$ are proportional to Δg_A and Δg_B (see Chap. 2), they are expected to be smaller in the Cu–Mn than in the Cu–Cu pair because of $g_{Mn} \simeq g_e$ [3.3]. It must be noted that the Cu–Mn couple is noncentrosymmetric and antisymmetric exchange interactions can be operative. Since the overall symmetry of the complex is not far from C_s, the σ plane including the copper(II) and manganese(II) ions and the oxygens of the water molecules, we expect a $\mathbf{d_{CuMn}}$ vector perpendicular to σ. The $\mathbf{g_{Mn}}$ tensor can be assumed as isotropic so that $\mathbf{g_{Cu}}$ and $\mathbf{g_{Mn}}$ are collinear and Eq. (3.28), which expresses the effect of $\mathbf{d}$ on the $S = 2$ state, can be applied. Using (3.21), (3.28) and the values of the coefficients reported in Table 3.3 we can decompose the $\mathbf{D}_2$ tensor according to

$$\mathbf{D}_2 = 8/6\,\mathbf{D_{Mn}} - 1/6\,\mathbf{D_{CuMn}} + \mathbf{D_a}, \tag{3.37}$$

65

where D_a is an axial tensor with the largest component parallel to d_{CuMn} whose magnitude is given by (3.28). Equation (3.37) shows that the observed D_{2z} direction, which is expected to be parallel to D_{az}, can be explained assuming a sizeable contribution of D_a. Unfortunately, we cannot use (3.37) to measure D_a from the EPR spectra since we cannot independently determine D_{Mn} and D_{CuMn}. The order of magnitude of D_a can, however, be estimated through the following assumptions: an axial D_{Mn} tensor with $D_{Mn} = 0.05$ cm^{-1}, according to the literature reports; the exchange contribution to D_{CuMn} taken to be the same as in the Cu–Cu couple, which overestimates it; computing the dipolar contribution to D_{CuMn} using the Cu–Cu direction seen in the crystal structure. With the above assumptions. Eq. (3.37) gives $D_a = -0.03$ cm^{-1}, $E_a/D_a = 0$, using for D_a the sign anticipated from Eq. (3.28), which corresponds to $\|d_{CuMn}\|^2/J_{CuMn}| = 0.48$ cm^{-1}.

3.4 Spin Levels in the Weak Exchange Limit

When the assumptions made in Sect 3.2 are no longer valid, the eigenvectors of S^2 and S_Z do not give a good description of the spin states and correct eigenvalues and eigenvectors must be obtained by diagonalization of the full hamiltonian matrix. The formalism developed in the previous sections can, however, be retained. It is sufficient to add to the spin hamiltonian (3.19), which has no nonzero matrix elements within states with different S, the additional terms:

$$H_{S \neq S'} = \mu_B (B \cdot g_- \cdot V)/2 + [\Sigma_k I^k \cdot (A_A^k - A_B^k) \cdot V]/2 + \tag{3.38}$$

$$+ [V \cdot (D_- - D_{AB}) \cdot V + 2S \cdot D_- \cdot V]/4. \tag{3.39}$$

In (3.38–39) we have neglected the antisymmetric term (3.25) for the sake of simplicity. By application of the Wigner–Eckart theorem (3.17) one can easily recognize the S states which are connected by (3.38–39). In fact, due to the triangular relationships of the 3j symbols, the first-rank irreducible tensor operators in (3.38) and the second-rank irreducible tensor operators in (3.39) can connect only states with $\Delta S = \pm 1, \pm 2$. For couples in which $S_A = S_B$ a further restriction applies to the matrix elements of the operator $S \cdot D_- \cdot V$ which are nonzero only when $\Delta S = \pm 1$. It must be also remembered that when an inversion center connects A and B, $D_- = 0$. General expressions for the matrix elements of (3.38–39) are reported in Appendix B. Some matrix elements of operators containing V have been analytically expressed as a function of S_A, S_B, S, and S' [3.4] and are reported in Table 3.6.

Perturbation expressions [3.17] have been derived for the eigenvalues and the eigenvectors of a two $S = 1/2$ spin system when the Zeeman effect is dominant over the anisotropic terms appearing in (3.19, 3.38–39). This situation has been experimentally encountered in a wide variety of spin-labeled transition metal complexes containing copper(II), silver(II), and oxovanadium(IV), in

Table 3.6. Analytical expression of matrix elements of operators containing V for $S' \neq S^a$

Matrix elements of operator V

$\langle S+1M\pm1|V_x|SM\rangle = \mp 1/2\{(S\pm M+1)(S\pm M+2)/[(2S+1)(S+1)]\}^{\frac{1}{2}} \times C$

$\langle S+1M\pm1|V_y|SM\rangle = i/2\{(S\pm M+1)(S\pm M+2)/[(2S+1)(S+1)]\}^{\frac{1}{2}} \times C$

$\langle S+1M|V_z|SM\rangle = \{(S-M+1)(S+M+1)/[(2S+1)(S+1)]\}^{\frac{1}{2}} \times C$

$C = \{(S_A+S_B+S+2)(S_A-S_B+S+1)(S_B-S_A+S+1)(S_A+S_B-S)/[(2S+3)(S+1)]\}^{\frac{1}{2}}$

Matrix elements of operator $S \cdot D_- \cdot V$

$\langle S+1M\pm1|S_xV_z+S_zV_x|SM\rangle = (S\mp2M)\{(S\pm M+2)(S\pm M+1)/[2S(S+1)(2S+1)(S+2)]\}^{\frac{1}{2}} \times C$

$\langle S+1M\pm1|S_yV_z+S_zV_y|SM\rangle = \mp i(S\mp2M)\{(S\pm M+2)(S\pm M+1)/[2S(S+1)(2S+1)(S+2)]\}^{\frac{1}{2}} \times C$

$\langle S+1M|3S_zV_z-S\cdot V|SM\rangle = M\sqrt{6}\{3(S-M+1)(S+M+1)/[S(S+1)(2S+1)(S+2)]\}^{\frac{1}{2}} \times C$

$\langle S+1M\pm2|S_xV_x-S_yV_y|SM\rangle = \mp \{(S\pm M+1)(S\pm M+2)(S\pm M+3)(S\mp M)/$
$\qquad\qquad [2S(S+1)(2S+1)(S+2)]\}^{\frac{1}{2}} \times C$

$\langle S+1M\pm2|S_xV_y+S_yV_x|SM\rangle = i\{(S\pm M+1)(S\pm M+2)(S\pm M+3)(S\mp M)/$
$\qquad\qquad [2S(S+1)(2S+1)(S+2)]\}^{\frac{1}{2}} \times C$

$C = \{(S_A+S_B+S+2)(S_A+S_B-S)(S_A-S_B+S+1)(S_B-S_A+S+1)S(S+2)/[(2S+3)(2S+2)]\}^{\frac{1}{2}}$

Matrix elements of operator $V \cdot O \cdot V$

$\langle S+1M\pm1|V_xV_z+V_zV_x|SM\rangle = (S\mp2M)\{(S\pm M+2)(S\pm M+1)/[2S(S+1)(2S+1)(S+2)]\}^{\frac{1}{2}} \times C$

$\langle S+1M\pm1|V_yV_z+V_zV_y|SM\rangle = \mp i(S\mp2M)\{(S\pm M+2)(S\pm M+1)/[2S(S+1)(2S+1)(S+2)]\}^{\frac{1}{2}} \times C$

$\langle S+1M|3V_zV_z-V\cdot V|SM\rangle = M\sqrt{6}\{3(S-M+1)(S+M+1)/[S(S+1)(2S+1)(S+2)]\}^{\frac{1}{2}} \times C$

$\langle S+1M\pm2|V_xV_x-V_yV_y|SM\rangle = \mp \{(S\pm M+1)(S\pm M+2)(S\pm M+3)(S\mp M)/$
$\qquad\qquad [2S(S+1)(2S+1)(S+2)]\}^{\frac{1}{2}} \times C$

$\langle S+1M\pm2|V_xV_y+V_yV_x|SM > = i\{(S\pm M+1)(S\pm M+2)(S\pm M+3)(S\mp M)/$
$\qquad\qquad [2S(S+1)(2S+1)(S+2)]\}^{\frac{1}{2}} \times C$

$C = 2\{(S_A+S_B+S+2)(S_A+S_B-S)(S_A-S_B+S+1)(S_B-S_A+S+1)/$
$\qquad [(2S+4)(2S+3)S(S+1)[S_A(S_A+1)-S_B(S_B+1)]]\}^{\frac{1}{2}}$

$\langle S+2M|3V_zV_z-V\cdot V|SM\rangle = \sqrt{6}\{3(S-M+2)(S-M+1)(S+M+2)(S+M+1)/$
$\qquad\qquad [(2S+1)(2S+2)(2S+3)(S+2)]\}^{\frac{1}{2}} \times C$

$\langle S+2M\pm1|V_xV_z+V_zV_x|SM\rangle = \mp \{(S\mp M+1)(S\pm M+3)(S\pm M+2)(S\pm M+1)/$
$\qquad\qquad [(2S+1)(2S+2)(2S+3)(S+2)]\}^{\frac{1}{2}} \times C$

$\langle S+2M\pm1|V_yV_z+V_zV_y|SM\rangle = \mp i\{(S\mp M+1)(S\pm M+3)(S\pm M+2)(S\pm M+1)/$
$\qquad\qquad [(2S+1)(2S+2)(2S+3)(S+2)]\}^{\frac{1}{2}} \times C$

$\langle S+2M\pm2|V_xV_x-V_yV_y|SM\rangle = \{(S\pm M+1)(S\pm M+3)(S\pm M+2)(S\pm M+4)/$
$\qquad\qquad [(2S+1)(2S+2)(2S+3)(2S+4)]\}^{\frac{1}{2}} \times C$

$\langle S+2M\pm2|V_xV_y+V_yV_x|SM\rangle = \mp i\{(S\pm M+1)(S\pm M+3)(S\pm M+2)(S\pm M+4)/$
$\qquad\qquad [(2S+1)(2S+2)(2S+3)(2S+4)]\}^{\frac{1}{2}} \times C$

$C = 2\{(S_A+S_B+S+3)(S_A+S_B+S+2)(S_A-S_B+S+2)(S_A-S_B+S+1)(S_B-S_A+S+2) \times$
$\qquad \times (S_B-S_A+S+1)(S_A+S_B-S)(S_A+S_B-S-1)/[(2S+5)(2S+4)(2S+3)(2S+2)]\}^{\frac{1}{2}}$

---

^{a}The operators are expressed in their cartesian components. If a matrix element becomes indeterminate when $S=0$, the matrix element is zero.

titanium(III), copper(II), and oxovanadium(IV) dinuclear complexes in which the metal centers are more than 8 Å apart, as well as in many systems of biological interest like B_{12}-dependent enzyme reactions. In these systems, moreover, anisotropic exchange interactions are usually assumed as zero and zero field splitting effects have been generally attributed only to dipolar interactions and used to estimate the distance between the spin centers.

A brief discussion and examples of these systems will be given in Sect. 3.5. We will report here relevant perturbation formulas for the calculation of the energy levels and EPR spectra of a couple of dissimilar centers $S_A = 1/2$ and $S_B = 1/2$. A complete treatment of the perturbation procedure can be found in Ref. [3.17a].

The hamiltonian we consider here has the form

$$H = \mu_B B \cdot g_A \cdot S_A + I^A \cdot A_A \cdot S_A + \tag{3.40}$$

$$+ \mu_B B \cdot g_B \cdot S_B + I^B \cdot A_B \cdot S_B + \tag{3.41}$$

$$+ J_{AB} S_A \cdot S_B + S_A \cdot D_{AB} \cdot S_B = \tag{3.42}$$

$$= H_A + H_B + H_{AB}. \tag{3.43}$$

In the single ion terms H_A and H_B we have considered only the hyperfine coupling with one nucleus. The H_{AB} term has been assumed to contain only isotropic exchange and anisotropic effects. These latter effects have been generally considered to be due to the magnetic dipolar interaction between the S_A and S_B spins. In order to work out a perturbation procedure all the spin hamiltonians must be referred to the same axes. It is generally used to refer (3.40–43) to the g_A principal axes system, (x_A, y_A, z_A). The general orientation of spin B with respect to spin A is shown in Fig. 3.11. In addition to the definition of the $A - B$ distance, r_{AB}, and to the polar angles, ε and η, of the r_{AB} vector one needs to specify the three Euler angles, α, β, and γ, defining the orientation of the g_B principal axes, (x_B, y_B, z_B), with respect to (x_A, y_A, z_A). In general, the g_B tensor will not be diagonal in the (x_A, y_A, z_A) reference system. Explicit expressions for the dipolar part of the D_{AB} term, D_{AB}^{dip}, have been given as a function of the principal values of g_A and g_B, the spin-spin distance r_{AB}, and the angular variables $\varepsilon, \eta, \alpha, \beta, \gamma$. A simple expression was derived by Pilbrow using the principal values

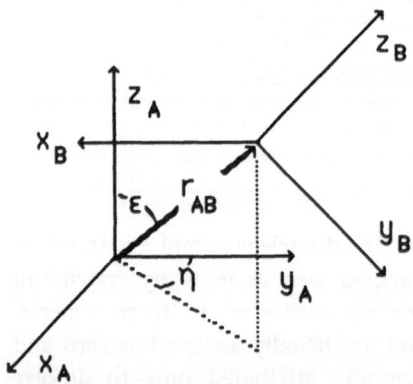

Fig. 3.11. General orientation of reference frame centered on spin center B with respect to the reference frame on spin center A. The orientation of the z axis of the dipolar tensor is specified by the angles ε and η

of the **g** tensors and the direction cosines of (x_B, y_B, z_B) and r_{AB} with respect to the (x_A, y_A, z_A) axes:

$$D_{ABij} = g_{Ai}\{\Sigma_\lambda g_{B\lambda} l_{j\lambda}(l_{i\lambda} - 3 l d_{ri}\Sigma_\mu l_{\mu\lambda} l_{r\mu})\}\mu_B^2/r_{AB}^3. \tag{3.44}$$

In (3.44) $\lambda \in \{x_B, y_B, z_B\}$, $\mu, i, j \in \{x_A, y_A, z_A\}$ and l_{ij} are the cosines of the angles between the i and j directions. The hamiltonian (3.40–43) can be written in the diagonal Zeeman representation using standard techniques as:

$$H_i = \mu_B B \cdot g_i \cdot S_{iz'} + K_i S_{iz'} I_{iz''} + \tau_{i1} S_{ix'} I_{ix''} + \tau_{i2} S_{iy'} I_{iy''} + \tau_{i3} S_{iy'} I_{ix''} + \\ + \tau_{i4} S_{ix'} I_{ix''} + \tau_{i5} S_{iy'} I_{iz''} \quad \{i = A, B\}, \tag{3.45}$$

where electron spin axes are denoted by ′ and nuclear spin axes by ″. The various quantities appearing in (3.45) are defined as follows:

$$g_i^2 = \Sigma_\mu g_\mu^2 l_\mu^2 \quad \{\mu = x_i, y_i, z_i, i = A, B\} \tag{3.46}$$

$$K_i^2 g_i^2 = \Sigma_\mu A_\mu^2 g_\mu^2 l_\mu^2 \tag{3.47}$$

$$\tau_{i1} = g_i A_z[(a_x^2 + a_y^2)/(g_x^2 l_x^2 + g_y^2 l_y^2)]^{\frac{1}{2}} \tag{3.48}$$

$$\tau_{i2} = A_x A_y[(g_x^2 l_x^2 + g_y^2 l_y^2)/(a_x^2 + a_y^2)]^{\frac{1}{2}}/(K_i g_i) \tag{3.49}$$

$$\tau_{i3} = a_x g_x l_x g_y l_y[A_y^2 - A_x^2]/\{[(a_x^2 + a_y^2)(g_x^2 l_x^2 + g_y^2 l_y^2)]^{\frac{1}{2}} K_i g_i\} \tag{3.50}$$

$$\tau_{i4} = g_z l_z(K_i^2 - A_z^2)/[K_i(g_x^2 l_x^2 + g_y^2 l_y^2)^{\frac{1}{2}}] \tag{3.51}$$

$$\tau_{i5} = [A_y^2 - A_x^2]g_x l_x g_y l_y/[K_i(g_x^2 l^2 + g_y^2 l_y^2)^{\frac{1}{2}}] \tag{3.52}$$

$$a_\mu = A_\mu g_\mu l_\mu/K_i g_i \quad \{\mu = x_i, y_i, z_i; i = A, B\}, \tag{3.53}$$

where l_x, l_y, and l_z are the direction cosines of **B** with respect to the x_i, y_i, and z_i axes. It must be stressed at this point that when ions A and B are dissimilar, the Zeeman quantization axes are in general different and can also be largely misaligned. The interaction term H_{AB} in the Zeeman representation becomes:

$$H_{AB} = \Sigma_{\mu'\lambda'}(D_{\mu'\lambda'} + J_{\mu'\lambda'})S_{A\mu'} S_{B\lambda'} \quad \{\mu = x_A, y_A, z_A; \lambda = x_B, y_B, z_B\} \tag{3.54}$$

with

$$D_{\mu'\lambda'} = \Sigma_{\mu\gamma\rho} D_{AB\mu\gamma} l_{\mu\mu'} l_{\rho\lambda'} l_{\gamma\rho} \quad \{\rho = x_B, y_B, z_B\} \tag{3.55}$$

$$J_{\mu'\lambda'} = J \Sigma_{\mu\lambda} l_{\mu\mu'} l_{\lambda\lambda'} l_{\mu\lambda}. \tag{3.56}$$

Equation (3.56) reduces to an isotropic interaction only for similar paramagnetic centers where $l_{\mu\lambda}=\delta_{\mu\lambda}$. In the general case the isotropic exchange interaction cannot be separated by the anisotropic interaction.

For centrosymmetric pairs $\{x_A, y_A, z_A\} \equiv \{x_B, y_B, z_B\}$ and (3.54) becomes:

$$H_{AB} = JS'_A \cdot S'_B + \Sigma_{\mu'\gamma'} D_{\mu'\gamma'} S_{A\mu'} S_{B\gamma'}. \tag{3.57}$$

Perturbation theory is conveniently carried out in the coupled representation defined by:

$$\begin{aligned}
|11\rangle &= |\tfrac{1}{2}\,\tfrac{1}{2}m_A m_B\rangle \\
|1-1\rangle &= |-\tfrac{1}{2}-\tfrac{1}{2}m_A m_B\rangle \\
|10\rangle &= (U+iV)|\tfrac{1}{2}-\tfrac{1}{2}m_A m_B\rangle + b|-\tfrac{1}{2}\tfrac{1}{2}m_A m_B\rangle
\end{aligned} \tag{3.58}$$

$$|00\rangle = (W+iX)|\tfrac{1}{2}-\tfrac{1}{2}m_A m_B\rangle + d|-\tfrac{1}{2}\tfrac{1}{2}m_A m_B\rangle, \tag{3.59}$$

where the kets on the right are product kets labeled as $|M_{S_A} M_{S_B} M_{I_A} M_{I_B}\rangle$ and the other symbols are defined as:

$$\begin{aligned}
&U = [(D_{xx}+D_{yy})+2J]/4f_1; \quad V = -(D_{yx}-D_{xy})/4f_1; \\
&W = [(D_{xx}+D_{yy})+2J]/4f_2; \quad X = -(D_{yx}-D_{xy})/4f_2; \\
&b = [2\varphi - K(m_A-m_B)]/2f_1; \quad d = [-2\varphi - K(m_A-m_B)]/2f_2 \\
&f_1 = \{1/16[(D_{xx}+D_{yy})+2J]^2 + 1/16(D_{yx}-D_{xy})^2 + [\varphi - 1/2K(m_A-m_B)]^2\}^{\frac{1}{2}}; \\
&f_2 = \{1/16[(D_{xx}+D_{yy})+2J]^2 + 1/16(D_{yx}-D_{xy})^2 \\
&\qquad + [-\varphi - 1/2K(m_A-m_B)]^2\}^{\frac{1}{2}};
\end{aligned} \tag{3.60}$$

$$\varphi = \{1/16[(D_{xx}+D_{yy})+2J]^2 + 1/16(D_{yx}-D_{xy})^2 + 1/4K^2(m_A-m_B)^2\}^{\frac{1}{2}}. \tag{3.61}$$

The hamiltonian (3.45) is not diagonal in the basis (3.58–59). Its representation matrix is shown in Table 3.7 where we have used the following definitions:

$$\begin{aligned}
S_1 &= (UD_{zx}+VD_{zy}+bD_{xz})/4 + ((U\tau_4+V\tau_5)m_B + b\tau_4 m_A)/2; \\
S_2 &= (VD_{zx}-UD_{zy}-bD_{yz})/4 + ((-U\tau_5+V\tau_4)m_B - b\tau_5 m_A)/2; \\
S_3 &= (D_{xx}-D_{yy})/4;
\end{aligned} \tag{3.62}$$

Table 3.7. Spin hamiltonian matrix for two interacting $S=\tfrac{1}{2}$ spins in the coupled representation[a]

	$|11\rangle$	$|10\rangle$	$|1-1\rangle$	$|00\rangle$
$|11\rangle$	$g\mu_B B+(D_{zz}+J)/4+$ $1/2\,K(m_A+m_B)$	S_1+iS_2	S_3+iS_4	S_7+iS_8
$|10\rangle$	S_1-iS_2	$-(D_{zz}+J)/4+\varphi$	S_5+iS_6	0
$|1-1\rangle$	S_3-iS_4	S_5+iS_6	$g\mu_B B+(D_{zz}+J)/4$ $-1/2\,K(m_A+m_B)$	S_9+iS_{10}
$|00\rangle$	S_7-iS_8	0	S_9-iS_{10}	$-(D_{zz}+J)/4-\varphi$

[a]The basis functions and the other symbols are defined in the text.

$S_4 = -(D_{xy} + D_{yx})/4;$
$S_5 = (-UD_{xz} + VD_{yz} - bD_{zx})/4 + ((U\tau_4 - V\tau_5)m_A + b\tau_4 m_B)/2;$
$S_6 = (VD_{xz} + UD_{yz} + bD_{zy})/4 - ((V\tau_4 + U\tau_5)m_A + b\tau_5 m_B)/2;$
$S_7 = (WD_{zx} + XD_{zy} + dD_{xz})/4 + ((W\tau_4 + X\tau_5)m_B + d\tau_4 m_A)/2;$
$S_8 = (XD_{zx} - WD_{zy} - dD_{yz})/4 + ((-W\tau_5 + X\tau_4)m_B - d\tau_5 m_A)/2;$
$S_9 = (-WD_{xz} + XD_{yz} - dD_{zx})/4 + ((W\tau_4 - X\tau_5)m_A + d\tau_4 m_B)/2;$
$S_{10} = (-XD_{xz} - WD_{yz} - dD_{zy})/4 + ((W\tau_5 + X\tau_4)m_A + d\tau_5 m_B)/2;$

Perturbation solutions can be obtained to the second order in energy yielding:

$$E(11) = g\mu_B B + (D_{zz} + J)/4 + K(m_A + m_B)/2 + (S_1^2 + S_2^2)/(E^0(11) - E^0(10)) +$$
$$+ (S_3^2 + S_4^2)/(E^0(11) - E^0(1-1)) + (S_7^2 + S_8^2)/(E^0(11) - E^0(00));$$
$$E(10) = -(D_{zz} + J)/4 + \varphi +$$
$$- (S_1^2 + S_2^2)/(E^0(11) - E^0(10)) + (S_5^2 + S_6^2)/(E^0(10) - E^0(1-1));$$
$$E(1-1) = -g\mu_B B + (D_{zz} + J)/4 - K(m_A + m_B)/2 - (S_3^2 + S_4^2)/(E^0(11) - E^0(1-1)) +$$
$$- (S_5^2 + S_6^2)/(E^0(10) - E^0(1-1)) - (S_9^2 + S_{10}^2)/(E^0(00) - E^0(10));$$
$$E(00) = -(D_{zz} + J)/4 - \varphi +$$
$$- (S_7^2 + S_8^2)/(E^0(11) - E^0(00)) + (S_9^2 + S_{10}^2)/(E^0(00) - E^0(1-1)).$$

(3.63)

The corresponding wave functions corrected at first order are given by (3.64):

$$|\psi_{11}\rangle = |11\rangle + (S_1 - iS_2)/(E^0(11) - E^0(10))|10\rangle + (S_3 - iS_4)/(E^0(11)$$
$$- E^0(1-1))|1-1\rangle + (S_7 - iS_8)/(E^0(11) - E^0(00))|00\rangle;$$
$$|\psi_{10}\rangle = |10\rangle - (S_1 + iS_2)/(E^0(11) - E^0(10))|11\rangle + (S_5 - iS_6)/$$
$$(E^0(10) - E^0(1-1))|1-1\rangle;$$

(3.64)

$$|\psi_{1-1}\rangle = |1-1\rangle - (S_3 + iS_4)/(E^0(11) - E^0(1-1))|11\rangle - (S_5 + iS_6)/$$
$$(E^0(10) - E^0(1-1))|10> - (S_9 - iS_{10})/(E^0(00) - E^0(1-1))|00\rangle;$$
$$|\psi_{00}\rangle = |00\rangle - (S_7 + iS_8)/(E^0(11) - 0(E^0(00))|11\rangle + (S_9 + iS_{10})/(E^0(00) -$$
$$E^0(1-1))|1-1\rangle.$$

Equations (3.63) and (3.64) can be used to obtain analytical expressions for the transition fields and intensities in the limit $J \ll h\nu$. In this limit the spin states (3.64) are not pure spin states and in general five transitions are expected corresponding to four $\Delta M_S = \pm 1$ transitions at zero order, and one $\Delta M_S = \pm 2$ transition. The first-order expressions for the transition fields are:

$$B_i = h\nu/g\mu_B \pm [(D_{zz} + J) - K(m_A + m_B)]/2g\mu_B \pm \varphi/g\mu_B \quad (i = 1-4) \tag{3.65}$$

for the four $\Delta M_S = \pm 1$ transitions and

$$B_5 = h\nu/2g\mu_B - K(m_A + m_B)/2g\mu_B. \tag{3.66}$$

for the $\Delta M_S = \pm 2$ transition.

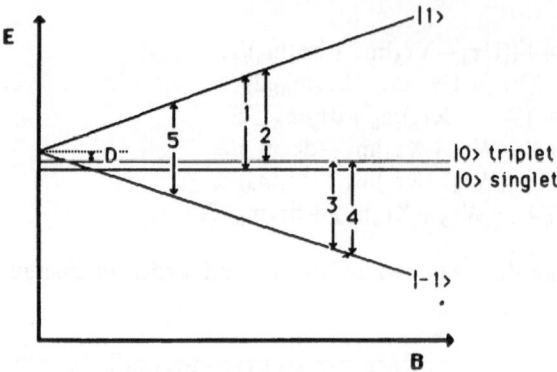

Fig. 3.12. Schematic representation of the spin levels and EPR transitions for an $S = 1$ system in the weak exchange limit. The levels are plotted for the static magnetic field parallel to the z axis of the D_{AB}^{dip} tensor. The graph represents $J_{AB} > 0$

Energy levels and transition fields are schematically shown in Fig. 3.12 for B parallel to the z axis of the D_{AB}^{dip} tensor. When the anisotropic interactions become vanishingly small, the φ term in (3.65) and (3.66) tends to the value $[J + K(m_A - m_B)]/2$ and the four $\Delta M_S = \pm 1$ transition fields reduce to three: two of them at $h\nu/g\mu_B$ and the other two separated at $(h\nu \pm J)/g\mu_B$; each transition is split into hyperfine lines separated by $K/2$ which is half the value observed in the mononuclear species as already encountered in the strong exchange limit. When the anisotropic interaction has sizeable values, the hyperfine splitting can be largely different from this value. It must be noted, however, that since B_5 in (3.66) does not depend on φ the hyperfine splitting is not affected at first order by the value of the anisotropic interaction. The transition probabilities strongly depend on the value of the anisotropic interaction and J. We will not work out here general expressions for the transition probabilities which can be found in [3.17a]. Particularly simple expressions are obtained at first order for the $\Delta M_S = \pm 1$ transitions in axial symmetry assuming a dipolar form for the anisotropic interaction:

$$P_{1,2} = (a+b)^2 \{1 \pm \{-\mu_B^2[g_\perp^2(g_\perp^2 + 2g_\parallel^2)\sin^2\theta/g^2]/(4\,r_{AB}^3)\}/h\nu\}/4; \qquad (3.67)$$

$$P_{3,4} = (a+b)^2 \{1 \mp \{-\mu_B{}^2[g^2{}_\perp(g_\perp^2 + 2g_\parallel^2)\sin^2\theta/g^2]/(4\,r_{AB}^3)\}/h\nu\}/4, \qquad (3.68)$$

where the numbers refer to the transitions shown in Fig. 3.12 and $a^2 = U^2 + V^2$, and $b^2 = W^2 + X^2$. The $\Delta M_S = \pm 2$ transition probability has no first-order contributions and the second-order ones can be expressed as:

$$P_5 \simeq \{\{\mu_B^2[((g_\perp^2 + 2g_\parallel^2)g_\parallel g_\perp \sin\theta\cos\theta)g^2]/(4\,r_{AB}^3)\}^2\}/(h\nu)^2 \cdot \qquad (3.69)$$

The only contribution to the $\Delta M_S = \pm 2$ transition comes from the anisotropic interaction and the intensity of this transition is expected to decrease on decreasing the value of the anisotropic interaction (increasing r_{AB} values) being zero for null anisotropic interaction.

The above formalism can still be applied for $J \gg h\nu$ by properly modifying the coefficients in (3.62) provided that anisotropic interactions are still smaller than $h\nu$. The above perturbation procedure becomes inadequate for $J \simeq h\nu$ since $|10\rangle$ and $|00\rangle$ are no longer close to the eigenstates of the system and a more general approach based on the diagonalization of the hamiltonian matrix appears more appropriate [3.18].

In the more general case of two different interacting centers $\{x_A, y_A, z_A\} \neq \{x_B, y_B, z_B\}$ perturbation theory can still be developed using the general form of the interaction hamiltonian (3.54–56). In this formalism the Zeeman terms of the two centers forming the pair are separately reduced to diagonal form and the isotropic exchange behaves as an anisotropic interaction. In this sense it is indistinguishable from the dipolar interaction and its value can be derived only by computer simulation of the spectra through (3.54–56). In order to apply the relevant equations derived above, it is sufficient to substitute the following relationships to Eq. (3.60):

$$
\begin{aligned}
&U = [(D_{xx} + D_{yy})]/4f_1; \; V = -(D_{yx} - D_{xy})/4f_1; \\
&W = [(D_{xx} + D_{yy})]/4f_2; \; X = -(D_{yx} - D_{xy})/4f_2; \\
&b = [2\varphi - (K_A m_A - K_B m_B) - B\mu_B(g_A - g_B)]/2f_1; \\
&d = [-2\varphi - (K_A m_A - K_B m_B) - B\mu_B(g_A - g_B)]/2f_2; \\
&f_1 = \{1/16[(D_{xx} + D_{yy})]^2 + 1/16(D_{yx} - D_{xy})^2 + [\varphi - 1/2(K_A m_A - K_B m_B) + \\
&\qquad - B\mu_B/2(g_A - g_B)]^2\}^{\frac{1}{2}}; \\
&f_2 = \{1/16[(D_{xx} + D_{yy})]^2 + 1/16(D_{yx} - D_{xy})^2 + [-\varphi - 1/2(K_A m_A - K_B m_B) + \\
&\qquad - B\mu_B/2(g_A - g_B)]^2\}^{\frac{1}{2}}; \\
&\varphi = \{1/16[(D_{xx} + D_{yy})]^2 + 1/16(D_{yx} - D_{xy})^2 + 1/4[(K_A m_A - K_B m_B) + \\
&\qquad + B\mu_B(g_A - g_B)]^2\}^{\frac{1}{2}}.
\end{aligned}
\tag{3.70}
$$

In Eqs. (3.62–63) the following changes should be made:

$$
\begin{aligned}
&\tau_4 m_B, \tau_5 m_B \to \tau_{24} m_B, \tau_{25} m_B; \; \tau_4 m_A, \tau_5 m_A \to \tau_{14} m_A, \tau_{15} m_A \\
&g = (g_A + g_B)/2; \; K(m_A + m_B) = K_A m_A + K_B m_B.
\end{aligned}
$$

The first-order equations for the transition fields become:

$$
B_i = h\nu/g\mu_B \pm [D_{zz} - (K_A m_A + K_B m_B)]/2g\mu_B \pm \varphi/g\mu_B \; (i = 1 - 4);
\tag{3.71}
$$

$$
B_5 = h\nu/2g\mu_B - (K_A m_A + K_B m_B)/2g\mu_B.
$$

In the limit situation of zero exchange and dipolar interactions φ is $1/2[(K_A m_A - K_B m_B) + B\mu_B(g_A - g_B)]$ and the EPR spectrum consists of two separate resonances at g_A and g_B with hyperfine splittings K_A and K_B, respectively.

3.5 Spectra of Pairs in the Weak Exchange Limit

In the weak exchange limit, and more generally when isotropic and anisotropic interactions have comparable magnitude, the singlet and triplet states are no longer pure spin states and the EPR spectra depend on both J_{AB} and D_{AB}. Weak anisotropic interactions between inequivalent centers result in EPR spectra which show separate resonances near the g values of the individual ions. These spectra have been referred to as AB patterns, in analogy with high resolution NMR spectra.

Following the angular dependence of the transition fields on single crystals, it is possible to measure isotropic and anisotropic contributions as well as hyperfine interactions. An example of the relative influence of isotropic, anisotropic, and hyperfine interactions on the appearance of single crystal EPR spectra is shown in Fig. 3.13. The combined isotropic and anisotropic interactions result in a further splitting of the hyperfine lines, which in the case of two $S = 1/2$ interacting spins are split into doublets.

In the literature only a few examples of single crystal spectra have been reported and much effort has been exerted to measure the spin hamiltonian parameters by computer simulation of polycrystalline powder or frozen solution spectra and a number of simulation procedures have been suggested [3.17–18]. In fluid solution the anisotropic interactions, which are represented by a traceless second-rank tensor, are averaged to zero by the tumbling motion of the molecule. Provided that the dinuclear structure is retained in the fluid solution, the EPR spectra can be used to estimate the J_{AB} value. This procedure has been applied to obtain the J_{AB} values of a wide variety of spin-labeled transition metal complexes of copper(II), silver(II), and oxovanadium(IV). The spectra generally show AB patterns, whose lines can be referred to as metal or radical according to the nature of the transition (g value) as $J_{AB} \to 0$. For J_{AB} values smaller than the g-value difference between the metal and radical resonances both the metal and radical resonances are split into doublets. On increasing the value of J_{AB} the intensity of the outer lines of the AB spectrum decreases and the positions of the inner metal and radical resonances tend to their average value. With this technique J_{AB} values in the range $10^{-4} - 10^{-1}$ cm^{-1} have been measured [3.17c].

The analysis of polycrystalline powder or frozen solution spectra is generally complicated by the fact that not all the $\Delta M_S = \pm 1$ transitions can be observed since they merge into a central broad line, because of the low value of D_{AB}, or can be obscured by the presence of monomeric impurities. Some information can, however, be extracted from the observation of the $\Delta M_S = \pm 2$ transition. The transition field in the polycrystalline powder spectrum is at

$$B_{min} = 1/(2\mu_B g_{min}) \, [(h\nu)^2 - 4(D_{AB}^2/3 + E_{AB}^2)]^{\frac{1}{2}}, \tag{3.72}$$

where $g_{min} = (g_x g_y \sin^2\alpha + g_z^2 \cos^2\alpha)^{\frac{1}{2}}$ and $\alpha^2 = \cos^{-1}\{[9 - 4(D_{AB}/h\nu)^2]/[27 - 36 (D_{AB}/h\nu)^2]\}$. In axial symmetry ($E_{AB} = 0$) it is possible to measure D_{AB} directly

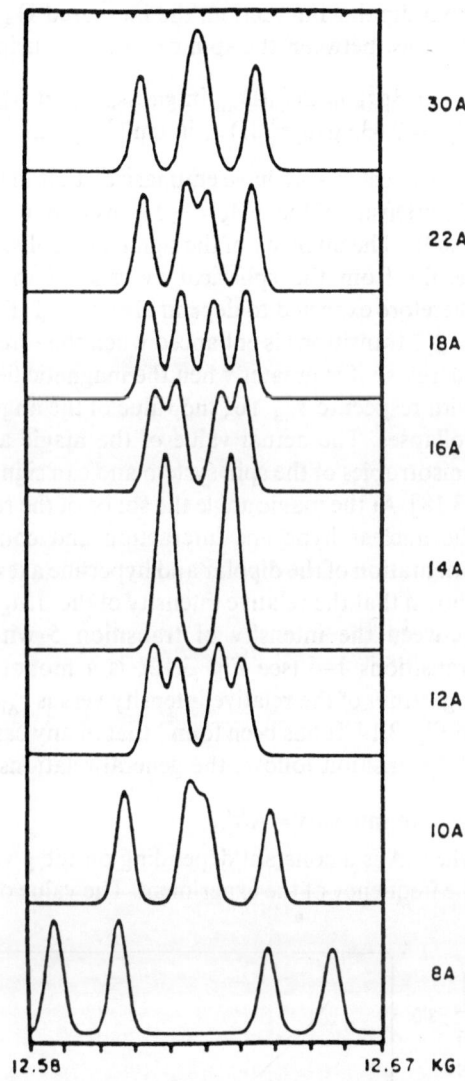

Fig. 3.13. Computed spectrum for the interaction of $S_A = 1/2$, $I^A = 1/2$ with $S_B = 1/2$ and no nuclear spin. The spectrum is computed along the z axis and r_{AB} is taken to be parallel to the z axis ($\varepsilon = 0$, $\eta = 0$). The g values are 1.956 and 1.899 for center A and B, respectively, $J_{AB} = 0.0013$ cm^{-1}. The D values are computed as a function of the distance indicated on the right-hand side of the figure [3.18]

from the position of the $\Delta M_S = \pm 2$ transition once one has measured independently the g values. Equation (3.72) can be applied only when singlet and triplet states are well separated in energy with respect to $h\nu$ ($D_{AB} \ll J_{AB} > h\nu$) and for $D_{AB} \gg A_A$, A_B. In this case the g values can be expressed to a good approximation through Eq. (3.20). In the hypothesis that all of the anisotropic interactions come

75

from dipolar interaction, the measured D_{AB} values can be used to estimate the distance between the spin centers, r_{AB}, using the relationship:

$$D_{AB} = 3g_{AB}^2\mu_B/2r_{AB}^3 (D_{AB} \text{ in gauss, } r_{AB} \text{ in Å});$$
$$D_{AB} = 0.433\, g_{AB}^2/r_{AB}^3 (D_{AB} \text{ in cm}^{-1}, r_{AB} \text{ in Å}). \tag{3.73}$$

In Sect. 3.4 we have emphasized that in the hypothesis of dipolar interaction, the intensity of the $\Delta M_S = \pm 2$ transition in single crystals depends on r_{AB}^{-6} [see e.g. (3.69)]. The intensity of the signal in a polycrystalline sample or a frozen solution results from the spherical average of the single crystal intensities and it is therefore expected to depend also on r_{AB}^{-6}. Furthermore, the intensity of the $\Delta M_S = \pm 2$ transitions is enhanced when the three spin levels are equally separated in energy so it is greater when the magnetic field is at about 54.7°, the magic angle with respect to r_{AB}, i.e., the value of the angle at which the dipolar fine structure collapses. The actual value of the magic angle is, however, dependent on the anisotropies of the spin system and can significantly deviate from the 54.7° value [3.18]. At the magic angle the shape of the resonance line is greatly influenced by the nuclear hyperfine interaction and could be used to monitor the relative orientation of the dipolar and hyperfine axes. By computer simulation it has been shown that the relative intensity of the $\Delta M_S = \pm 2$ transition, defined as the ratio between the intensity of transition 5 with respect to the total intensity of transitions 1–4 (see Fig. 3.12), is a monotonic function of r_{AB}. A plot of the logarithm of the relative intensity versus r_{AB}, computed for $g_A \simeq g_B = 2$, is shown in Fig. 3.14. It has been found that in any case the relative intensity of the $\Delta M_S = \pm 2$ transition follows the general relationship [3.20]:

$$\text{relative intensity} = A/r_{AB}^6, \tag{3.74}$$

where A is a constant depending on the g values of the two spin centers and on the frequency of the experiment. The value of the constant A at a given frequency

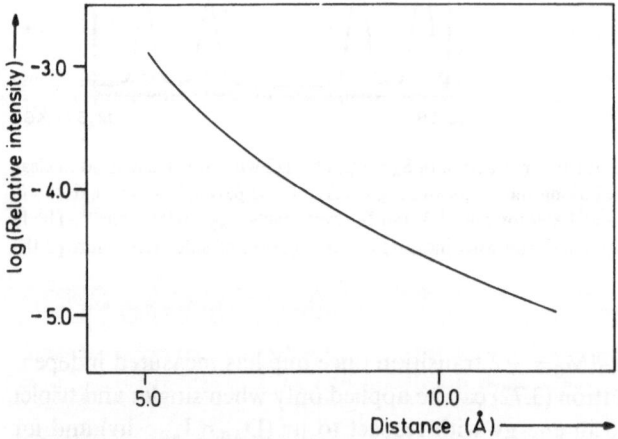

Fig. 3.14. Plot of the logarithm of the relative intensity of the $\Delta M_S = \pm 2$ transition versus r_{AB}. The curve is computed for $g_A \simeq g_B = 2$ [3.19]

Table 3.8. Comparison between distances[a] measured from EPR spectra and X-ray techniques

Compound	g_x	g_y	g_z	r_{AB}	r_{AB}(X-ray)	J_{AB}(cm^{-1})	Ref.
Zinc(II) doped bis(pyridine--N-oxide)copper(II) nitrate)	2.04	2.04	2.30	3.46(5)	3.46	-30	[3.17]
Bis(salicylaldeidato) copper(II) in chloroform	2.02	2.02	2.26	4.05(5)	4.05		[3.17]
Zinc(II) doped copper(II) diethyldithiocarbamate	2.02	2.01	2.07	3.85(5)	3.59	-24	[3.17]
Copper(II) d,l-tartrate in water-glycol solution	2.06	2.06	2.28	3.77(5)	2.99	18	[3.17]
Copper(II) salen in chloro-form-toluene solution	2.05	2.05	2.15	4.55(5)	3.18	20	[3.17]
1				4.48	4.45		[3.19]
2				5.7	≈ 6		[3.19]

[a]J values from magnetic susceptibility measurements. Compounds:

1
Dimer

2

and for given g_A and g_B can be determined by a least squares fit of the computed relative intensity of the transition with (3.74). This procedure can be followed for any value of hν and g and allows one to determine the A value appropriate to the system under examination. In the case of only one anisotropic g_i tensor the A constant takes the form:

$$A = (19.5 + 10.9 \Delta g)(9.100/\nu)^2, \tag{3.75}$$

where $\Delta g = |g_{max} - g_{min}|$. Equation (3.75) has been found to follow accurately the frequency dependence of A up to Q band frequency.

In order to make a comparison between the different methods to obtain r_{AB} from the EPR spectra we report in Table 3.8 the r_{AB} values measured in a number of compounds whose X-ray structure was determined.

3.6 Intermediate Exchange

In a number of experimental situations two or more terms in (3.1-3) can have comparable magnitudes and in this case no simple perturbation procedure can be applied. The eigensolutions of (3.1–3) are no longer pure spin states and the exact knowledge of the energies and eigenvectors is required to interpret the EPR spectra. This situation is expected to occur in principle in heterobimetallic couples in which at least one of the ions has $S > 1/2$ as well as in couples where $S_A = S_B > 1/2$. In fact, all these systems single ion anisotropies can be as large as or larger than J_{AB}.

A general interpretation of the EPR spectra is still possible for all those systems with an odd number of electrons, such as $S_A = \frac{1}{2}$, $S_B = 1$ [3.22]. In this case, in fact, the ground and the next excited state are always Kramers doublets and, if zero field splitting effects cause the excited states to occur at energies larger than $h\nu$ above the ground Kramers doublet, the observed EPR transitions are always within states belonging to the same Kramers doublet and the EPR spectra can be interpreted using an effective $S' = 1/2$ spin hamiltonian. Neglecting superhyperfine couplings a convenient form of the effective spin hamiltonian is

$$H_{eff} = \mu_B \mathbf{B} \cdot \mathbf{g}_{eff} \cdot \mathbf{S}' + \mathbf{I}^A \cdot \mathbf{A}_{eff}^A \cdot \mathbf{S}' + \mathbf{I}^B \cdot \mathbf{A}_{eff}^B \cdot \mathbf{S}'. \tag{3.76}$$

Since $\mathbf{S}'$ is an effective spin, its quantization axes can largely deviate from the laboratory system (to which $\mathbf{B}$ is referred) and from nuclear spin axes; this means that $\mathbf{g}_{eff}$ and $\mathbf{A}_{eff}^k$ in general do not transform like second-rank tensors upon rotation of the reference frame.

Let $|m\rangle$ and $|m'\rangle$ indicate the components of the ground Kramers doublet. A formal relationship between the effective spin $\mathbf{S}'$ and $\mathbf{S}_A$ and $\mathbf{S}_B$ spin operators is

$$\mathbf{S}_k = \mathbf{M}_k \cdot \mathbf{S}' \quad (k = A, B), \tag{3.77}$$

where the elements of the $\mathbf{M}_k$ matrices can be evaluated from the real and imaginary parts of the expectation value of the components of $\mathbf{S}_k$ over $|m\rangle$ and $|m'\rangle$ as shown in Table 3.9. The actual form of the $|m\rangle$ and $|m'\rangle$ states can be

Table 3.9. General expressions for the $\mathbf{M}_k$ matrix elements[a]

M_{kxx}	$2\,\mathrm{Re}\langle m|S_{kx}|m'\rangle$
M_{kxy}	$-2\,\mathrm{Im}\langle m|S_{kx}|m'\rangle$
M_{kxz}	$2\ \ \langle m|S_{kx}|m\rangle$
M_{kyx}	$2\,\mathrm{Re}\langle m|S_{ky}|m'\rangle$
M_{kyy}	$-2\,\mathrm{Im}\langle m|S_{ky}|m'\rangle$
M_{kyz}	$2\ \ \langle m|S_{ky}|m\rangle$
M_{kzx}	$2\,\mathrm{Re}\langle m|S_{kz}|m'\rangle$
M_{kzy}	$-2\,\mathrm{Im}\langle m|S_{kz}|m'\rangle$
M_{kzz}	$2\ \ \langle m|S_{kz}|m\rangle$

[a]$|m\rangle$ and $|m'\rangle$ are the components of the ground Kramers doublet. $k = A$ or B. Im and Re indicate the imaginary and real part of the matrix element, respectively.

obtained by diagonalization of the spin hamiltonian including isotropic and anisotropic interactions

$$H = J_{AB}S_A \cdot S_B + S_A \cdot D_{AB} \cdot S_B + d_{AB} \cdot S_A \times S_B + S_A \cdot D_A \cdot S_A + S_B \cdot D_B \cdot S_B, \tag{3.78}$$

By substituting (3.77) into the Zeeman and hyperfine terms in (3.1–2) we obtain the effective spin hamiltonian

$$H_{eff} = \mu_B B \cdot (g_A \cdot M_A + g_B \cdot M_B) \cdot S' + I^A \cdot (A_A^A \cdot M_A) \cdot S' + I^B \cdot (A_B^B M_B) \cdot S' \tag{3.79}$$

which, by comparison with (3.76), yields

$$g_{eff} = g_A \cdot M_A + g_B \cdot M_B; \tag{3.80}$$

$$A_{eff}^k = A_k^k \cdot M_k \quad (k = A, B) \tag{3.81}$$

Equations (3.80–81) provide a generalization of (3.20) and (3.22–23) where we have substituted general M_k matrices $(k = A, B)$ to constant factors.

The above formalism has been applied to interpret the EPR spectra of copper (II)-nickel (II) pairs [3.22] and the EPR spectra of doublet-triplet organometallic molecular pairs in diamagnetic host lattices [3.23]. In the following part of this section we will consider in some detail the EPR spectra of $(S_A = 1/2, S_B = 1)$ couples as an example.

From the isotropic exchange interaction two total spin states $S = 3/2$ and $S = 1/2$ arise. In general, the $S = 3/2$ state splits in zero field into two Kramers doublets so that a total of three Kramers doublets arise from the combined effect of the isotropic and nonzero anisotropic interactions in (3.78).

The energies of the three Kramers doublets are shown in Fig. 3.15 as a function of the ratio $R = J_{AB}/D_B$. For simplicity the curves have been computed from (3.78) for an axial D_B tensor and for $D_{AB} = 0$ and $d_{AB} = 0$ (D_A being always zero in the present case). The doublets in Fig. 3.15 are labeled in the limit $J_{AB} \gg D_B$. For an antiferromagnetic exchange interaction, $J_{AB} > 0$, the ground Kramers doublet is always $|1/2 \pm 1/2\rangle$ irrespective of the sign of D_B, while when the exchange interaction is ferromagnetic, $J_{AB} < 0$, the ground doublet is $|3/2 \pm 1/2\rangle$ for $D_B > 0$ and $|3/2 \pm 3/2\rangle$ for $D_B < 0$. Therefore, when the isotropic exchange interaction is ferromagnetic, the sign of the local zero field splitting of nickel (II) determines the nature of the ground doublet and from the EPR spectra it is thus possible to determine the sign of both J_{AB} and D_B.

Under the above assumptions the M_k matrices are diagonal and axial and expressions for their matrix elements can be easily obtained as a function of R since the eigenvalue problem can be solved analytically:

$$M_{A, zz} = \pm 1/\sqrt{(1 + e^2)}; \quad M_{A, xx} = M_{A, yy} = 1/2[\pm 1 + 1/\sqrt{(1 + e^2)}]; \tag{3.82}$$

$$M_{B, zz} = 1 \mp 1/\sqrt{(1 + e^2)}; \quad M_{B, xx} = M_{B, yy} = -\sqrt{2}e/\sqrt{(1 + e^2)} \tag{3.83}$$

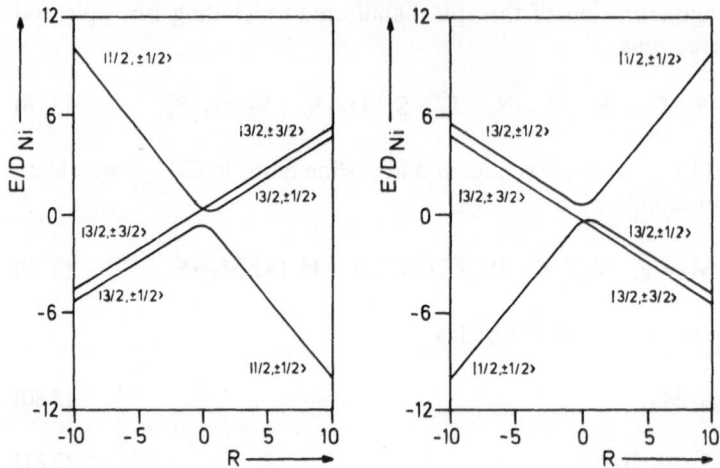

Fig. 3.15. Reduced energy (E/D_B) of the low lying Kramers doublets arising from the interaction between $S_A = 1/2$ and $S_B = 1$ spins as a function of the $R = J_{AB}/D_B$ ratio. The curves have been computed with $g_{Ax} = g_{By} = 2.06$, $g_{Az} = 2.35$, $g_B = 2.2$, $E_B/D_B = 0$. The levels are labeled in the $J_{AB} \gg D_B$ limit. *left* $D_B > 0$; *right* $D_B < 0$ [3.22]

for $D_B > 0$, and

$$M_{A,zz} = 1; \quad M_{A,xx} = M_{A,yy} = 0; \tag{3.84}$$

$$M_{B,zz} = 2; \quad M_{B,xx} = M_{B,yy} = 0 \tag{3.85}$$

for $D_B < 0$ and $R > 0$, and

$$M_{A,zz} = 1/\sqrt{(1+e)^2}; \quad M_{A,xx} = M_{A,yy} = 1/2[1 + 1/\sqrt{(1+e^2)}]; \tag{3.86}$$

$$M_{B,zz} = 1 - 1/\sqrt{(1+e^2)}; \quad M_{B,xx} = M_{B,yy} = -\sqrt{2}e/\sqrt{(1+e^2)} \tag{3.87}$$

for $D_B < 0$ and $R < 0$. The upper and lower sign in (3.82–87) holds for $R < 2$ and $R > 2$, respectively, and $e = \tan|\sqrt{8R/(2-R)}|$.

In more general cases than that considered above the solution of the eigenvalue problem requires numerical diagonalization of the hamiltonian matrix and no analytical expression for M_k is possible. The two matrices must be obtained from Table 3.9 using the eigenvectors obtained from the diagonalization of (3.78).

As an example we show in Figs. 3.16 and 3.17 the effective **g** and **A** tensors computed including only the B single ion anisotropy, in the axial case, and for $\lambda_B = E_B/D_B = 0.13$, for D_B positive and D_B negative, respectively. It is apparent that the largest deviations from the limiting values computed using (3.20–22) are computed on the A values. In fact, the computed g values closely approach the

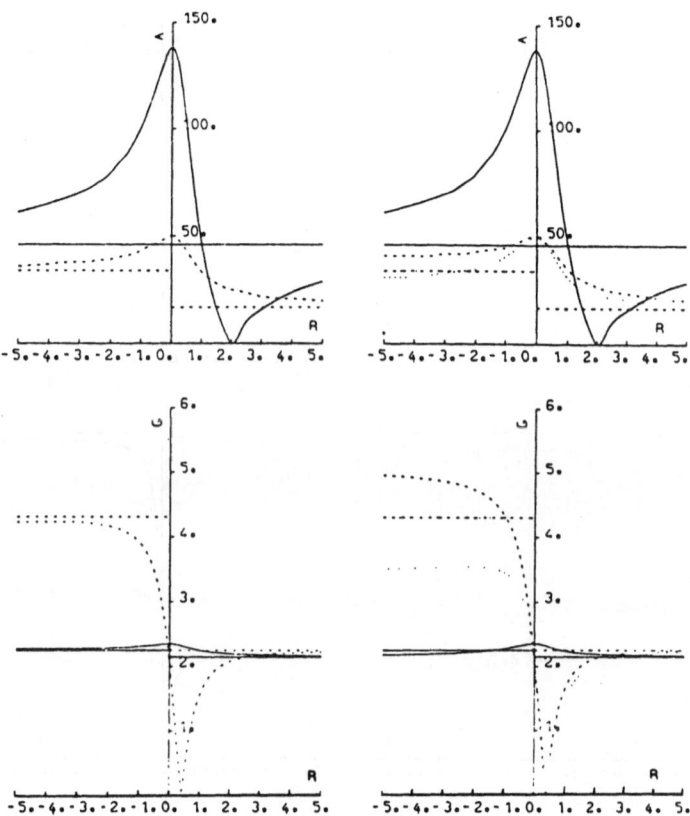

Fig. 3.16. Computed eigenvalues of the effective **g** (*lower*) and **A** (*upper*) tensors as a function of the R $=J_{AB}/D_B$ ratio with $D_B > 0$, $g_{Ax} = g_{By} = 2.06$, $g_{Az} = 2.35$, $g_B = 2.2$, $A_{Ax} = A_{Ay} = 50 \times 10^{-4}$ cm^{-1}, $A_{Az} = 138 \times 10^{-4}$ cm^{-1}. Figures on the *left* correspond to $E_B/D_B = 0$, and on the *right* to E_B/D_B $= 0.13$ *Full lines* represent the z component, *broken lines* the y component, and *dotted lines* the x component. *Horizontal lines* correspond to the strong exchange limiting values computed by (3.20–23) [3.22]

limiting values for $R > 3$, while the A values are still significantly different from the limiting values at $R = 5$. This is clearly a consequence of eqs (3.80) and (3.81) since the A_{eff}^k tensor depends on the M_k matrix only and even a small deviation of M_k from the constant value required by (3.23) is important. These calculations suggest the use of the deviation of A_{eff}^k from the limiting value of $1/3A_{Cu}^{Cu}$ as a sensitive probe for the estimation of the R ratio.

In the above discussion we have neglected both anisotropic and antisymmetric exchange interactions. Their effect is easily taken into account using the exact eigenvectors and Table 3.9. In general, their inclusion will cause the M_k matrices to have a general form and the effective tensors will thus be largely misaligned with respect to the single ions reference frames. The effect of anisotropic and antisymmetric terms has been considered with some detail in the

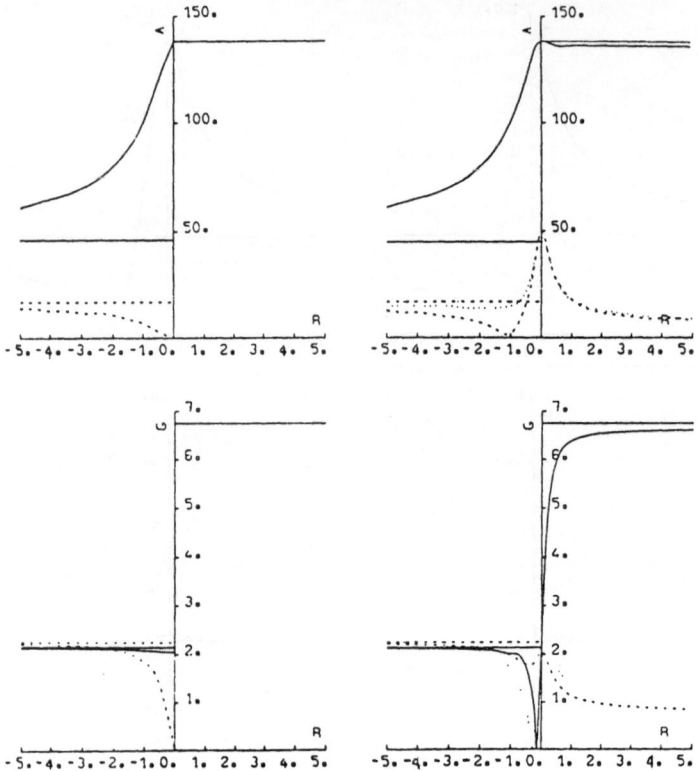

Fig. 3.17. Computed eigenvalues of the effective **g** (*lower*) and **A** (*upper*) tensors as a function of the $R = J_{AB}/D_B$ ratio with $D_B < 0$, $g_{Ax} = g_{By} = 2.06$, $g_{Az} = 2.35$, $g_B = 2.2$, $A_{Ax} = A_{Ay} = 50 \times 10^{-4}$ cm^{-1}, $A_{Az} = 138 \times 10^{-4}$ cm^{-1}. Figures on the *left* correspond to $E_B/D_B = 0$, and on the *right* to $E_B/D_B = 0.13$. *Full lines* represent the z component, *broken lines* the y component, and *dotted lines* the x component. *Horizontal lines* correspond to the strong exchange limiting values computed by (3.20-23) [3.22]

case when D_B is the leading term. In this situation it was found that at first order in perturbation theory the effective **g** tensor is given by $\mathbf{g}_{\text{eff}} = \mathbf{g}_A + \mathbf{g}_B M_B$ and it is no longer a symmetric matrix, while the **A** tensor is not affected by the interaction.

To conclude this section we will report an example where the above formalism was applied. In Fig. 3.18 the polycrystalline powder EPR spectra of nickel(II) doped tetra(μ-benzoato-O,O')bis(quinoline)dicopper(II) recorded at 4.2 K are shown [3.24]. A sketch of the structure of the pure copper complex is shown in Fig. 3.19 [3.25]. The single crystal spectra show transitions within one Kramers doublet and were interpreted with an effective $S' = 1/2$ spin hamiltonian with $g_x = 4.51$, $g_y = 3.44$, $g_z = 2.24$, and $A_z = 94 \times 10^{-4}$ cm^{-1}. Since both the pure copper(II) and the pure nickel(II) complexes are antiferromagnetically coupled with isotropic exchange constants of the order of 300 cm^{-1}, they cannot

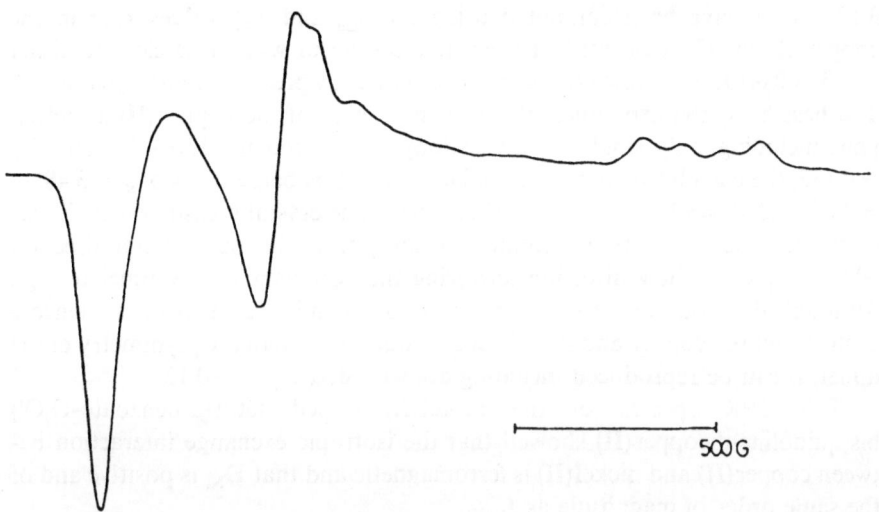

Fig. 3.18. Polycrystalline powder EPR spectra of nickel(II) doped tetra(μ-benzoato-O,O′)bis(quinoline)dicopper(II) recorded at X-band frequency and 4.2 K. The spectrum is plotted between 0 and 5000 Gauss

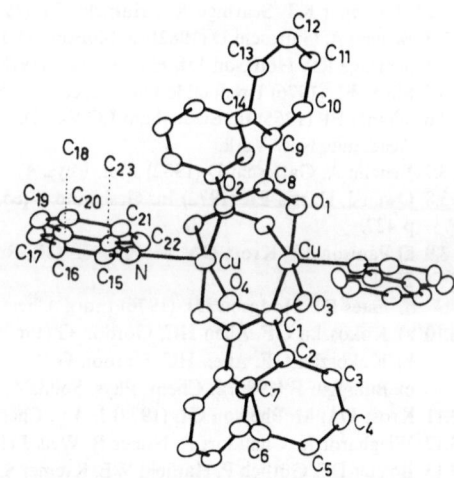

Fig. 3.19. ORTEP view of the dinuclear complex tetra(μ-benzoato-O,O′) bis(quinoline)dicopper(II) [3.25]

contribute to the EPR spectra at 4.2 K, which should arise only from copper(II)-nickel(II) couples. Using for A_{Cuz}^{Cu} the value seen in the spectra of the copper(II)–zinc(II), $A_{Cuz}^{Cu} = 138 \times 10^{-4}$ cm^{-1}, Eq. (3.81) gives $M_{Cuzz} \simeq 2/3$ instead of the value 1/3 expected in the strong exchange limit. From Figs. 3.16 and

3.17, which have been computed using the g_{Cu} and A_{Cu}^{Cu} values seen in the copper(II)-zinc(II) pair, we learn that this can occur when R is close to either $-1/3$ or 0.6. Only the first choice, however, correctly predicts g_x and g_y close to 4. The best fit to the experimental spin hamiltonian of the copper(II)-nickel(II) pair, including only single ion anisotropy, was found for $R = -1.2$ and $\lambda_{Ni} = 0.133$, the calculated spin hamiltonian parameters being $g_x = 4.5$, $g_y = 3.48$, $g_z = 2.26$, and $A_z = 93 \times 10^{-4}$ cm^{-1}. From the single crystal measurements it was found that the g_z direction is rotated 14° away from the metal–metal direction which would be the z direction idealizing the symmetry of the dimer to C_{4v}. Although this rotation can be ascribed to slight misalignment of the tensors centered on the copper and nickel centers, due to the actual C_i symmetry of the dimer, it can be reproduced including $\mathbf{d}_{AB}$ with $\mathbf{d}_{ABx}/J_{AB} = -0.12$.

The EPR spectra of the nickel(II) doped tetra(μ-benzoato-O,O') bis(quinoline)dicopper(II) showed that the isotropic exchange interaction between copper(II) and nickel(II) is ferromagnetic and that D_{Ni} is positive and of the same order of magnitude as J_{CuNi}.

References

3.1 a) Banwell CN, Primas H (1963) Mol. Phys. 6: 225.
   b) Belford GG, Belford RL, Burkhalter JF (1973) J. Magn. Reson. 11: 251.
3.2 McGregor KT, Scaringe RP, Hatfield WE (1975) Mol. Phys. 30: 1925
3.3 Bencini A, Gatteschi D (1982) In: Melson GA Figgis BN (eds) Transition metal chemistry. 8: 1
3.4 Scaringe RP, Hodgson DJ, Hatfield WE (1978) Mol. Phys. 35: 701.
3.5 Silver BL (1976) Irreducible tensor methods, Academic, New York.
3.6 Wigner EP (1965) In: Biedenham LC Van Dam (eds) Quantum theory of angular momentum. Academic, New York.
3.7 Bencini A, Gatteschi D (1982) Mol. Phys. 47: 161.
3.8 Owen J, Harris EA (1972) In: Geshwind S (ed.) Electron spin resonance Plenum New York, p 427.
3.9 a) Paulson JA, Krost DA, McPherson GL, Rogers LD, Atwood JL (1980) Inorg. Chem. 19: 2519.
   b) Estes DE Hodgson DJ, (1976) Inorg. Chem. 15: 348.
3.10 a) Kokoszka GF, Allen HC, Gordon G (1967) J. Chem. Phys. 46: 3020.
   b) Kokoszka GF, Allen HC, Gordon G (1967) J. Chem. Phys. 46: 3013.
   c) Buluggiu E (1980) J. Chem. Phys. Solids 41: 43.
3.11 Krost DA, McPherson GL (1978) J. Am. Chem. Soc. 100: 987.
3.12 Wieghardt K, Chauduri P, Nuber B, Weis J (1982) Inorg. Chem., 21: 3086.
3.13 Bolster DE, Gütlich P, Hatfield WE, Kremer S, Müller EW, Wieghardt K (1983) Inorg. Chem. 22: 1725.
3.14 Kremer S (1985) Inorg. Chem. 24: 887
3.15 Al'tshuler SA, Kozyrev BM (1974) Electron paramagentic resonance in compounds of transition elements, Wiley, New York.
3.16 Banci L, Bencini A, Gatteschi D (1981) Inorg. Chem. 20: 2734.
3.17 a) Smith TD, Pilbrow JR (1974) Coord. Chem. Rev. 13: 173.
   b) Carr SG, Smith TD, Pilbrow JR (1974) J. Chem. Soc. Faraday Trans. 2: 497
   c) Eaton SS, More KM, Sawant BM, Boymel PM, Eaton GR (1983) J. Magn. Reson. 52: 435.
3.18 Coffman RE, Buettner GR (1979) J. Chem. Phys. 83: 2392.
3.19 Eaton SS, More KM, Sawant BM, Eaton G R (1983) J. Am. Chem. Soc. 105: 6560.

3.20 Coffman RE, Pezeshk A (1986) J. Magn. Reson. 70: 21.
3.21 Boyd PWD, Toy AD, Smith TD, Pilbrow JR (1973) J. Chem. Soc. Dalton Trans., 1549.
3.22 Bencini A Gatteschi D (1985) Mol. Phys. 54: 969.
3.23 a) Hulliger J (1984) Ph.D. Thesis, University of Zürich, Zürich.
      b) Hulliger J (1987) Mol. Phys. 60: 97
3.24 Bencini A, Benelli C, Gatteschi D, Zanchini C (1980) J. Am. Chem. Soc. 102: 5820.
3.25 Bencini A, Gatteschi D, Mealli C (1979) Cryst. Struct. Commun. 8: 305.

85

4 Spectra of Clusters

4.1 The Spin Hamiltonian for Oligonuclear Systems of Interacting Spins

The spin hamiltonian appropriate to describe the EPR spectra of clusters of three or more weakly interacting paramagnetic centers can be written as an obvious extension of (3.1–3) and takes the form

$$H = \Sigma_i H_i + \Sigma_{i<j} H_{ij}. \tag{4.1}$$

In (4.1) H_i is the spin hamiltonian (3.1) of the individual spin center i and H_{ij} is the interaction hamiltonian (3.3) between couples of spins assumed to be symmetric with respect to the i and j indices. The sums in (4.1) extend over all the paramagnetic centers forming the cluster. It must be stressed here that (4.1) is not the most general hamiltonian to describe the magnetic and exchange inter-actions in a cluster. Such a hamiltonian, in fact, has the only requisite to be invariant with respect to the symmetry operations of the cluster and should include *n*-center interactions as well as spin operators of the form $j_{ij}(S_i \cdot S_j)^2$. Hamiltonian (4.1) is, however, the most widely used to interpret the EPR spectra and we will use this operator in the following. A number of general equations, which we will report in this chapter, can be, however, easily modified to include the j_{ij} terms.

The most commonly investigated clusters are the ones in which the isotropic exchange interaction is the dominant term in (4.1). In this case the eigenstates of (4.1) are most conveniently found among the eigenstates of S_z and S^2 where $S = \Sigma_i S_i$ is the total spin operator. This situation is completely analogous to that found for the dinuclear case and we will call this the strong exchange limit. In this situation the EPR transitions occur within a given S manifold or within states arising from a given S manifold and the EPR spectra can be interpreted using spin hamiltonians containing parameters which can be expressed as linear combinations of terms including single ion spin hamiltonian parameters and interaction terms.

When the above assumptions are no longer valid, the interpretation of the EPR spectra requires the knowledge of the eigenstates of (4.1), which must be obtained through matrix diagonalization. These eigenstates can then be used as zero-order functions to which one can apply the Zeeman perturbation. This approach, which we already noted to be hardly applicable to the dinuclear case, becomes practically unworkable when applied to higher nuclearity clusters.

In the literature a number of trinuclear clusters have been carefully investigated through EPR spectroscopy, only a few tetranuclear clusters have been studied, and very little has been done on larger clusters. In almost all the cases the strong exchange limit was applied. Thus, in the following we will develop the necessary formalism to compute the energy levels of trinuclear clusters in the strong exchange limit and we will give the necessary information to extend the formalism to larger clusters. Later we will derive the equations for interpreting the EPR spectra of trinuclear systems in the strong exchange limit.

4.2 Spin Levels of Exchange Coupled Clusters in the Strong Exchange Limit

In the strong exchange limit we can label the spin levels of the cluster according to the eigenvalues of S^2 and S_Z. These operators, in fact, commute with the isotropic part of (4.1). Since we are dealing with more than two spins, the eigenstates of S_Z and S^2 are not uniquely determined by the spin quantum numbers S and M_S, but a number of additional quantum numbers are required according to the different "coupling schemes" of the interacting spins. Consider, for example, the simplest case of the coupling of three spins $S_i (i = A, B, C)$. The total spin can be obtained by coupling spins S_A and S_B first to give $S_{AB} = S_A + S_B$ and then S_{AB} and S_C to give $S = S_{AB} + S_C$. The resulting states will be conveniently labeled using the eigenvalues of the commuting observables $\{S_A^2, S_B^2, S_{AB}^2, S_C^2, S^2, S_Z\}$ as $|S_A S_B S_{AB} S_C, SM\rangle$. In the case of three spins $\frac{1}{2}$ the allowed spin states are:

$$
\begin{array}{lll}
|\tfrac{1}{2}\tfrac{1}{2} 1 \tfrac{1}{2}, 3/2\, M\rangle & \text{4 states} - 3/2 \le M \le 3/2; & \\
|\tfrac{1}{2}\tfrac{1}{2} 1 \tfrac{1}{2}, \tfrac{1}{2} M\rangle & \text{2 states} -\tfrac{1}{2} \le M \le \tfrac{1}{2}; & (4.2) \\
|\tfrac{1}{2}\tfrac{1}{2} 0 \tfrac{1}{2}, \tfrac{1}{2} M\rangle & \text{2 states} -\tfrac{1}{2} \le M \le \tfrac{1}{2}; &
\end{array}
$$

corresponding to one quartet, $S = 3/2$, and two doublet, $S = 1/2$, states. We might as well couple S_B and S_C first to give $S_{BC} = S_B + S_C$ and then S_A and S_{BC} to give $S = S_A + S_{BC}$. With this coupling scheme we obtain states like $|S_A S_B S_C S_{BC}, SM\rangle$, corresponding to the same eigenvalues of S^2 and S_Z as the states $|S_A S_B S_{AB} S_C, SM\rangle$. From the coupling of three spins $\frac{1}{2}$ we can obtain a maximum of eight spin functions orthogonal to each other which span the $|SM\rangle$ space; therefore the two eightfold orthonormal sets $|S_A S_B S_{AB} S_C, SM\rangle$ and $|S_A S_B S_C S_{BC}, SM\rangle$ are not linearly independent and must be transformed one into the other by a unitary transformation. In general, this means that any $|S_A S_B S_{AB} S_C, SM\rangle$ ket can be expressed as a linear combination of $|S_A S_B S_C S_{BC}, SM\rangle$ kets according to:

$$|S_A S_B S_{AB} S_C, SM\rangle = \Sigma \langle S_A S_B S_{AB} S_C, SM | S_A S_B S_C S_{BC}, SM\rangle |S_A S_B S_C S_{BC}, SM\rangle, \quad (4.3)$$

where the sum is over all the allowed values of the intermediate spin coupling quantum number S_{BC}. The expansion coefficients in (4.3) are just the scalar products between kets taken from the two coupling schemes and are independent of M (scalar products do not depend on the orientation of the coordinate system). These coefficients, which are also called recoupling coefficients, are most commonly written using a 6-j symbol, so called for obvious reasons, to which they are related by a phase factor according to

$$
\begin{Bmatrix} S_A S_B S_{AB} \\ S_C S S_{BC} \end{Bmatrix} = (-1)^{S_A + S_B + S_C + S} [(2S_{AB} + 1)(2S_{BC} + 1)]^{-\frac{1}{2}} \times \tag{4.4}
$$
$$
\langle S_A S_B S_{AB} S_C, SM | S_A S_B S_C S_{BC}, SM \rangle,
$$

The above arguments can be extended to the coupling of four spins to give states of the form $|S_A S_B S_{AB} S_C S_D S_{CD}, SM\rangle$. All these kets are related by unitary transformations using 9-j symbols

$$
\begin{Bmatrix} S_A & S_B & S_{AB} \\ S_C & S_D & S_{CD} \\ S_{AC} & S_{BD} & S \end{Bmatrix} = [(2S_{AB} + 1)(2S_{CD} + 1)(2S_{AC} + 1)(2S_{BD} + 1)]^{-\frac{1}{2}} \times \tag{4.5}
$$
$$
\langle S_A S_B S_{AB} S_C S_D S_{CD}, SM | S_A S_C S_{AC} S_B S_D S_{BD}, SM \rangle.
$$

The most relevant properties of the 6-j and 9-j symbols are given in Appendix B. Numerical tables of 6-j and 9-j symbols are given in Refs. [4.1–3].

Recoupling of five, six, and seven spins are possible through 12-, 15-, and 18-j symbols [4.4] which will not be used in this book.

In the most general case the isotropic exchange hamiltonian for a rn-nuclear cluster

$$
H_{ex}^{(n)} = \Sigma_{i < j} J_{ij} S_i \cdot S_j \tag{4.6}
$$

commutes with S^2 and S_Z, but does not commute with all of the intermediate spin coupling operators. Since we use a representation whose states are eigenstates of the squares of the intermediate spin coupling operators and of S^2 and S_Z, the representation matrix of (4.6) will be block diagonal, each block corresponding to states with the same value of S independent of the eigenvalues of the other operators forming the commuting set of observables of our representation. The full representation matrix of (4.6) can be built up by recurrence starting from one pair of spins and adding one spin at a time, thus reducing the problem always to the coupling of two spins, or using the general vector coupling scheme. The first approach is slowly convergent especially when dealing with $S_i > \frac{1}{2}$, due to the large number of spin states to be considered, while the second method becomes unworkable for clusters with more than four atoms, since it requires the use of 12-j and higher multiplicity symbols which are difficult to handle without the help of a fast computer. Vector coupling techniques are, however, very helpful in dealing with trinuclear systems and coupling this technique with the recurrence method for larger clusters is probably the best way to operate.

Let us first examine the case of three interacting spins. It is convenient to rewrite the interaction hamiltonian in (4.1) in a generalized form as

$$H_{int}^{(3)} = \Sigma_k [2k+1]^{\ddagger} O_k(k_1 k_2 k_{12} k_3) \cdot \{\{T_{k1}(S_1) \otimes T_{k2}(S_2)\}_{k12} \otimes T_{k3}(S_3)\}_k =$$
$$= \Sigma_k [2k+1]^{\ddagger} O_k(k_1 k_2 k_{12} k_3) \cdot X_k(k_1 k_2 k_{12} k_3), \tag{4.7}$$

where $T_k(S_i)$ is a k-rank irreducible tensor operator of the operator variable S_i [4.5] and the spin centers are labeled with numbers instead of letters for the sake of clarity. We will adopt this notation in the rest of the chapter. O_k is a k-rank irreducible tensor operator which contains the spin hamiltonian parameters. Operator (4.7) is equivalent to the interaction spin hamiltonian H_{ij} in (4.1). Let us consider, for example, the $J_{12}S_1 \cdot S_2$ part of the isotropic exchange interaction. In (4.7) it is obtained by putting $T_{k3}(S_3) = 1$, which implies $k_3 = 0$ since the identity operator is zero rank, $k = 0$, and $k_{12} = 0$ as

$$H_{int}(12) = O_0(1100) \cdot X_0(1100) = O_0(1100)\{T_1(S_1) \otimes T_1(S_2)\}_0. \tag{4.8}$$

Since

$$S_1 \cdot S_2 = -\sqrt{3}\{T_1(S_1) \otimes T_1(S_2)\}_{00}, \tag{4.9}$$

we get

$$H_{int}(12) = -1\sqrt{3} O_{00}(1100)S_1 \cdot S_2 = J_{12}S_1 \cdot S_2, \tag{4.10}$$

where we have defined

$$O_{00}(1100) = -\sqrt{3} J_{12}. \tag{4.11}$$

In a similar way we can define the O_k tensors to express the interaction hamiltonian in the usual way. It must be noted that in (4.7) we included also terms of the form $S_i^2 \cdot S_j^2$ as well as terms with higher spin powers and multicenter interactions.

The matrix elements of X_k can be efficiently computed using the Wigner-Eckart theorem [4.6]:

$$\langle S_1 S_2 S_{12} S_3 SM | X_{kq} | S_1' S_2' S_{12}' S_3' S'M' \rangle = (-1)^{S-M} \begin{pmatrix} S & k & S' \\ -M & q & M' \end{pmatrix}$$
$$\times \langle S_1 S_2 S_{12} S_3 S \| X_k \| S_1' S_2' S_{12}' S_3' S' \rangle, \tag{4.12}$$

where the reduced matrix element is given by the general equation

$$\langle S_1 S_2 S_{12} S_3 S \| X_k \| S'_1 S'_2 S'_{12} S'_3 S' \rangle = [(2S+1)(2S'+1)(2k+1)(2S_{12}+1)$$
$$(2S'_{12}+1)(2k_{12}+1)]^{\ddagger}$$

$$\begin{Bmatrix} S_{12} & S'_{12} & k_{12} \\ S_3 & S'_3 & k_3 \\ S & S' & k \end{Bmatrix} \begin{Bmatrix} S_1 & S'_1 & k_1 \\ S_2 & S'_2 & k_2 \\ S_{12} & S'_{12} & k_{12} \end{Bmatrix} \langle S_1 \| T_{k1}(S_1) \| S'_1 \rangle \times \tag{4.13}$$

$$\times \langle S_2 \| T_{k2}(S_2) \| S'_2 \rangle \langle S_3 \| T_{k3}(S_3) \| S'_3 \rangle.$$

General expressions for the reduced matrix elements of zero-, first-, and second-rank tensor operators appearing in the right-hand side of (4.13) are given in Table 3.2.

One of the possible forms of the generalized interaction spin hamiltonian for a tetranuclear cluster is:

$$H_{int}^{(4)} = \Sigma_k [2k+1]^{\ddagger} O_k(k_1 k_2 k_{12} k_3 k_{123} k_4 k) \cdot$$
$$\{\{\{T_{k1}(S_1) \otimes T_{k2}(S_2)\}_{k12} \otimes T_{k3}(S_3)\}_{k123} \otimes T_{k4}(S_4)\}_k =$$
$$= \Sigma_k [2k+1]^{\ddagger} O_k(k_1 k_2 k_{12} k_3 k_{123} k_4 k) \cdot X_k(k_1 k_2 k_{12} k_3 k_{123} k_4 k), \tag{4.14}$$

where we have coupled a tensor operator of rank k_1 to a tensor operator of rank k_2 to give a tensor operator of rank k_{12} ($|k_1 - k_2| \leq k_{12} \leq k_1 + k_2$); then we coupled this operator to a tensor operator of rank k_3 to get a tensor operator of rank k_{123} ($|k_{12} - k_3| \leq k_{123} \leq k_{12} + k_3$) which finally couples with a fourth tensor operator of rank k_4 to give a total tensor operator of rank k ($0 \leq |k_{123} - k_4| \leq k \leq k_{123} + k_3 \leq 2$). Since the tensor operator of rank k_{12}, $T_{k12}(S_{12})$ $= \{T_{k1}(S_1) \otimes T_{k2}(S_2)\}_{k12}$, is always obtained, in this coupling scheme, by coupling the same two tensor operators $T_{k1}(S_1)$ and $T_{k2}(S_2)$ we can rewrite (4.14) in a way formally similar to (4.7) as

$$H_{int}^{(4)} = \Sigma_k [2k+1]^{\ddagger} O_k(k_{12} k_3 k_{123} k_4 k) \cdot \{\{T_{k12}(S_{12}) \otimes T_{k3}(S_3)\}_{k123} \otimes T_{k4}(S_4)\}_k =$$
$$= \Sigma_k [2k+1]^{\ddagger} O_k(k_{12} k_3 k_{123} k_4 k) \cdot X_k(k_{12} k_3 k_{123} k_4 k). \tag{4.15}$$

The matrix elements of (4.15) can be computed through (4.12) and (4.13). The reduced matrix element of $T_{k12}(S_{12})$ needed in (4.13) can be computed as

$$\langle S_1 S_2 S_{12} \| T_{k12}(S_{12}) \| S_1 S_2 S'_{12} \rangle = [(2S_{12}+1)(2S'_{12}+1)(2k_{12}+1)]^{\ddagger} \times$$

$$\begin{Bmatrix} S_1 & S_1 & k_1 \\ S_2 & S_2 & k_2 \\ S_{12} & S'_{12} & k_{12} \end{Bmatrix} \times$$

$$\times \langle S_1 \| T_{k1}(S_1) \| S_1 \rangle \langle \| S_2 \| T_{k2}(S_2) \rangle. \tag{4.16}$$

Repeated applications of (4.12), (4.13), and (4.15) allow one to compute the matrix elements of any interaction operator for clusters containing more than four spin centers.

It must be noted here that although the above equations seem rather complicated to be handled, they offer the great advantage to be close equations which can be easily included in a general computer program which deals with any spin system.

4.2.1 Spin Levels of Trinuclear Clusters

The most general arrangement of three centers is on the vertex of a general triangle (Fig. 4.1).

Each spin interacts with the adjacent ones and the strength of these interactions determines the magnetic geometry of the triad. When the three coupling constants are equal, we have a regular triad (represented by an equilateral triangle). When two of them, e.g. J_{13} and J_{23}, are equal but different from the other, J_{12}, we have a symmetric triad (represented by an isosceles triangle) and when all the coupling constants are different we have a general triad (represented by a scalene triangle). A limiting situation can occur when one of the coupling constants, e.g. J_{13}, is zero. This corresponds to a linear arrangement of the three spin centers (linear triad) when only nearest neighbor interactions are considered. This approximation is the most widely used in the treatment of spin clusters and most often only interactions between adjacent spins are included in (4.6). The linear triad is symmetric when the two coupling constants J_{12} and J_{23} have the same value.

The spin hamiltonian (4.6) describing the isotropic exchange interaction between three paramagnetic centers in a general triad takes the form:

$$H_{int}^{(3)} = J_{12}S_1 \cdot S_2 + J_{13}S_1 \cdot S_3 + J_{23}S_2 \cdot S_3. \tag{4.17}$$

It can be rewritten in the generalized form taking the $k=0$ term in (4.7) and summing over all the $k_i k_j (i \le j)$ pairs as

$$H_{int}^{(3)} = O_{00}(1100)X_{00}(1100) + O_{00}(1011)X_{00}(1011) + O_{00}(0111)X_{00}(0111), \tag{4.18}$$

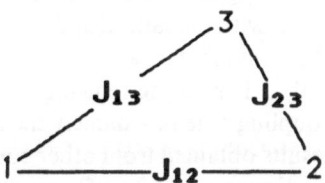

Fig 4.1. Exchange coupling constants for the coupling of three spins

where
$$O_{00}(1100) = -\sqrt{3}J_{12};$$
$$O_{00}(1011) = -\sqrt{3}J_{13};$$
$$O_{00}(0111) = -\sqrt{3}J_{23}. \tag{4.19}$$

Application of (4.13) gives the following reduced matrix elements:

$$\langle S_1S_2S_{12}S_3S\|X_0(1100)\|S_1S_2S'_{12}S_3S'\rangle = -\delta_{SS'}\delta_{S_{12}S'_{12}}\sqrt{(2S+1)}/\sqrt{3} \times$$
$$\times [S_{12}(S_{12}+1) - S_1(S_1+1) - S_2(S_2+1)] = -\delta_{SS'}\delta_{S_{12}S'_{12}}\sqrt{(2S+1)}/\sqrt{3} \times A_{12} \tag{4.20}$$

$$\langle S_1S_2S_{12}S_3S\|X_0(1011)\|S_1S_2S'_{12}S_3S'\rangle = (-1)^{S_1+S_2+S_3+S}\delta_{SS'}\sqrt{(2S+1)}/\sqrt{3} \times$$
$$[(2S_{12}+1)(2S'_{12}+1)(2S_1+1)(S_1+1)S_1(2S_3+1)(S_3+1)S_3]^{\frac{1}{2}} \times$$
$$\times \begin{Bmatrix} S_{12} & S'_{12} & 1 \\ S_3 & S_3 & S \end{Bmatrix} \times \begin{Bmatrix} S_1 & S_1 & 1 \\ S'_{12} & S_{12} & S_2 \end{Bmatrix} = (-)^{S_1+S_2+S_3+S}\delta_{SS'}\sqrt{(2S+1)}/\sqrt{3} \times A_{13} \tag{4.21}$$

$$\langle S_1S_2S_{12}S_3S\|X_0(0111)\|S_1S_2S'_{12}S_3S'\rangle = (-1)^{S'_{12}+S_{12}+S_1+S_2+S_3+S}$$
$$\delta_{SS'}\sqrt{(2S+1)}/\sqrt{3} \times$$
$$[(2S_{12}+1)(2S'_{12}+1)(2S_1+1)(S_1+1)S_1(2S_3+1)(S_3+1)S_3]^{1/2}$$
$$\times \begin{Bmatrix} S_{12} & S'_{12} & 1 \\ S_3 & S_3 & S \end{Bmatrix} \begin{Bmatrix} S_{12} & S'_{12} & 1 \\ S_2 & S_2 & S_1 \end{Bmatrix} =$$
$$= (-1)^{S'_{12}+S_{12}+S_1+S_2+S_3+S}\delta_{SS'}\sqrt{(2S+1)}/\sqrt{3} \times A_{23} \tag{4.22}$$

where the definition of A_{ij} is obvious.

By using (4.12), (4.18), and (4.19) one easily obtains the following non-zero matrix elements:

$$\langle S_1S_2S_{12}S_3SM|J_{12}S_1 \cdot S_2|S_1S_2S_{12}S_3SM'\rangle = \delta_{MM'}J_{12}/2\,A_{12} \tag{4.23}$$

$$\langle S_1S_2S_{12}S_3SM|J_{13}S_1 \cdot S_3|S_1S_2S'_{12}S_3SM'\rangle = \delta_{MM'}(-1)^{S_1+S_2+S_3+S+1}J_{13}A_{13} \tag{4.24}$$

$$\langle S_1S_2S_{12}S_3SM|J_{23}S_2 \cdot S_3|S_1S_2S'_{12}S_3SM'\rangle = \delta_{MM'}$$
$$(-1)^{S'_{12}+S_{12}+S_1+S_2+S_3+S+1}J_{23}A_{23} \tag{4.25}$$

We thus found from (4.20) and (4.23) that the operator $J_{12}S_1 \cdot S_2$ couples states with the same S and S_{12} and from (4.24–25) and the symmetry properties of the 6-j symbols that the other matrix elements are zero unless $\Delta S_{12} = |S_{12} - S'_{12}| = 0, 1$. It must be remembered here that we are now using one of the possible coupling schemes, namely the $S_{12} + S_3$, and care must be taken when comparing results obtained from other coupling schemes in the labeling of the energy levels.

Equations (4.23–25) can also be expressed in an analytical form [4.7] using algebraic formulae of the 6-j symbols [4.8–9] and we obtain the following

92

expressions for the diagonal and nondiagonal matrix elements of (4.17):

$$\langle S_{12}SM|H^{(3)}_{int}|S_{12}SM\rangle = \tfrac{1}{2}J_{12}[S_{12}(S_{12}+1)-S_1(S_1+1)-S_2(S_2+1)]+$$
$$+\{(J_{13}+J_{23})+(J_{13}-J_{23})[S_1(S_1+1)-S_2(S_2+1)]/[S_{12}(S_{12}+1)]\}\times$$
$$\times[S(S+1)-S_3(S_3+1)-S_{12}(S_{12}+1)]/4; \qquad (4.26)$$

$$\langle S_{12}SM|H^{(3)}_{int}|(S_{12}-1)SM\rangle = (J_{23}-J_{13})/(4S_{12})[(S+S_3+S_{12}+1)(S_{12}+S_3-S)\times$$
$$\times(S+S_{12}-S_3)(S+S_3-S_{12}+1)(S_{12}+S_1+S_2+1)(S_{12}+S_2-S_1)\times$$
$$\times(S_{12}+S_1-S_2)(S_1+S_2-S_{12}+1)/(4S^2_{12}-1)]^{\frac{1}{2}}. \qquad (4.27)$$

It is apparent from (4.27) that the matrix elements connecting states with $\Delta S_{12}=\pm1$ are zero when $J_{13}=J_{23}$ and the hamiltonian matrix is diagonal. This is the case of a symmetric triad. For a regular triad all the coupling constants are equal and spin states with the same S are at the same energy:

$$\langle S_{12}SM|H^{(3)}_{int}|S_{12}SM\rangle = J[S(S+1)-S_1(S_1+1)-S_2(S_2+1)-S_3(S_3+1)]. \qquad (4.28)$$

For a symmetric linear triad the hamiltonian matrix is diagonal and the diagonal elements have the form:

$$\langle S_{12}SM|H^{(3)}_{int}|S_{12}SM\rangle = J[S(S+1)-S_3(S_3+1)-S_{12}(S_{12}+1)]/2. \qquad (4.29)$$

Explicit expressions for the hamiltonian matrix of a number of general triples are given in Table 4.1.

The reduced energies of the spin levels for trinuclear systems with $1/2$ spins are reported in Fig. 4.2 as a function of the $r=J_{13}/J_{23}$ and $R=J_{12}/J_{23}$ ratios. From the coupling the following spin states arise: $|1\ 3/2\rangle$, $|1\ 1/2\rangle$, and $|0\ 1/2\rangle$. Figure 4.2a shows the energy levels for a symmetric triple ($r=1$). In this case the hamiltonian matrix is always diagonal independently of the R value. When $R=1$ we have a regular triad and the $S=1/2$ states are degenerate. The energy levels for a linear triad are shown in Fig. 4.2b. When $R=1$ we have a symmetric linear triad. In Fig. 4.2c and d the energy levels for general triads with $r=0.5$ and -0.5, respectively, are shown. The ground state of the triad is always a doublet state when r is negative and smaller than $\simeq -0.5$ independently of the sign of J_{23}. The quartet state is the ground state only for $R>0$ and $J_{23}<0$, i.e., when all the pairwise interactions are ferromagnetic.

4.2.2 Spin Levels of Tetranuclear Clusters

On increasing the number of interacting nuclei the topology of the magnetic interaction becomes more complicate since the number of possible geometrical arrangements of the spin centers is larger. For clusters with more than three spins, in fact, we pass from a plane topology to a space topology. Fortunately not all the possible topological structures for an *n*-vertex cluster can be found in

Table 4.1. Explicit expression of the nonzero matrix elements[a] of the isotropic exchange operator for general triads with $\frac{1}{2} \leq S_1 = S_2 = S_3 \leq 5/2$

S_1	S_{12}	S'_{12}	S	J_{12}	$J_{23}+J_{13}$	$J_{23}-J_{13}$
1/2	1	1	3/2	1/4	1/4	0
	1	1	1/2	1/4	−1/2	0
	0	0	1/2	−3/4	0	0
	1	0	1/2	0	0	$\sqrt{3}/4$
1	2	2	3	1	1	0
	2	2	2	1	−1/2	0
	1	1	2	−1	1/2	0
	2	1	2	0	0	$\sqrt{3}/2$
	2	2	1	1	−3/2	0
	1	1	1	−1	−1/2	0
	0	0	1	−2	0	0
	2	1	1	0	0	$\sqrt{5}/(2\sqrt{3})$
	1	0	1	0	0	$2/\sqrt{3}$
	1	1	0	−1	−1	0
3/2	3	3	9/2	9/4	9/4	0
	3	3	7/2	9/4	0	0
	2	2	7/2	−3/4	3/2	0
	3	2	7/2	0	0	$3\sqrt{3}/4$
	3	3	5/2	9/4	−7/4	0
	2	2	5/2	−3/4	−1/4	0
	1	1	5/2	−11/4	3/4	0
	2	3	5/2	0	0	$2\sqrt{2}/\sqrt{5}$
	1	2	5/2	0	0	$3\sqrt{7}/(2\sqrt{5})$
	3	3	3/2	9/4	−3	0
	2	2	3/2	−3/4	−3/2	0
	1	1	3/2	−11/4	−1/2	0
	0	0	3/2	−15/4	0	0
	3	2	3/2	0	0	$3\sqrt{7}/(4\sqrt{5})$
	2	1	3/2	0	0	$2\sqrt{15}/5$
	1	0	3/2	0	0	$5\sqrt{3}/4$
	2	2	1/2	−3/4	−9/4	0
	1	1	1/2	−11/4	−5/4	0
	2	1	1/2	0	0	$\sqrt{3}/2$
2	4	4	6	4	4	0
	4	4	5	4	1	0
	3	3	5	0	3	0
	4	3	5	0	0	$\sqrt{3}$
	4	4	4	4	−3/2	0
	3	3	4	0	1/2	0
	2	2	4	−3	2	0
	4	3	4	0	0	$3\sqrt{11}/(2\sqrt{7})$
	3	2	4	0	0	$\sqrt{40}/\sqrt{7}$
	4	4	3	4	−7/2	0
	3	3	3	0	−3/2	0
	2	2	3	−3	0	0
	1	1	3	−5	1	0
	4	3	3	0	0	$5\sqrt{3}/(2\sqrt{7})$
	3	2	3	0	0	$6\sqrt{6}/\sqrt{35}$

94

Table 4.1 (*continued*)

S_1	S_{12}	S'_{12}	S	J_{12}	$J_{23} + J_{13}$	$J_{23} - J_{13}$
	2	1	3	0	0	$\sqrt{42}/\sqrt{5}$
	4	4	2	4	-5	0
	3	3	2	0	-3	0
	2	2	2	-3	$-3/2$	0
	1	1	2	-5	$-1/2$	0
	0	0	2	-6	0	0
	4	3	2	0	0	$3/\sqrt{7}$
	3	2	2	0	0	$12/\sqrt{35}$
	2	1	2	0	0	$21/(2\sqrt{15})$
	1	0	2	0	0	$\sqrt{12}$
	3	3	1	0	-4	0
	2	2	1	-3	$-5/2$	0
	1	1	1	-5	$-3/2$	0
	3	2	1	0	0	$2\sqrt{2}/\sqrt{5}$
	2	1	1	0	0	$3\sqrt{7}/(2\sqrt{5})$
5/2	5	5	15/2	25/4	25/4	0
	5	5	13/2	25/4	5/2	0
	4	4	13/2	5/4	5	0
	5	4	13/2	0	0	$\sqrt{75}/4$
	5	5	11/2	25/4	$-3/4$	0
	4	4	11/2	5/4	7/4	0
	3	3	11/2	$-11/4$	15/4	0
	5	4	11/2	0	0	$\sqrt{56}/3$
	4	3	11/2	0	0	$5\sqrt{13}/6$
	5	5	9/2	25/4	$-7/2$	0
	4	4	9/2	5/4	-1	0
	3	3	9/2	$-11/4$	1	0
	2	2	9/2	$-23/4$	5/2	0
	5	4	9/2	0	0	$\sqrt{91}/4$
	4	3	9/2	0	0	$4\sqrt{5}/\sqrt{7}$
	3	2	9/2	0	0	$3\sqrt{165}/(4\sqrt{7})$
	5	5	7/2	25/4	$-23/4$	0
	4	4	7/2	5/4	$-13/4$	0
	3	3	7/2	$-11/4$	$-5/4$	0
	2	2	7/2	$-23/4$	1/4	0
	1	1	7/2	$-31/4$	5/4	0
	5	4	7/2	0	0	2
	4	3	7/2	0	0	$5\sqrt{11}/(2\sqrt{7})$
	3	2	7/2	0	0	$6\sqrt{3}/\sqrt{7}$
	2	1	7/2	0	0	$3\sqrt{2}$
	5	5	5/2	25/4	$-15/2$	0
	4	4	5/2	5/4	-5	0
	3	3	5/2	$-11/4$	-3	0
	2	2	5/2	$-23/4$	$-3/2$	0
	1	1	5/2	$-31/4$	$-1/2$	0
	0	0	5/2	$-35/4$	0	0
	5	4	5/2	0	0	$5\sqrt{11}/12$
	4	3	5/2	0	0	$20/(3\sqrt{7})$
	3	2	5/2	0	0	$81/(4\sqrt{35})$

Table 4.1 (*continued*)

S_1	S_{12}	S'_{12}	S	J_{12}	$J_{23}+J_{13}$	$J_{23}-J_{13}$
	2	1	5/2	0	0	$16/\sqrt{15}$
	1	0	5/2	0	0	$35/(4\sqrt{3})$
	4	4	3/2	5/4	$-25/4$	0
	3	3	3/2	$-11/4$	$-17/4$	0
	2	2	3/2	$-23/4$	$-11/4$	0
	1	1	3/2	$-31/4$	$-7/4$	0
	4	3	3/2	0	0	$5\sqrt{3}/(2\sqrt{7})$
	3	2	3/2	0	0	$6\sqrt{6}/\sqrt{35}$
	2	1	3/2	0	0	$\sqrt{42}/\sqrt{5}$
	3	3	1/2	$-11/4$	-5	0
	2	2	1/2	$-23/4$	$-7/2$	0
	3	2	1/2	0	0	$3\sqrt{3}/4$

[a]The matrix elements are labeled as in Eqs. (4.23–25). The last three columns are the coefficients which multiply the specified J_{ij} values. The final form of the matrix elements is given as the sum of the contributions along each line.

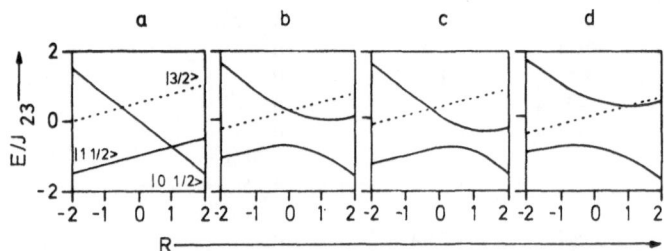

Fig. 4.2. Dependence of the reduced energies, E/J_{23}, for a trinuclear cluster with 1/2 spins on the $R = J_{12}/J_{23}$ ratio with (from *left* to *right*) $r = J_{13}/J_{23} = 1, 0, 0.5, -0.5$, respectively

nature and we can thus avoid the complication of a general treatment of the exchange interaction in tetranuclear clusters to a few cases which have been studied. Any interested reader can of course perform the calculation of the spin levels of any cluster following the equations of Sect. 4.1.

The most common arrangements of four spins arise from the tetrahedral geometry shown in Fig. 4.3.

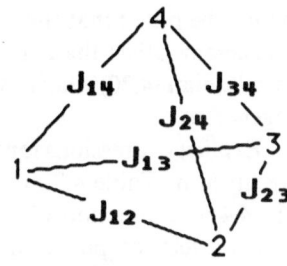

Fig. 4.3. Exchange coupling constants for the coupling of four spins

In the most general arrangement six exchange coupling constants are needed to describe the magnetic interaction. In a regular tetrahedral geometry all the spin centers are equivalent and the exchange interaction can be described by the hamiltonian:

$$H_{int}^{(T_d)} = J\Sigma_{i<j} S_i \cdot S_j. \tag{4.30}$$

Only one exchange coupling constant, $J = J_{12} = J_{13} = J_{14} = J_{23} = J_{24} = J_{34}$, can in this case account for the exchange interaction.

Two common distortions from the tetrahedral geometry will be now considered, one preserving an S_4 symmetry like a symmetric tetragonal deformation and the other preserving a C_{2v} symmetry corresponding to unequal, pairwise distortions. The spin hamiltonians relevant to describe the exchange interaction in these cases are:

$$H_{int}^{S_4} = J(S_1 \cdot S_2 + S_3 \cdot S_4) + J'(S_1 \cdot S_3 + S_1 \cdot S_4 + S_2 \cdot S_3 + S_2 \cdot S_4); \tag{4.31}$$

$$H_{int}^{C_{2v}} = J_{12}S_1 \cdot S_2 + J_{34}S_3 \cdot S_4 + J''(S_1 \cdot S_3 + S_1 \cdot S_4 + S_2 \cdot S_3 + S_2 \cdot S_4). \tag{4.32}$$

In (4.31) $J = J_{12} = J_{34}$ and $J' = J_{13} = J_{14} = J_{23} = J_{34}$, and in (4.32) $J'' = J_{13} = J_{14} = J_{23} = J_{34}$.

The eigenvalues of (4.32) can be easily found in the coupling scheme $|S_1 S_2 S_{12} S_3 S_4 S_{34} S\rangle$ since (4.32) commutes with both S_{12}^2, S_{23}^2, and S^2. It is convenient to rewrite (4.32) as

$$H_{int}^{C_{2v}} = \tfrac{1}{2}J_{12}(S_{12}^2 - S_1^2 - S_2^2) + \tfrac{1}{2}J_{34}(S_{34}^2 - S_3^2 - S_4^2) + J''(S_{12} \cdot S_{34}), \tag{4.33}$$

which gives:

$$\langle S_{12}S_{34}SM|H_{int}^{C_{2v}}|S_{12}S_{34}SM\rangle = \tfrac{1}{2}J_{12}[S_{12}(S_{12}+1) - S_1(S_1+1) - S_2(S_2+1)] + \\ + \tfrac{1}{2}J_{34}[S_{34}(S_{34}+1) - S_3(S_3+1) - S_4(S_4+1)] + \\ + \tfrac{1}{2}J''[S(S+1) - S_{12}(S_{12}+1) - S_{34}(S_{34}+1)]. \tag{4.34}$$

It must be noted that the symmetry of the hamiltonian (4.33) does not allow for degeneracy other than the spin degeneracy itself. This is not the case for the hamiltonian (4.30) whose symmetry requires all the states with the same S to be degenerate.

Explicit expressions for the energy of the spin states for tetrads of spin $\frac{1}{2}$ and 1 are given in Table 4.2. The energy levels of a tetrad of $\frac{1}{2}$ spins are plotted in Fig. 4.4 as a function of the $R = J_{12}/J_{34}$ and $r = J''/J_{34}$ ratios. In Fig. 4.4a the energy levels computed for a tetrad of S_4 symmetry ($R = 1$) are shown. It should be observed that the $|101\rangle$ and $|011\rangle$ triplet levels are accidentally degenerate. The quintet state is expected to be the ground state for pairwise ferromagnetic interactions ($r > 0$) and for strong ferromagnetic J'' interaction ($r < -2$, i.e. $J'' < -2J_{34}$). The point at $r = 1$ corresponds to the tetrahedral cluster in which the coupling constants are equal; in this case all the levels with the same S are degenerate. On passing to tetrads of C_{2v} symmetry (Fig. 4.4b, c) removes the accidental degeneracy of the $|101\rangle$ and $|011\rangle$ triplets.

It is appropriate to report here the general equation for computing the matrix elements of the generalized interaction hamiltonian in the $|S_1S_2S_{12}S_3S_4S_{34}S\rangle$ coupling scheme which can be helpful in evaluating the matrix elements of

Table 4.2. Energies of the spin states[a] for C_{2v} tetrads with 1/2 and 1 spins

S_1	S_{12}	S_{34}	S	J_{12}	J'_{34}	J''	S_1	S_{12}	S_{34}	S	J_{12}	J_{34}	J''
1/2	1	1	2	1/4	1/4	1	1	2	2	4	1	1	4
	1	1	1	1/4	1/4	−1		2	2	3	1	1	0
	1	1	0	1/4	1/4	−2		2	2	2	1	1	−3
	1	0	1	1/4	−3/4	0		2	2	1	1	1	−5
	0	1	1	−3/4	1/4	0		2	2	0	1	1	−6
	0	0	0	−3/4	−3/4	0		2	1	3	1	−1	2
								2	1	2	1	−1	−1
								2	1	1	1	−1	−3
								2	0	2	1	−2	−3
								1	2	3	−1	1	−3
								1	2	2	−1	1	2
								1	2	1	−1	1	−3
								1	1	2	−1	−1	1
								1	1	1	−1	−1	−1
								1	1	0	−1	−1	−2
								1	0	1	−1	−2	−1
								0	2	2	−2	1	−3
								0	1	1	−2	1	−1
								0	0	0	−2	−2	0

[a]For any spin state the appropriate energy is obtained by adding together the coefficients times the spin coupling constants along each line.

98

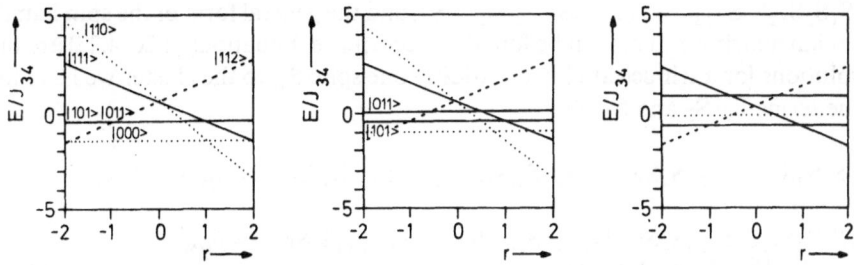

Fig. 4.4. Dependence of the reduced energies, E/J_{34}, for a tetranuclear cluster with 1/2 spins on the $r = J''/J_{34}$ ratio with (from *left* to *right*) $R = J_{12}/J_{34} = 1, 0.5, -0.5$, respectively

operators of first and second rank. The generalized hamiltonian in the present coupling scheme takes the form:

$$H_{int}^{(4)} = \Sigma_k [2k+1]^{\frac{1}{2}} O_k(k_1 k_2 k_{12} k_3 k_4 k_{34} k) \cdot \{\{T_{k1}(S_1) \otimes T_{k2}(S_2)\}_{k12} \otimes$$
$$\otimes \{T_{k3}(S_3) \otimes T_{k4}(S_4)\}_{k34}\}_k = \Sigma_k [2k+1]^{\frac{1}{2}} O_k(k_1 k_2 k_{12} k_3 k_4 k_{34} k) \cdot$$
$$\cdot X_k(k_1 k_2 k_{12} k_3 k_4 k_{34} k). \tag{4.35}$$

The reduced matrix element of X_k can be easily computed through standard tensor operator techniques as:

$$\langle S_1 S_2 S_{12} S_3 S_4 S_{34} S \| X_k \| S_1 S_2 S'_{12} S_3 S_4 S'_{34} S' \rangle =$$
$$= [(2S+1)(2S'+1)(2k+1)(2S_{12}+1)(2S'_{12}+1)(2k_{12}+1)(2S_{34}+1)$$
$$(2S'_{34}+1)(2k_{34}+1)]^{\frac{1}{2}} \times$$

$$\times \begin{Bmatrix} S_{12} & S'_{12} & k_{12} \\ S_{34} & S'_{34} & k_{34} \\ S & S' & k \end{Bmatrix} \begin{Bmatrix} S_1 & S_1 & k_1 \\ S_2 & S_2 & k_2 \\ S_{12} & S'_{12} & k_{12} \end{Bmatrix} \begin{Bmatrix} S_3 & S_3 & k_3 \\ S_4 & S_4 & k_4 \\ S_{34} & S'_{34} & k_{34} \end{Bmatrix} \times$$

$$\times \langle S_1 \| T_{k1} \| S_1 \rangle \langle S_2 \| T_{k2} \| S_2 \rangle \langle S_3 \| T_{k3} \| S_3 \rangle \langle S_4 \| T_{k4} \| S_4 \rangle. \tag{4.36}$$

4.2.3 Spin Levels of Linear Clusters

In Sect. 4.2.1 we have found the energy levels of a linear trinuclear cluster as a particular case of the general trinuclear cluster. In the following we will give a general solution for the energy levels of a linear N atom cluster. The appropriate spin hamiltonian is

$$H_{int}^{(l)} = \Sigma_i J_{ii+1} S_i \cdot S_{i+1}, \tag{4.37}$$

where the summation is extended to N-1 centers. The appropriate states can be built up adding the N-th spin to the N-1 cluster, i.e.,

$|S_1 S_2 S_{12} S_3 S_{123} \ldots \ldots \ldots S \ldots_{n-2\,n-1} S\rangle$, and the general form of the spin hamiltonian matrix elements can be found by recurrence. Equations (4.20, 4.22) are the solutions for a trinuclear cluster. Adding one spin, S_4, to the cluster we have for the terms $S_1 \cdot S_2$ and $S_2 \cdot S_3$:

$$\langle S_1 S_2 S_{12} S_3 S_{123} S_4 SM | J_{12} S_1 \cdot S_2 | S_1 S_2 S_{12} S_3 S_{123} S_4 SM' \rangle = \delta_{MM'} J_{12}/2 \; A_{12} \quad (4.38)$$

$$\langle S_1 S_2 S_{12} S_3 S_{123} S_4 SM | J_{23} S_2 \cdot S_3 | S_1 S_2 S'_{12} S_3 S'_{123} S_4 SM' \rangle = \delta_{MM'}$$
$$(-1)^{S'_{12}+S_{12}+S_1+S_2+S_3+S+1} J_{23} A_{23} \quad (4.39)$$

with A_{12} and A_{23} defined in (4.20, 4.22) since both operators commute with S_4. For the $S_3 \cdot S_4$ term we note that Eq. (4.13) can be applied since the operator does not involve S_1 and S_2. This gives:

$$\langle S_1 S_2 S_{12} S_3 S_{123} S_4 SM | J_{34} S_3 \cdot S_4 | S_1 S_2 S_{12} S_3 S'_{123} S_4 SM' \rangle = \delta_{MM'}$$
$$(-1)^{S'_{123}+S_{123}+S_{12}+S_3+S_4+S+1} J_{34} A_{34} \quad (4.40)$$

where
$$A_{34} = [(2S_{123}+1)(2S'_{123}+1)(2S_3+1)(S_3+1)S_3(2S_4+1)(S_4+1)S_4 \times$$
$$\times \begin{Bmatrix} S_{123} & S'_{123} & 1 \\ S_4 & S_4 & S \end{Bmatrix} \begin{Bmatrix} S_{123} & S'_{123} & 1 \\ S_3 & S_3 & S_{12} \end{Bmatrix}. \quad (4.41)$$

One can easily prove on this ground that adding the center n to a chain of $n-1$ requires the addition of matrix elements of the type:

$$\langle S_1 S_2 S_{12} \ldots S \ldots_{n-2\,n-1} S_n SM | J_{n-1\,n} S_{n-1} \cdot S_n | S_1 S_2 S_{12} \ldots S'_{\ldots n-2\,n-1} S_n SM' \rangle$$
$$= \delta_{MM'} (-1)^{S'_{\ldots n-2\,n-1}+S_{\ldots n-2\,n-1}+S_{n-2}+S_{n-1}+S_n+S+1} J_{n-1\,n} A_{n-1\,n} \quad (4.42)$$

with
$$A_{n-1\,n} = [(2S_{\ldots n-2\,n-1}+1)(2S'_{\ldots n-2\,n-1}+1)(2S_{n-1}+1)$$
$$(S_{n-1}+1)S_{n-1}(2S_n+1)(S_n+1)S_n]^{\frac{1}{2}}$$
$$\times \begin{Bmatrix} S_{\ldots n-2\,n-1} & S'_{\ldots n-2\,n-1} & 1 \\ S_n & S_n & S \end{Bmatrix} \begin{Bmatrix} S_{\ldots n-2\,n-1} & S'_{\ldots n-2\,n-1} & 1 \\ S_{n-1} & S_{n-1} & S_{n-2} \end{Bmatrix}. \quad (4.43)$$

Equation (4.43) has been used to compute the energy matrix and the eigenvalues and eigenvectors for chains with $\frac{1}{2} \leq S_i \leq 5/2$ and $5 \leq N \leq 11$ [4.10].

4.3 EPR Spectra of Exchange Coupled Clusters in the Strong Exchange Limit

In order to interpret the EPR spectra of clusters of exchange coupled centers we must add to the spin hamiltonian (4.6) describing the isotropic exchange interaction the other terms in (4.1), which describe the Zeeman, nuclear hyperfine, and zero field splitting interactions. They have the form:

$$H_i = \mu_B B \cdot g_i \cdot S_i + \Sigma_k I^k \cdot A^k_i \cdot S_i \quad (i = 1, 2, \ldots, n); \tag{4.44}$$

$$H_{ij} = \Sigma_i S_i \cdot D_i \cdot S_i + \Sigma_{i < j} S_i \cdot D_{ij} \cdot S_j, \tag{4.45}$$

where the i and j indices run over all the paramagnetic centers forming the cluster and k runs over all the nuclei with nonzero nuclear spin.

In the strong exchange limit the EPR transitions occur within the S levels which are thermally populated at the temperature of the experience. The EPR spectra can thus be interpreted using one or more effective S spin hamiltonians of the from:

$$H_{(S=S')} = \mu_B B \cdot g_S \cdot S + \Sigma_k I^k \cdot A^k_S \cdot S + S \cdot D_S \cdot S. \tag{4.46}$$

In the particular case in which the isotropic exchange hamiltonian for a n-nuclear cluster, $H^{(n)}_{ex}$ of Eq. (4.6), commutes with all of the intermediate spin coupling operators, i.e., its representation matrix is diagonal, a relationship between the spin hamiltonian parameters in (4.44–45) and the spin hamiltonian parameters in the operator equivalent (4.46) can be easily set up. This situation occurs, for example, in a symmetric triad or in a C_{2v} tetrad, as it has been shown in the preceding section.

In the following part of this section we will explicitly consider trinuclear clusters, the extension to larger clusters being only a matter of more algebra.

In the case of a symmetric trinuclear cluster the spin hamiltonian parameters in (4.44–45) and (4.46) are related by:

$$g_S = c_1 g_1 + c_2 g_2 + c_3 g_3; \tag{4.47}$$

$$A^k_S = c_1 A^k_1 + c_2 A^k_2 + c_3 A^k_3; \tag{4.48}$$

$$D_S = d_1 D_1 + d_2 D_2 + d_3 D_3 + d_{12} D_{12} + d_{13} D_{13} + d_{23} D_{23}. \tag{4.49}$$

In (4.47–49), due to the symmetry of the system, we also have $T_2 = T_3$ and $D_{13} = D_{23}$, where T_i is any single ion second rank tensor. The c and d coefficients can be expressed according to (3.18) as

$$c_i = \langle S_{12} S \| T_1(S_i) \| S_{12} S \rangle / \langle S \| T_1(S) \| S \rangle; \tag{4.50}$$

$$d_i = \langle S_{12} S \| T_2(S_i) \| S_{12} S \rangle / \langle S \| T_2(S) \| S \rangle; \tag{4.51}$$

$$d_{ij} = \langle S_{12} S \| T_2(S_i S_j) \| S_{12} S \rangle / \langle S \| T_2(S) \| S \rangle. \tag{4.52}$$

It must be noted now that the actual form of the c and d coefficients depends on the coupling scheme one chooses since the reduced matrix elements in the numerator in (4.50–52) are actually dependent on the coupling scheme [see Eq. (4.13)].

Relevant expressions for the c and d coefficients obtained using Eq. (4.13) are shown in Table 4.3. In Table 4.4 we report the numerical values of these coefficients for a number of symmetric triples. In any case we used the $\{S_{12}S_3S\}$ coupling scheme.

Since $T_1(S) = T_1(S_1) + T_1(S_2) + T_1(S_3)$ and $T_2(S) = T_2(S_1) + T_2(S_2) + T_2(S_3) + 2\Sigma_{i<j}T_2(S_iS_j)$ the c and d coefficients are not linearly independent, but from (4.50–52) it follows that

$$c_1 + c_2 + c_3 = 1 \tag{4.53}$$

$$d_1 + d_2 + d_3 + 2d_{12} + 2d_{13} + 2d_{23} = 1 \tag{4.54}$$

with obvious extensions to larger clusters. It is also easy to show that for the

Table 4.3. Relevant expressions[a] for the c and d coefficients of a symmetric triad

$$c_{1,2} = (-1)^{2(S_1 + S_2 + S_3 + 2S_{12} + S)}[S' + S'_{12} - S'_3][S'_{12} + S'_{1,2} - S'_{2,1}]/[4S'_{12}S(S+1)]$$

$$c_3 = (-1)^{2(S_3 + S_{12} + S)}[S' + S'_3 - S'_{12}]/[2S(S+1)]$$

$$d_{1,2} = (-1)^{2S_{12} + S_1 + S_2 + S_3 + S}\{S''S''_{12}[(2S_{1,2} + 3)S''_{1,2}(2S_{1,2} - 1)S_{1,2}(S_{1,2} + 1)]^{\frac{1}{2}}/$$
$$[(2S+3)S''(2S-1)S(S+1)]^{\frac{1}{2}}\} \times$$
$$\begin{Bmatrix} S_{12} & S_{12} & 2 \\ S & S & S_3 \end{Bmatrix} \begin{Bmatrix} S_{12} & S_{12} & 2 \\ S_{1,2} & S_{1,2} & S_{2,1} \end{Bmatrix}$$

$$d_3 = (-1)^{2S_1 + 2S_2 + S_3 + 3S_{12} + S}\{S''[(2S_3 + 3)S''_3(2S_3 - 1)S_3(S_3 + 1)]^{\frac{1}{2}}/$$
$$[(2S+3)S''(2S-1)S(S+1)]^{\frac{1}{2}} \quad \begin{Bmatrix} S & S & 2 \\ S_3 & S_3 & S_{12} \end{Bmatrix}$$

$$d_{12} = (-1)^{S_3 + S + S_{12}}\{\sqrt{(30)}S''S''_{12}[S_1(S_1 + 1)S''_1 S_2(S_2 + 1)S''_2]^{\frac{1}{2}}/$$
$$[(2S+3)S''(2S-1)S(S+1)]^{\frac{1}{2}}\} \quad \begin{Bmatrix} S_{12} & S_{12} & 2 \\ S & S & S_3 \end{Bmatrix} \begin{Bmatrix} S_1 & S_1 & 1 \\ S_2 & S_2 & 1 \\ S_{12} & S_{12} & 2 \end{Bmatrix}$$

$$d_{1,23} = (-1)^{S_1 + S_2 + S_{12} + 1}\{\sqrt{(30)}S''S''_{12}[S_{1,2}(S_{1,2} + 1)S''_{1,2}S_3(S_3 + 1)S''_3]^{\frac{1}{2}}/$$
$$[(2S+3)S''(2S-1)S(S+1)]^{\frac{1}{2}}\} \quad \begin{Bmatrix} S_{12} & S_{12} & 1 \\ S_{1,2} & S_{1,2} & S_{2,1} \end{Bmatrix} \begin{Bmatrix} S_{12} & S_{12} & 1 \\ S_3 & S_3 & 1 \\ S & S & 2 \end{Bmatrix}$$

[a] The shorthand notations $S' = S(S+1)$ and $S'' = 2S+1$ have been used. The c coefficients are zero when the numerator is zero; when the denominator is zero they are equal to 1.

Table 4.4. Numerical values of the c and d coefficients for selected symmetric triads

S_1	S_2	S_3	S_{12}	S	c_1	c_2	c_3	d_1	d_2	d_3	d_{12}	d_{13}	d_{23}
1/2	1/2	1/2	1	3/2	1/3	1/3	1/3	0	0	0	1/6	1/6	1/6
1/2	1/2	1/2	1	1/2	2/3	2/3	−1/3	0	0	0	0	0	0
1/2	1/2	1/2	0	1/2	0	0	1	0	0	0	0	0	0
1/2	1/2	1	1	2	1/4	1/4	1/2	0	0	1/6	1/12	1/6	1/6
1/2	1/2	1	1	1	1/4	1/4	1/2	0	0	−1/2	−1/4	1/2	1/2
1/2	1/2	1	1	0	0	0	0	0	0	0	0	0	0
1/2	1/2	1	0	1	0	0	1	0	0	1	0	0	0
1/2	1/2	3/2	1	5/2	1/5	1/5	3/5	0	0	3/10	1/20	3/20	3/20
1/2	1/2	3/2	1	3/2	2/15	2/15	11/15	0	0	1/5	−2/15	4/15	4/15
1/2	1/2	3/2	1	1/2	−1/3	−1/3	5/3	0	0	0	0	0	0
1/2	1/2	3/2	0	3/2	0	0	1	0	0	1	0	0	0
1/2	1/2	2	1	3	1/6	1/6	2/3	0	0	2/5	1/30	4/30	4/30
1/2	1/2	2	1	2	1/12	1/12	10/12	0	0	1/2	−1/12	1/6	1/6
1/2	1/2	2	1	1	−1/4	−1/4	3/2	0	0	21/10	1/20	−3/10	−3/10
1/2	1/2	2	0	2	0	0	1	0	0	1	0	0	0
1/2	1/2	5/2	1	7/2	1/7	1/7	5/7	0	0	10/21	1/42	5/42	5/42
1/2	1/2	5/2	1	5/2	2/35	2/35	31/35	0	0	23/35	−2/35	4/35	4/35
1/2	1/2	5/2	1	3/2	−1/5	−1/5	7/5	0	0	28/15	1/30	−7/30	−7/30
1/2	1/2	5/2	0	5/2	0	0	1	0	0	1	0	0	0

highest spin multiplicity state $S = (S_1 + S_2) + S_3$ the ratio between the c coefficients is

$$c_1 : c_2 : c_3 = S_1 : S_2 : S_3, \tag{4.55}$$

and since S_i is proportional to the number of unpaired electrons n_i, it follows that

$$c_1 : c_2 : c_3 = n_1 : n_2 : n_3, \tag{4.56}$$

as already obtained for two interacting spins.

In the case of a general triad the isotropic exchange spin hamiltonian (4.6) no longer commutes with S_{12}^2, and states with the same S and S_{12} are mixed together. The hamiltonian matrix will thus be block diagonal, each block containing the states having the same S and S_{12} values. A general expression for the eigenstates of (4.6) is:

$$|\alpha SM\rangle = \Sigma_{S_{12}} C_{SS_{12}} |S_1 S_2 S_{12} S_3 SM\rangle, \tag{4.57}$$

where the $C_{SS_{12}}$ coefficients must be obtained from the diagonalization of the blocks of the hamiltonian matrix corresponding to a given S state or from a perturbative solution of the eigenvalue problem. α is any additional quantum number, or set of quantum numbers, added to conveniently label the eigenstates. When the eigenstates obtained from this procedure are well separated in energy, so that only EPR transitions within each spin state are observed, the spin hamiltonian (4.44–45) describing the EPR experiment can be transformed into operator equivalents containing $\mathbf{S}$ and $\mathbf{S}^2$ like (4.46). This transformation can be obtained using (3.18) and yields

$$\mathbf{g}_s = c_1^\alpha \mathbf{g}_1 + c_2^\alpha \mathbf{g}_2 + c_3^\alpha \mathbf{g}_{31}; \tag{4.58}$$

$$\mathbf{A}_s^k = c_1^\alpha \mathbf{A}_1^k + c_2^\alpha \mathbf{A}_2^k + c_3^\alpha \mathbf{A}_3^k; \tag{4.59}$$

$$\mathbf{D}_s = d_1^\alpha \mathbf{D}_1 + d_2^\alpha \mathbf{D}_2 + d_3^\alpha \mathbf{D}_3 + d_{12}^\alpha \mathbf{D}_{12} + d_{13}^\alpha \mathbf{D}_{13} + d_{23}^\alpha \mathbf{D}_{23}; \tag{4.60}$$

where
$$c_i^\alpha = \langle \alpha S \| T_1(S_i) \| \alpha S \rangle / \langle S \| T_1(S) \| S \rangle = \Sigma_{12}^S C_{SS_{12}}^2 c_i^2 + \Sigma_{S_{12} \neq S'_{12}} c_{SS_{12}} c_{SS'_{12}}$$
$$\langle S_{12} S \| T_1(S_i) \| S_{12} \cdot S \rangle \langle S \| T_1(S) \| S \rangle \tag{4.61}$$

$$d_i^\alpha = \langle \alpha S \| T_2(S_i) \| \alpha S \rangle / \langle S \| T_2(S) \| S \rangle = \Sigma_{S_{12}} c_{SS_{12}}^2 d_i^2 + \Sigma_{S_{12} \neq S'_{12}} c_{SS_{12}} c_{SS'_{12}}$$
$$\langle S_{12} S \| T_2(S_i) \| S_{12} \cdot S \rangle / \langle S \| T_2(S) \| S \rangle \tag{4.62}$$

$$d_{ij}^\alpha = \langle \alpha S \| T_2(S_i S_j) \| \alpha S \rangle / \langle S \| T_2(S) \| S \rangle = \Sigma_{S_{12}} c_{SS_{12}}^2 d_{ij}^2 + \Sigma_{S_{12} \neq S'_{12}} c_{SS_{12}} c_{SS'_{12}}$$
$$\langle S_{12} S \| T_2(S_i S_j) \| S_{12} \cdot S \rangle / \langle S \| T_2(S) \| S \rangle \tag{4.63}$$

Equations (4.58–60) show that also for general triads the spin hamiltonian parameters can be expressed as a linear combination of one and two center contributions.

For clusters having more than three or four atoms application of (4.50–52) can be tedious. It can be helpful, in these cases, to use recurrence relationships which reduce the calculation of the c and d coefficients to products of coefficients relative to the coupling of two spins. Let us consider the trinuclear case first. It can be easily shown that the coefficients given in Table 4.4 can be obtained by coupling the two spins $S_{12} = S_1 + S_2$ and S_3 using the coupling coefficients in Table 3.2 or 3.3. The relevant equations are given in Table 4.5. It can be easily verified that conditions (4.53–54) are fulfilled.

With the use of Table 4.5 any spin system can in fact be handled. A tetramer, for example, can be derived adding a new spin S_4 to a trimer and Table 4.5 can be used to evaluate the c and d coefficients provided that one substitutes for the $c(S_1 S_2 S_{12})$ and $d(S_1 S_2 S_{12})$ coefficients in Table 4.5 the $c(S_1 S_2 S_{12} S_3 S_{123})$ and $d(S_1 S_2 S_{12} S_3 S_{123})$ coefficients obtained for the trinuclear case. With the coefficients obtained for the tetramer one can use Table 4.5 again to compute the relevant coefficients for a pentamer and so on.

Table 4.5. Recurrence relationships for calculating the c and d coefficients of a trinuclear cluster[a]

$$c_1(S_1S_2S_{12}S_3S) = c_1(S_{12}S_3S)c_1(S_1S_2S_{12})$$
$$c_2(S_1S_2S_{12}S_3S) = c_1(S_{12}S_3S)c_2(S_1S_2S_{12})$$
$$c_3(S_1S_2S_{12}S_3S) = c_2(S_{12}S_3S)$$

$$d_1(S_1S_2S_{12}S_3S) = d_1(S_{12}S_3S)\,d_1(S_1S_2S_{12})$$
$$d_2(S_1S_2S_{12}S_3S) = d_1(S_{12}S_3S)\,d_2(S_1S_2S_{12})$$
$$d_3(S_1S_2S_{12}S_3S) = d_2(S_{12}S_3S)$$

$$d_{12}(S_1S_2S_{12}S_3S) = d_1(S_{12}S_3S)d_{12}(S_1S_2S_{12})$$
$$d_{13}(S_1S_2S_{12}S_3S) = d_{12}(S_{12}S_3S)c_1(S_1S_2S_{12})$$
$$d_{23}(S_1S_2S_{12}S_3S) = d_{12}(S_{12}S_3S)c_2(S_1S_2S_{12})$$

[a]The spin states of the trimer are labeled according to the coupling scheme $\{S_1S_2S_{12}S_3S\}$. The coefficients for the dinuclear case appearing on the right-hand side of each equation are given in Tables 3.2 and 3.3.

Table 4.6. Recurrence relationship for calculating the c and d coefficients for a tetranuclear cluster[a]

$$c_1(S_1S_2S_{12}S_3S_4S_{34}S) = c_1(S_{12}S_{34}S)c_1(S_1S_2S_{12})$$
$$c_2(S_1S_2S_{12}S_3S_4S_{34}S) = c_1(S_{12}S_{34}S)c_2(S_1S_2S_{12})$$
$$c_3(S_1S_2S_{12}S_3S_4S_{34}S) = c_2(S_{12}S_{34}S)c_1(S_3S_4S_{34})$$
$$c_4(S_1S_2S_{12}S_3S_4S_{34}S) = c_2(S_{12}S_{34}S)c_2(S_3S_4S_{34})$$

$$d_1(S_1S_2S_{12}S_3S_4S_{34}S) = d_1(S_{12}S_{34}S)d_1(S_1S_2S_{12})$$
$$d_2(S_1S_2S_{12}S_3S_4S_{34}S) = d_1(S_{12}S_{34}S)d_2(S_1S_2S_{12})$$
$$d_3(S_1S_2S_{12}S_3S_4S_{34}S) = d_2(S_{12}S_{34}S)d_1(S_3S_4S_{34})$$
$$d_4(S_1S_2S_{12}S_3S_4S_{34}S) = d_2(S_{12}S_{34}S)d_2(S_3S_4S_{34})$$

$$d_{12}(S_1S_2S_{12}S_3S_4S_{34}S) = d_1(S_{12}S_{34}S)d_{12}(S_1S_2S_{12})$$
$$d_{34}(S_1S_2S_{12}S_3S_4S_{34}S) = d_2(S_{12}S_{34}S)d_{12}(S_3S_4S_{34})$$
$$d_{13}(S_1S_2S_{12}S_3S_4S_{34}S) = d_{12}(S_{12}S_{34}S)c_1(S_1S_2S_{12})c_1(S_3S_4S_{34})$$
$$d_{14}(S_1S_2S_{12}S_3S_4S_{34}S) = d_{12}(S_{12}S_{34}S)c_1(S_1S_2S_{12})c_2(S_3S_4S_{34})$$
$$d_{23}(S_1S_2S_{12}S_3S_4S_{34}S) = d_{12}(S_{12}S_{34}S)c_2(S_1S_2S_{12})c_1(S_3S_4S_{34})$$
$$d_{24}(S_1S_2S_{12}S_3S_4S_{34}S) = d_{12}(S_{12}S_{34}S)c_2(S_1S_2S_{12})c_2(S_3S_4S_{34})$$

[a]The spin levels of the tetramer are labeled according to the coupling scheme $\{S_1S_2S_{12}S_3S_{34}S_{34}S\}$. The coefficients on the right-hand side of each equation refer to the coupling of the two indicated spins and are given in Tables 3.2 and 3.3.

For tetramers, or any other cluster with an even number of spins, however, it can be convenient to split the cluster into two moieties and to look at the coupling of these two smaller clusters. The spin levels of a tetramer, for example, can be obtained, as already done in the preceding part of the chapter, by coupling the two couples of spins, $S_1 + S_2$ and $S_3 + S_4$, together. The recurrence expressions to be used for the c and d coefficients in this coupling scheme are easily worked out and are reported in Table 4.6.

The combined use of Tables 3.2, 4.3, 4.5, and 4.6 allows one to compute the c and d coefficients for any spin cluster.

4.3.1 EPR Spectra of Trinuclear Clusters

The investigation of the magnetic properties of trinuclear clusters has been generally based on magnetic susceptibility data [4.11]. The interpretation of these data was, however, not always unambiguous since three coupling constants should be extracted by fitting the temperature dependence of the bulk magnetic susceptibility measured on polycrystalline samples. For example, the interpretation of the magnetic susceptibility data of iron (III) basic acetate $[Fe_3O(CH_3COO)_6(H_2O)_3]Cl \cdot 5H_2O$ requires the use of at least two different coupling constants, while the iron atoms are at the vertex of an almost equilateral triangle with the oxygen atom at the center, and the acetato groups bridge six pairs of vertices of the FeO_6 octahedra [4.12], as shown in Fig. 4.5. The magnetic data have been in fact fitted with $J_{12} = 76\,cm^{-1}$ and $J_{13} = J_{23} = 58\,cm^{-1}$ [4.13], a poorer fit being obtained with the regular triad model with $J_{12} = J_{13} = J_{23} \simeq 58\,cm^{-1}$ [4.14]. This fact can reflect either the inadequacy of the Heisenberg-Dirac-van Vleck spin hamiltonian or the intrinsic imprecision of the bulk magnetic susceptibility measurements which generally results in large correlations between the parameters used in the fitting procedure.

EPR spectroscopy can play an important role in the study of the magnetic properties of small clusters and in principle can give information which is complementary to the magnetic susceptibility data. The correlation of the measured spin hamiltonian parameters $(\mathbf{g}, \mathbf{D}, \mathbf{A}, \ldots)$ to the spin hamiltonian parameters of the individual spins forming the cluster can depend on the J_{ij}

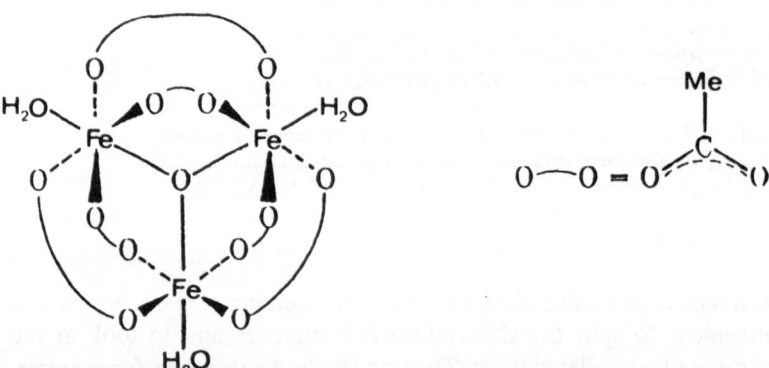

Fig. 4.5. Schematic view of the structure of iron (III) basic acetate $[Fe_3O(CH_3COO)_6(H_2O)_3]Cl \cdot 5H_2O$

values, and information on both their sign and their relative values can be obtained from EPR spectroscopy. The maximum information is of course obtained by measuring the spin hamiltonian parameters for all the spin states arising from the isotropic exchange interaction. This situation has never been met experimentally. The complete analysis of the spectra requires the knowledge of the single ion spin hamiltonian parameters as we will show with some examples. In the following part of this section we will report some examples of analysis of EPR spectra of trinuclear systems with $S_i = 1/2$ and $S_i = 5/2$, respectively. Owing to the small number of trinuclear clusters investigated so far, we cannot fully explore all of the subtleties of the spin hamiltonian formalism; in particular we will interpret the spectra in the strong exchange limit and no formal theory of the single ion anisotropies will be developed.

Let us consider a triad of $S_i = 1/2$ interacting spins. From the isotropic exchange interaction three total spin states arise (see Table 4.1): one quartet ($|1\,3/2\rangle$) and two doublets ($|1\,1/2\rangle$, $|0\,1/2\rangle$). In the case of antiferromagnetic interactions a doublet state is the ground state. In symmetric triads, $J_{13} = J_{23}$, and the $|0\,1/2\rangle$ state is the ground state for $r = J_{12}/J_{23} > 1$ (see Fig. 4.2.). For a general triad the energies of the two doublet states are easily computed as eigenvalues of the 2×2 hamiltonian matrix given in Table 4.1. The energies are:

$$E(a1/2) = -1/4(J_{12} + J_{13} + J_{23}) + 1/2[J_{12} - J_{13}/2 - J_{23}/2)^2 + 3/4$$
$$(J_{23} - J_{13})^2]^{\frac{1}{2}} \tag{4.64}$$

$$E(b1/2) = -1/4(J_{12} + J_{13} + J_{23}) - 1/2[(J_{12} - J_{13}/2 - J_{23}/2)^2 + 3/4$$
$$(J_{23} - J_{13})^2]^{\frac{1}{2}}$$

where the two states are labeled as $|\alpha S\rangle$ according to Eq. (4.57). The eigenvectors corresponding to (4.64) are given by:

$$|a1/2\rangle = \cos\lambda|1\,1/2\rangle - \sin\lambda|0\,1/2\rangle$$
$$|b\,1/2\rangle = \sin\lambda|1\,1/2\rangle + \cos\lambda|0\,1/2\rangle \tag{4.65}$$

with
$$\lambda = 1/2\,\mathrm{tg}^{-1}[\sqrt{3}(J_{23} - J_{13})/(2J_{12} - J_{13} - J_{23})]. \tag{4.66}$$

When $\lambda = 0$ the triad is symmetric. Using (4.65–66) in (4.61) we get:
$$c_1^a = 2/3 - 2/3\sin^2\lambda + 2/\sqrt{3}\,\sin\lambda\cos\lambda$$
$$c_2^a = 2/3 - 2/3\sin^2\lambda - 2/\sqrt{3}\,\sin\lambda\cos\lambda \tag{4.67}$$
$$c_3^a = -1/3 + 4/3\sin^2\lambda$$

and
$$c_1^b = 2/3\sin^2\lambda - 2/\sqrt{3}\,\sin\lambda\cos\lambda$$
$$c_2^b = 2/3\sin^2\lambda + 2/\sqrt{3}\,\sin\lambda\cos\lambda \tag{4.68}$$
$$c_3^b = 1 - 4/3\sin^2\lambda.$$

107

Owing to the ternary symmetry of the system independent solutions are obtained only for λ values in the range 0-$\pi/6$. λ values larger than $\pi/6$ correspond to a different numbering of the spin centers. The dependence of the c_i^α coeffecients on λ is graphically shown in Fig. 4.6. It is apparent that these coefficients are strongly dependent on the λ values; this suggests that the g values themselves depend on λ and that they can be used to estimate it. In order to compute the λ value from the experimental g values Eq. (4.58) should be used. Once the single ion $\mathbf{g}_i$ tensors are known, in fact, Eq. (4.58) depends directly on λ. In a symmetric triad $J_{12} = 0$ and Eq. (4.66) becomes:

$$\lambda = 1/2 \, tg^{-1}[\sqrt{3}(r-1)/(r+1)], \tag{4.69}$$

and the ratio $r = J_{13}/J_{23}$ between the two coupling constants can be directly evaluated from the measure of λ. For a general triad Eq. (4.66) should be used to relate λ to the J_{ij} values and ranges of relative values of the J_{ij} values can be derived from the measure of λ.

Let us consider the EPR spectra [4.15] of bis[N,N'-ethylenebis(o-hydroxyacetophenone iminato)copper(II)] copper(II) diperchlorate dihydrate. (CuHAPen)$_2$Cu. The trimer is formed by two CuHAPen complexes (Fig. 4.7) which bind to the central copper atom through the oxygen atom of the ligand. The central copper atom achieves a five-coordinate structure binding to a water molecule as is shown in Fig. 4.8 [4.16].

From Fig. 4.8 it is expected that the exchange interaction between Cu$_1$ and Cu$_2$ is negligible, and magnetic susceptibility data have shown that a strong antiferromagnetic interaction ($\simeq 400 \text{ cm}^{-1}$) is operative between Cu$_3$ and Cu$_1$, Cu$_2$ [4.17]. Single crystal EPR spectra recorded at 4.2 K were attributed to the $|b\,1/2\rangle$ doublet state. The spectra showed evidence of intermolecular exchange

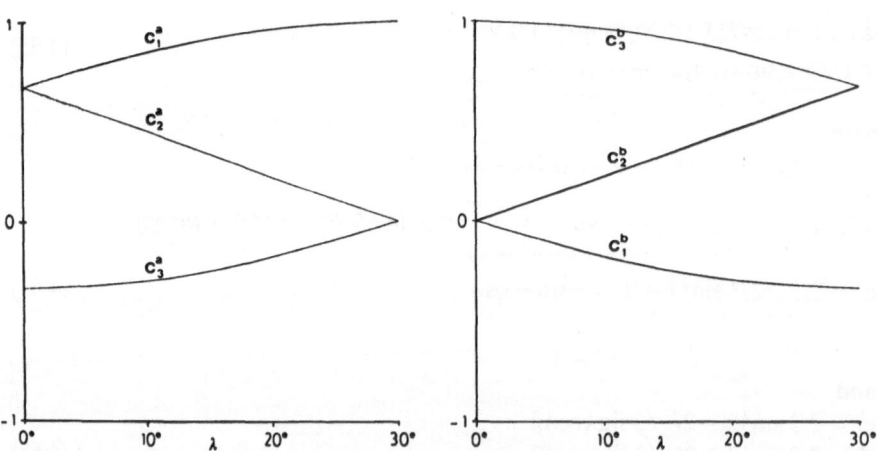

Fig. 4.6. The dependence of the c_i^α coefficients of Eqs. (4.67–68) on λ(see text)

interactions which averaged out the signal of the magnetically inequivalent molecules. Since the crystals are orthorhombic, space group Pbca, the principal g values were found along the crystallographic axes ($g_a = 2.139$, $g_b = 2.042$, $g_c = 2.044$). Using in (4.58) and (4.68) for g_1 and g_2 the tensor observed in the (CuHAPen)$_2$Zn complex and the experimentally observed g_s values, the g_3 values were computed as a function of λ. These values are shown in Fig. 4.9. Only the values computed for $0° < \lambda < 11°$ could be considered satisfactory for a distorted five-coordinated copper(II) ion [4.18]. Equation (4.69) shows that r $= J_{13}/J_{23}$ should be in the range 0–1.6.

In three-iron clusters containing iron(III) ions in the high spin states ($S_i = 5/2$) EPR spectra arising from a S $= 1/2$ state have been observed [4.19]. In the ferredoxin from *Azotobacter vinelandii* the three-iron site is [Fe$_3$S$_3$(S-Cys)$_5$(oxo)]. This complex contains a cyclic [Fe$_3(\mu_2 - $S)$_3$]$^{3+}$ moiety schematically shown in Fig. 4.10. The iron atoms have a S$_4$ pseudotetrahedral coordination. The average Fe–Fe separation is 4.1 Å [4.20]. More details on iron sulfur proteins may be found in Chapter 9.

Kent et al. [4.21], using magnetically perturbed Mössbauer spectroscopy, measured three different hyperfine coupling constants $A^1 = 47 \cdot 10^{-4}$ cm^{-1}, $|A^2| = 13 \cdot 10^{-4}$ cm^{-1}, and $A^3 = -106 \cdot 10^{-4}$ cm^{-1} for the three iron centers, respectively. From the isotropic exchange interaction only two doublets arise (Table

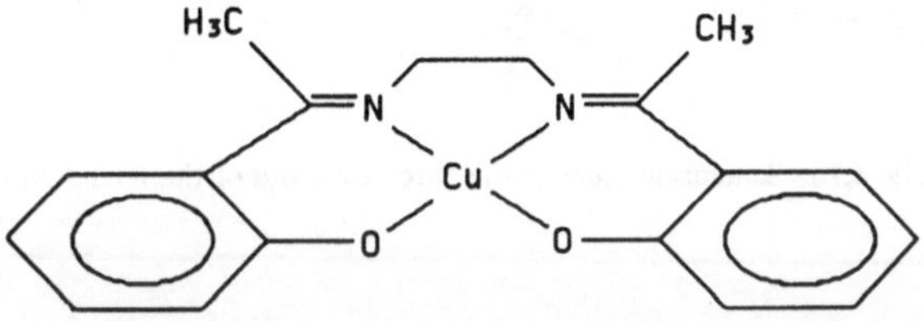

Fig. 4.7. Schematic structure of N, N'-ethylenebis(o-hydroxyacetophenone iminato)copper(II), (CuHAPen)

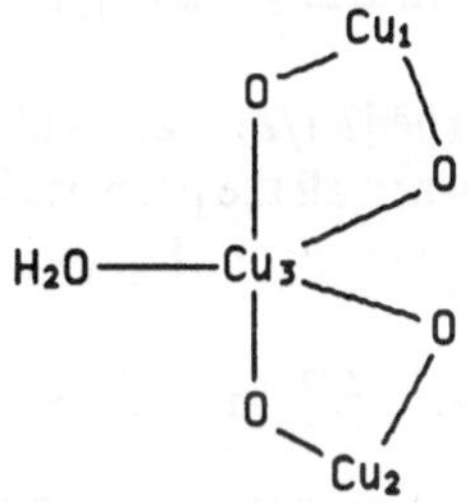

Fig. 4.8. The five coordinate structure of the central copper atom in the trinuclear cluster bis[N,N'-ethylenebis(o-hydroxyacetophenone iminato)copper (II)]copper (II) diperchlorate

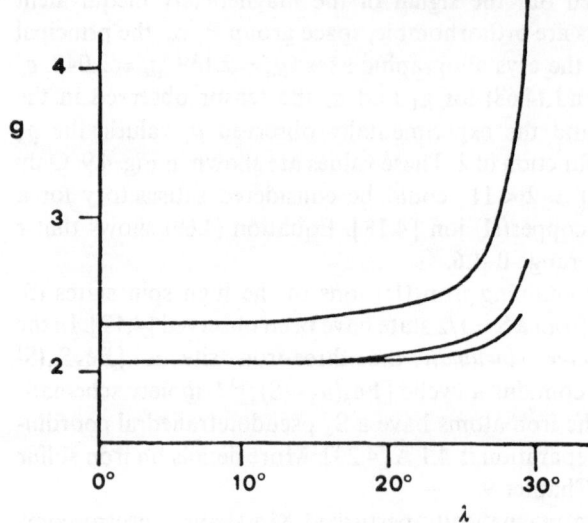

Fig. 4.9. The principal values of the g_3 tensor computed as a function of the λ angle of Eq. (4.69)

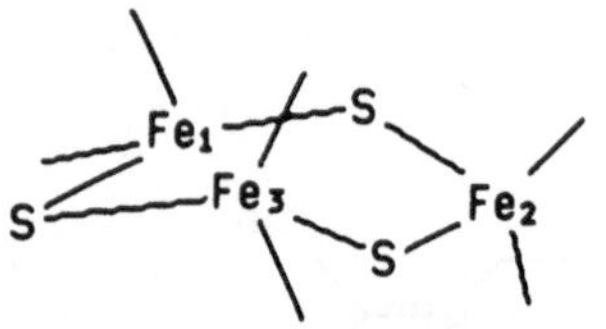

Fig. 4.10. Schematic view of the three-iron site of the ferredoxin from *Azotobacter vinelandii*

4.1), namely $|3\ 1/2\rangle$ and $|2\ 1/2\rangle$. In a symmetric triad ($J_{13} = J_{23}$) the energies of these states are given by:

$$E(3\ 1/2) = -11\ J_{12}/4 - 10\ J_{13};$$
$$E(2\ 1/2) = -23\ J_{12}/4 - 7\ J_{13}. \tag{4.70}$$

The $|2\ 1/2\rangle$ state will be the ground state for $J_{13} > 0$ and $4/7 < J_{12}/J_{13} < 1$, i.e., when all the pairwise interactions are antiferromagnetic and nearly equal to each other. From Eq. (4.50) the following c_i coefficients can be computed:

$$c_1 = 4/3, c_2 = 4/3, c_3 = -5/3 \qquad \text{for } |3\ 1/2\rangle \tag{4.71}$$

$$c_1 = -2/3, c_2 = -2/3, c_3 = 7/3 \qquad \text{for } |2\ 1/2\rangle \tag{4.72}$$

and using the value of $-52 \cdot 10^{-4}$ cm^{-1} for the hyperfine coupling constant of the single iron atoms we compute from (4.48, 4.71–72) the following values (cm^{-1}) of the hyperfine coupling constants:

$$A^1_{1/2} = -69 \cdot 10^{-4}, \; A^2_{1/2} = -69 \cdot 10^{-4}, \; A^3_{1/2} = 87 \cdot 10^{-4} \text{ for } |3\ 1/2\rangle; \tag{4.73}$$

$$A^1_{1/2} = 35 \cdot 10^{-4}, \; A^2_{1/2} = 35 \cdot 10^{-4}, \; A^3_{1/2} = -121 \cdot 10^{-4} \text{ for } |2\ 1/2\rangle, \tag{4.74}$$

It is apparent that the spin coupling model can explain why the hyperfine coupling constants are different for the three spin centers and the values in (4.74) reproduce also the observed signs of the hyperfine constants. The experimental values of A_1 and A_2 are, however, significantly different from each other contrary to (4.74).

If one allows for $J_{13} \neq J_{23}$ the two $S = 1/2$ states are admixed and from the diagonalization of the 2×2 matrix of Table 4.1 two states $|a\ 1/2\rangle$ and $|b\ 1/2\rangle$ are obtained. Their energies are:

$$E(a\ 1/2) = -17/4(J_{12} + J_{13} + J_{23}) + 3/2[(2J_{12} - J_{13} - J_{23})^2 + 3(J_{23} - J_{13})^2]^{\frac{1}{2}}$$
$$E(b\ 1/2) = -17/4(J_{12} + J_{13} + J_{23}) - 3/2[(2J_{12} - J_{13} - J_{23})^2 + 3(J_{23} - J_{13})^2]^{\frac{1}{2}}$$
$$\tag{4.75}$$

and their eigenvectors are:

$$|a\ 1/2\rangle = \cos\lambda |3\ 1/2\rangle - \sin\lambda |2\ 1/2\rangle$$
$$|b\ 1/2\rangle = \sin\lambda |3\ 1/2\rangle + \cos\lambda |2\ 1/2\rangle \tag{4.76}$$

where

$$\lambda = 1/2 \, \mathrm{tg}^{-1}[\sqrt{3}(J_{23} - J_{13})/(2J_{12} - J_{13} - J_{23})]. \tag{4.77}$$

For the symmetric triad $\lambda = 0$ and $|b\ 1/2\rangle = |2\ 1/2\rangle$.

Equation (4.6.1) gives for the $|b\ 1/2\rangle$ state:

$$c^b_1 = -2/3 + 2\sin^2\lambda - 2\sqrt{3}\sin\lambda\cos\lambda;$$
$$c^b_2 = -2/3 + 2\sin^2\lambda + 3\sqrt{3}\sin\lambda\cos\lambda; \tag{4.78}$$
$$c^b_3 = 7/3 - 4\sin^2\lambda.$$

In Fig. 4.11 the dependence of the c^b_i coefficients on λ is shown. Only λ values in the range $0 - \pi/6$ need be considered as already noted. It is apparent that significant variations of c^b_1 and c^b_2 from the $-2/3$ values are allowed even by small λ values. The experimental A^i values can be used to extract one value of λ from (4.78) and a reasonable fit was obtained for $\lambda = 6°$. This means [Eq. (4.77)] that the J_{ij} values should obey the following relationships: $J_{12} > J_{13} > J_{23} > 0$; $0.5 < J_{23}/J_{12} < 1$; $0.6 < J_{13}/J_{12} < 1$; $J_{13} - J_{23} \leq 0.2 \text{ cm}^{-1}$; i.e., the exchange interactions should all be negative and nearly equal.

As a last example we will consider the EPR spectra of quasi-linear trimer octachlorodiadeniniumtricopper(II) tetrahydrate, $Cu_3(ade)_2Cl_8$ (ade = adenin-

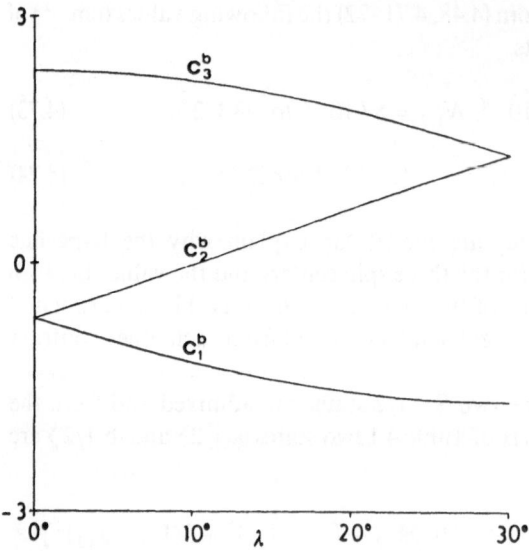

Fig. 4.11. The dependence of the c_i^b coefficients of Eq. (4.78) on λ (see text)

ium ion). The structure of the trimer is schematically shown in Fig. 4.12. The copper atoms lie on a straight line and are coupled by a double chlorine bridge and an adenine molecule acting as a bidentate ligand [4.22]. Since the adeninium bridge can create a coupling between Cu_1 and Cu_3 we call this triad a quasi-linear triad. The temperature dependences of the magnetic susceptibility were fitted with either $J_{12} = J_{23} = 32.2 \text{ cm}^{-1}$, $J_{13} = 0$, and $g = 2.14$ or $J_{12} = J_{23} = 33.2 \text{ cm}^{-1}$, $J_{13} = -16.1 \text{ cm}^{-1}$, and $g = 2.14$ [4.23]. An equivalent fit, however, was obtained also with $J_{12} = J_{23} = 28 \text{ cm}^{-1}$, $J_{13} = 18 \text{ cm}^{-1}$, and $g = 2.15$ [4.24]. These results show that the magnetic susceptibility alone cannot give a meaningful estimate of the J_{13} coupling constant. In any case the resulting ground state is the $|1 \ 1/2\rangle$ doublet, with the $|0 \ 1/2\rangle$ doublet and the $|1 \ 3/2\rangle$ state $J_{12} - J_{13}$ and $3/2 \ J_{12}$ above, respectively. The effect of the J_{13} coupling is therefore that of varying the spacing between the two doublet states. The EPR spectra of $Cu_3(ade)_2Cl_8$ were recorded in the temperature range $300 - 4.2$ K and were interpreted with an $S = 1/2$ spin hamiltonian with $g_1 = 2.059$, $g_2 = 2.140$, $g_3 = 2.189$ at 300 K; $g_1 = 2.049$, $g_2 = 2.133$, $g_3 = 2.194$ at 44 K, and $g_1 = 2.028$, $g_2 = 2.127$, $g_3 = 2.226$ at 4.2 K, and no signal attributable to the $S = 3/2$ state was observed [4.24]. The spectra showed a small variation of the g values between 300 and 77 K, and a much steeper variation between 40 and 10 K. Below 10 K the g values become constant. Unfortunately intermolecular exchange interactions are operative to average the molecular signals in the cyrstals, thus preventing the measurement of the molecular **g** tensor and of other spin hamiltonian parameters. The temperature dependence of the spectra alone can give, however, some information. If the temperature dependence is due to the

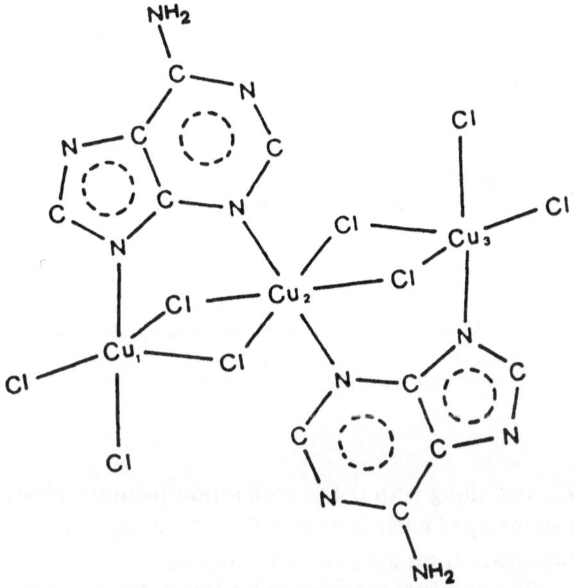

Fig. 4.12. Schematic structure of the trimer octachlorodiadeniniumtricopper(II) tetrahydrate

variation of the thermal population of the different spin levels it indicates that the $|0\ 1/2\rangle$ is almost depopulated only below $\simeq 10$ K. This requires a small energy difference between the $|1\ 1/2\rangle$ and $|0\ 1/2\rangle$ states. Using the three parameter sets obtained from the fitting of magnetic susceptibility data we compute an energy separation between the two doublets of $\simeq 32$ cm^{-1}, $\simeq 49$ cm^{-1}, and $\simeq 10$ cm^{-1}, respectively. The EPR experiment indicates that the last figure should be correct and shows that only the third set of parameters describes the exchange interaction in $Cu_3(ade)_2Cl_8$.

The above examples show that knowledge of the spin hamiltonian parameters of exchange coupled triads and measurement of the temperature dependence of EPR spectra can be fundamental in understanding the exchange interactions. In particular, very small differences in the exchange coupling constants, which could hardly be revealed by magnetic susceptibility measurements, can be evidenced in this way.

4.3.2 EPR Spectra of Tetranuclear Clusters

The cubic tetramer $Cu_4OCl_6(TPPO)_4$ where TPPO represents tetraphenylphosphine oxide is surely the most studied tetrameric copper system. A sketch of the structure is shown in Fig. 4.13. The crystals of the compound belong to the cubic system, $P43m$ space group [4.25]. The oxygens are located at the vertex of a regular tetrahedron with Cu-Cu distance of 311 pm and lie on a crystallographic

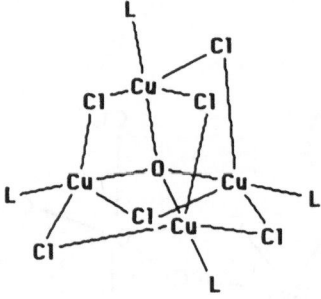

Fig. 4.13. Schematic structure of the cubic tetramer $Cu_4OCl_6(L)_4$ (L = TPPO = tetraphenylphosphine oxide)

C_3 axis along with the oxygen atoms from the phosphine oxide ligands. The site symmetry of each copper is C_{3v}. An oxygen atom lies at the center of the Cu_4 tetrahedron on a T_d crystallographic site.

The spin states arising from the isotropic exchange interaction are reported in Table 4.2. The temperature dependence of the magnetic susceptibility was measured by Lines et al. [4.26]. No reasonable fit of the data was obtained using the spin hamiltonian (4.30) for the regular Cu_4 tetrahedron. The existence of intercluster exchange interaction was claimed to explain the observed low temperature antiferromagnetism. Within this model the data were fitted with J = −29 cm^{-1} and z'J' = 7 cm^{-1}, where z'J' represents the intercluster exchange in a molecular field scheme. With these parameters the states with the same total spin S are degenerate: the S = 2 state is the ground state with three S = 1 states at $\simeq 58$ cm^{-1} and two singlet states at $\simeq 87$ cm^{-1}.

Black et al. [4.27–29] have measured the single crystal EPR spectra of $Cu_4OCl_6(TPPO)_4$. The 4.2-K spectra were interpreted using the cubic S = 2 spin hamiltonian:

$$H_{s=2} = \mu_B g \mathbf{B} \cdot \mathbf{S} + B_4(O_4^0 + 5 O_4^4), \tag{4.79}$$

where O_4^0 and O_4^4 are the fourth-order cubic fine structure operators [4.18]:

$$O_4^0 = 35 S_z^4 - 30 S(S+1)S_z^2 + 25 S_z^2 - 6 S(S+1) + 3 S^2(S+1)^2; \tag{4.80}$$

$$O_4^4 = 1/2(S_+^4 + S_-^4), \tag{4.81}$$

with g = 2.10 and $B_4 = 0.0044$ cm^{-1}. The $O_4^0 + 5 O_4^4$ operator causes a zero field splitting of the spin levels of the S = 2 manifold into a doublet (M = 0, 2) and a triplet (M = ± 1, −2). The energy separation Δ between the doublet and the triplet is $\Delta = 120 B_4$. Up to now we have neglected the fourth-order operators in the spin hamiltonian since they are usually negligible with respect to the second-

114

order ones. They became important, however, in cubic complexes since the second-order terms are zero by symmetry.

From the temperature dependence of the intensity of the EPR signals, it was found that the $S = 2$ state is not the ground state in $Cu_4OCl_6(TPPO)_4$, but it is $\simeq 14 \text{ cm}^{-1}$ above the ground state which should be diamagnetic. This level ordering is shown in Fig. 4.14 and compared to the level ordering obtained from magnetic susceptibility data. In fact, the temperature dependence of the magnetic susceptibility could be reproduced using the differences $\Delta_1 = E(1) - E(0)$ and $\Delta_2 = E(2) - E(0)$ as free parameters, the best fit being obtained for $\Delta_1 \simeq 85 \text{ cm}^{-1}$, $\Delta_2 \simeq 10 \text{ cm}^{-1}$, and $g = 2.10$. The Δ_2 value computed in this way is in agreement with the EPR results and confirms the EPR information about the level ordering. This level ordering demands a justification since it cannot be obtained from the regular tetrad spin hamiltonian (4.30) and the cubic symmetry of the EPR spectra demands that any low symmetry distortion from the tetrahedral cluster, if any, should be dynamically averaged. Kamase et al. [4.30] took into consideration four center interactions through a spin hamiltonian of the form

$$H_{int}^{(4)} = J \Sigma_{i<j} S_i \cdot S_j + J' \Sigma_{i<j} \Sigma_{k<l} (S_i \cdot S_j)(S_k \cdot S_l) \quad i \neq j \neq k \neq l, \tag{4.82}$$

which yields $\Delta_1 = J - 5J'/2$ and $\Delta_2 = 3(J - J'/2)$. The fitting of the magnetic data requires $J \simeq 17 \text{ cm}^{-1}$ and $J' \simeq -41 \text{ cm}^{-1}$. It is apparent that although this model can explain the observed level ordering it requires a quite large four-center interaction.

Another important feature of the EPR spectra of $Cu_4OCl_6(TPPO)_4$ is the absence of EPR signals attributable to transitions within the three $S = 1$ degenerate multiplets. This can be due to the fact that the triplet states are depopulated at 4.2 K, again in agreement with the level ordering of Fig. 4.14b.

The analysis of the zero field splitting Δ and its correlation to magnetic dipolar and anisotropic and antisymmetric exchange interactions was performed by Buluggiu [4.31] using second-order perturbation theory. He has shown that only anisotropic exchange and magnetic dipolar interactions contribute to the

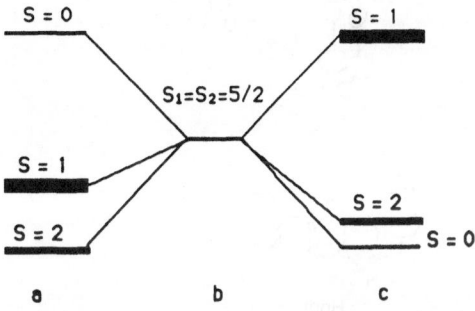

Fig. 4.14. Level ordering in $Cu_4OCl_6(TPPO)_4$ as computed from the temperature dependence of the magnetic susceptibility (a) and from EPR spectra (c)

115

zero field splitting of the $S = 2$ state, while both antisymmetric and anisotropic exchange contribute to the zero field splitting of the triplet states.

Another interesting example of a rather symmetric tetranuclear cluster is provided by the complex tetrakis [aqua(3-(pyridin-2-yl)-5-(pyrazin-2-yl)-1,2,4-triazolato)copper(II)] tetranitrate dodecahydrate, $[Cu(\mu\text{-ppt})(H_2O)]_4(NO_3)_4(H_2O)_{12}$, schematically shown in Fig. 4.15. The compound crystallizes in the tetragonal system, space group $I4_1/a$ [4.32]. The cluster consists of four copper(II) ions at the vertex of a tetragonally compressed tetrahedron with short copper-copper distances $Cu_1 - Cu_3 = Cu_2 - Cu_3 = Cu_3 - Cu_4 = Cu_1 - Cu_4$ $= 426.9$ pm, and long copper-copper distances $Cu_1 - Cu_2 = Cu_3 - Cu_4$ $= 432.0$ pm; four ppt$^-$ groups bridging the edges of the tetrahedron, and four coordinated water molecules. The coordination of each copper atom can be described as a slightly distorted $(N_3O)N$ square pyramid with three of the four nitrogens from two ppt molecules and one water molecule in the basal plane, the other nitrogen occupying the axial position.

The temperature dependence of the magnetic susceptibility was reproduced using the $H_{int}^{S_4}$ spin hamiltonian (4.31) with $J = -0.6$ cm^{-1}, $J' = 12.2$ cm^{-1}, g $= 2.17$ [4.33]. The energies of the total spin states arising from the isotropic exchange interaction can be obtained from Table 4.2. The ordering of the energy levels computed from the magnetic data is shown in Fig. 4.16. The $S = 2$ state is

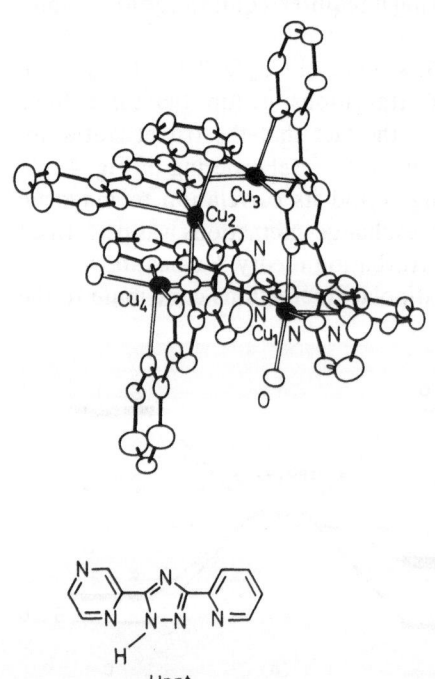

Hppt

Fig. 4.15. The structure of the tetranuclear cluster tetrakis[aqua(3-(pyridin-2-yl)-5-(pyrazin-2-yl)-1,2,4-triazolato) copper(II)] tetranitrate dodecahydrate, $[Cu(\mu\text{-ppt})(H_2O)]_4(NO_3)_4(H_2O)_{12}$

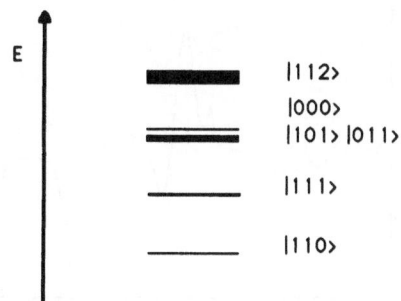

Fig. 4.16. Level ordering in $[Cu(\mu\text{-ppt})(H_2O)]_4(NO_3)_4(H_2O)_{12}$

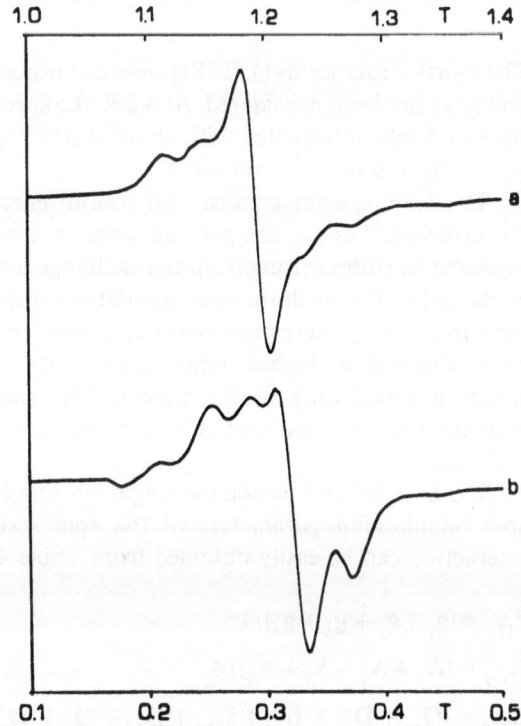

Fig. 4.17 a, b. Room temperature polycrystalline powder EPR spectra of $[Cu(\mu\text{-ppt})(H_2O)]_4(NO_3)_4(H_2O)_{12}$ measured at Q-band (35 GHz) (**a**) and X-band (9.1 GHz) (**b**)

$\simeq 37$ cm^{-1} above the ground singlet state and it is expected to be depopulated at 4.2 K. Two of the three triplet states are degenerate and close in energy to the $|000\rangle$ state, the third triplet lying $\simeq 13$ cm^{-1} below them and $\simeq 12$ cm^{-1} above the $|110\rangle$ ground singlet state. Polycrystalline powder EPR spectra recorded at room temperature and at 4.2 K are shown in Figs 4.17 and 4.18, respectively.

Single crystal spectra recorded at room temperature were analyzed using an axial S = 2 spin hamiltonian with $g_\parallel = 2.03$, $g_\perp = 2.16$, and $|D| = 0.0379$ cm^{-1}.

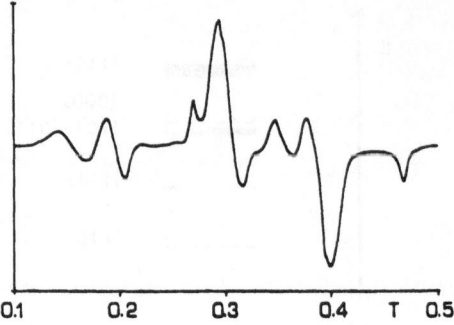

0.1	0.2	0.3	0.4	T	0.5

Fig. 4.18. Polycrystalline powder EPR spectrum of $[Cu(\mu\text{-ppt})(H_2O)]_4(NO_3)_4(H_2O)_{12}$ measured at X-band and 4.2 K

The fourth-order terms (4.79–81) were not necessary for the fitting of the spectra and have not been considered. At 4.2 K the spectra are characteristic of a triplet state and were interpreted with an axial S = 1 spin hamiltonian with $g_{//} = 2.03$, $g_\perp = 2.16$, and $|D| = 0.0906$ cm^{-1}.

The EPR spectra confirm the overall energy level ordering of Fig. 4.16. Nevertheless, they are also peculiar, since they show signals only from one of the three triplet states arising from the exchange interaction. While this fact can be explained by the small thermal population of the $|011\rangle$ and $|101\rangle$ states, since the triplet spectra were observed only at temperatures lower than 10 K, they will be unobserved at higher temperatures only assuming fast relaxation which causes a broadening of the signals. The spectra, however, show that the relaxation between the triplets is more effective than the relaxation within the S = 2 state.

The correlation between the single ion spin hamiltonian parameters and the spin hamiltonian parameters of the spin states arising from the exchange interaction can be easily obtained from Table 4.6 as:

$$g_{112} = (g_1 + g_2 + g_3 + g_4)/4$$
$$A_{112} = (A_1 + A_2 + A_3 + A_4)/4 \quad\quad (4.83)$$
$$D_{112} = (D_{12} + D_{13} + D_{14} + D_{23} + D_{24} + D_{34})/12$$

$$g_{011} = (g_3 + g_4)/2$$
$$A_{011} = (A_3 + A_4)/2 \quad\quad (4.84)$$
$$D_{011} = D_{34}/2$$

$$g_{101} = (g_1 + g_2)/2$$
$$A_{101} = (A_1 + A_2)/2 \quad\quad (4.85)$$
$$D_{101} = D_{12}/2$$

$$g_{111} = (g_1 + g_2 + g_3 + g_4)/4$$
$$A_{111} = (A_1 + A_2 + A_3 + A_4)/4 \qquad (4.86)$$
$$D_{111} = (-D_{12} + D_{13} + D_{14} + D_{23} + D_{24} - D_{34})$$

After referring all of the tensors in (4.83–4.86) to the tetragonal x, y, z axes and applying the S_4 symmetry relationships between the tensors, we obtain the relevant equations to interpret the EPR spectra:

$$g_{113,\parallel} = g_{1,zz}; g_{112,\perp} = (g_{1,xx} + g_{1,yy}) \qquad (4.87)$$

$$D_{112} = D_{13,zz}/2 + D_{12,zz}/4; \; E_{112} = 0 \qquad (4.88)$$

$$g_{111,\parallel} = g_{1,zz}; g_{111,\perp} = (g_{1,xx} + g_{1,yy}) \qquad (4.89)$$

$$D_{111} = 3\,D_{13,zz}/2 - 3\,D_{12,zz}/4; \; E_{111} = 0 \qquad (4.90)$$

where D and E are the usual zero field splitting parameters. Equations (4.83, 4.86) show that the **g** tensors of the $|1\,1\,2\rangle$ and $|1\,1\,1\rangle$ states should be equal, in excellent agreement with the experimental results. Assuming a regular C_{4v} symmetry around each copper ion, nice agreement with the observed **g** tensors is obtained using $g_{Cu,\parallel} = 2.34$ and $g_{Cu,\perp} = 2.01$ in (4.87, 4.88), which are in fair agreement with the values expected for a monomeric square pyramidal copper(II) complex. From (4.88, 4.90) the following zero field splitting parameters were obtained: $D_{13,zz} = 0.0681 \text{ cm}^{-1}$ and $D_{12,zz} = 0.0153 \text{ cm}^{-1}$.

The spectra at 4.2 K show a resolved copper hyperfine structure in the plane perpendicular to the crystallographic c axis. The observed maximum splitting into seven line components is $A_{111}^{Cu} = 0.0045 \times 10^{-4} \text{ cm}^{-1}$. This splitting is observed when the static magnetic field is parallel to the z molecular axes of two copper ions. In the same crystal orientation the other two copper ions forming the tetramer have their x molecular axes, parallel to the field, therefore, the expected splitting is very small. The observed value of A_{111}^{Cu} requires $A_{111}^{Cu} = 0.0180 \times 10^{-4} \text{ cm}^{-1}$ for the individual copper ion, in agreement with the value expected for a square pyramidal complex [4.18].

References

4.1 Rotenberg M, Bivens R, Metropolis N, Wooten JK Jr (1959) The 3-j and 6-j symbols, Technology, MIT, Cambridge, Massachussets

4.2 Smith K (1958) Table of Wigner 9-j symbols for integral and half-integral values of the parameters, ANL-5860, Argonne National Lab., Chicago

4.3 Howell KM (1959) Revised tables of 6-j symbols, Res. Rep. 59-1, Univ. of Southampton Maths. Dept., Southampton, England

4.4 Yutsis AP, Levinson IB, Vanagas VV (1962) Mathematical apparatus of the theory of angular momentum, Israel Program for Scientific Translations, Jerusalem

4.5 Silver BL (1976) Irreducible Tensor Methods, Academic, New York

4.6 Wigner EP (1965) In. Biedenham LC, VanDam H(eds) Quantum theory of angular momentum. Academic, New York

4.7 Griffith JS (1972) Structure and Bonding, 10: 87

4.8 Edmonds AR, (1968) Angular momentum in quantum mechanics, Princeton University Press, Princeton, New Jersey

4.9 Viehland LA, Curtis CF (1974) J. Chem. Phys., 60: 492

4.10 Rakitin YuV, Kalinnikov VT, Hatfield WE (1977) Phys. Stat. Sol. b 81: 379

4.11 O'Connor CJ (1982) Progr. Inorg. Chem, 29: 203

4.12 Figgis BN, Robertson GB (1965) Nature 205: 694

4.13 Blake AB, Yavari A, Hatfield WE, Sethulekshmi CN (1985) J. Chem. Soc. Dalton Trans., 2509

4.14 Earnshaw A, Figgis BN, Lewis J (1966) J. Chem. Soc. A, 1656

4.15 Banci L, Bencini A, Dei A, Gatteschi D (1983) Inorg. Chem. 22: 4018

4.16 Epstein JM, Figgis BN, White A, Willis AC (1974) J. Chem. Soc. Dalton Trans., 1954

4.17 Gruber SJ, Harris CM, Sinn E (1968) J. Chem. Phys., 49: 2183

4.18 Bencini A, Gatteschi D In. Melson GA, Figgis BN (eds) Transitional metal chemistry, Marcel Dekker, New York, 8: 1 (1982)

4.19 Guigliarelli B, Gayda JP, Bertrand P, More C (1986) Biochim. Biophys. Acta 871: 149

4.20 Beinert H, Emptage MH, Dreyer J-L, Scott RA, Hahn JE, Hodgson KA, Thompson AJ (1983) Proc Natl. Acad. Sci. USA, 80: 393

4.21 Kent TA, Huynh BH, Münck E (1980) Proc. Natl. Acad. Sci. USA, 77: 6574

4.22 De Meester P, Skapski AC (1972) J. Chem. Soc. Dalton Trans, 2400

4.23 Brown DB, Wasson JR, Hall JW, Hatfield WE (1977) Inorg. Chem. 16: 2526

4.24 Banci L, Bencini A, Gatteschi D (1983) Inorg. Chem. 22: 2681

4.25 Bertrand JA, (1967) Inorg. Chem 6: 495

4.26 Lines ME, Ginsberg AP, Martin RL, Sherwood RC (1972) J. Chem. Phys. 57: 1

4.27 Rubins RS, Black TD, Barak J (1986) J. Chem. Phys. 85: 3770

4.28 Dickinson RC, Baker WA, Jr, Black TD, Rubins RS (1983) J. Chem. Phys. 79: 2609

4.29 Black TD, Rubins RS, De DK, Dickinson RC, Baker WA, Jr (1984) J. Chem. Phys. 80: 4620

4.30 Kamase K, Osaki K, Uryu N, (1979) Phys. Letters A73: 241

4.31 Buluggiu E, (1986) J. Chem. Phys., 84: 1243

4.32 Prins R, de Graaf RAG, Haasnoot J G, Vaden C, Reedijk J (1986) J. Chem. Soc. Chem. Commun. 1430

4.33 Bencini A, Gatteschi D, Zanchini C, Haasnoot JG, Prins R Reedijk (1987) J. Am. Chem. Soc. 109: 2926

5 Relaxation in Oligonuclear Species

5.1 Introduction

Relaxation is extremely important in the EPR of magnetically coupled systems, exactly as it is important for the spectra of isolated spins. However, the study of relaxation properties of exchange coupled systems has not yet had the same attention which has been devoted to the other parameters in the EPR spectra of these compounds. Some work has been performed on pairs, much less is available for oligonuclear species, a little more has been worked out for extended systems, as will be shown in Chap. 6.

The main limitation in the study of EPR relaxation is the experimental difficulty in the determination of either T_1 and T_2, which are very short at high temperature. Under this respect also the spectra of isolated spins have not been fully investigated, suffering the same limitations. Therefore, only a few studies at very low temperature are available, while the high temperature range has been hardly investigated at all, except for some indirect investigations through NMR measurements [5.1].

In the following we will briefly outline the theoretical basis of electron spin-lattice relaxation in pairs, resuming in short the general theories and specializing them to couples. We will work out in detail some examples and then we will mention some results in clusters containing more than two electron spins.

5.2 Theoretical Basis of Spin Relaxation in Pairs

The theoretical model currently used to describe spin-lattice relaxation in pairs is analogous to that of mononuclear species [5.2–4]. This treatment is now well established and we refer to the quoted texts, also for the necessary references.

For pairs, like for mononuclear species, several different processes can be operative, namely the direct, the Raman, and the Orbach processes. The spin system is assumed to be immersed in a thermal bath, to which it can dissipate nonradiatively energy. Since the theory was first developed for solids, the environment in which the spin is immersed is called lattice, and the lattice vibrations, phonons, are responsible for the relaxation.

A spin system is characterized by a set of eigenvalues of the hamiltonian which describes it in the absence of dynamic phenomena. Turning on the interaction with the lattice determines a fluctuation of all the terms of the hamiltonian, which are then modulated by the phonon bath. All the components can be modulated, but the most important for relaxation in single spin systems are crystal field and zero field splitting modulation. In coupled spin systems also modulation of the exchange hamiltonian becomes important, as will be shown later. The modulated hamiltonians have the same form as the static ones, but the relative parameters reflect the strain dependence of the static parameters and the symmetry of the dynamic hamiltonian can be reduced compared to the static one. For instance, the hamiltonian appropriate to the modulation of the isotropic exchange can take the form:

$$H = J'S_A \cdot S_B,　\qquad (5.1)$$

where $J' = r_o \, \delta J/\delta r_{AB}$. r_{AB} is the distance between the two coupled centers, and r_o is their equilibrium distance. The matrix elements of the dynamic hamiltonians, therefore, follow the same selection rules as the corresponding static ones. A phonon can be either absorbed or emitted when two levels have matrix elements different from zero with the dynamic hamiltonian.

Considering two levels $|m\rangle$ and $|n\rangle$, a possible relaxation mechanism is operative when the lattice can induce a transition between the two, with the absorption of one phonon, whose energy corresponds to the energy difference between the two levels, as shown in Fig. 5.1. This is the so-called direct process, which, in order to be efficient requires that an elevated number of phonons of energy ω_{mn} is present in the lattice. Since ω_{mn} is small for an ordinary EPR experiment, only at very low temperature is this condition fulfilled. In fact, the energy density of phonons as a function of $x = \hbar\omega_{mn}/kT$ is shown in Fig. 5.2. At high temperature ($x \approx 0$) the number of phonons available is very small, but on decreasing T (increasing x), the number of phonons increases sharply. For instance, for a frequency of 9 GHz, the ratio of the energy densities of phonons at $T = 1$ K and $T = 10$ K is ≈ 82.

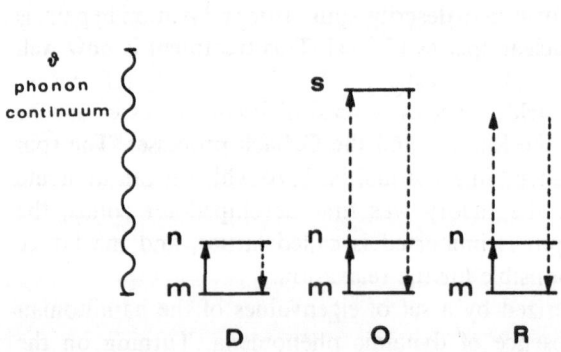

Fig. 5.1. Scheme of the energy levels responsible for the relaxation mechanism: *D* direct process; *O* Orbach process; *R* Raman process

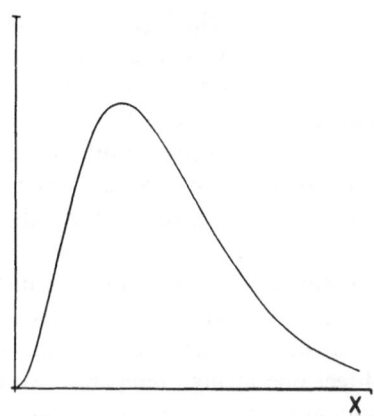

Fig. 5.2. Energy density of phonons as a function of $x = \hbar\omega_{mn}/kT$

The characteristic time T_1 in which the transitions bring the spin system into equilibrium is given by:

$$T_1^{-1} \propto \hbar^3 \omega_{mn}^3 \coth(\hbar\omega_{mn}/2kT) \Sigma_k |\langle m|H_k|n\rangle|^2, \tag{5.2}$$

where H_k is the dynamic hamiltonian responsible for the transition, and the summation is over the various modes of vibration. If $\hbar\omega_{mn} \ll kT$, (5.2) shows that T_1^{-1} is directly proportional to T.

The above expression has been obtained on the assumption that the transfer of energy of phonons to the bath is instantaneous. However, since the number of phonons available is much smaller than that of the spins in an undiluted material, each phonon must experience a very high number of collisions for every collision experienced by a spin. This has the effect of considerably reducing the mean lifetime for the phonon, and consequently to reduce the mean free path of phonons. At low temperature this may become shorter than the crystal size, thus making the process of exchange with the bath much longer. This situation is referred to as "phonon bottleneck", and it alters considerably the relaxation mechanism. It has been established that T_1^{-1} under these conditions is proportional to:

$$T_1^{-1} \propto \coth^2(\hbar\omega_{mn}/2kT). \tag{5.3}$$

If $\hbar\omega_{mn} \ll kT$, T_1^{-1} is proportional to T^2 rather than to T, as required by (5.2).

Beyond the direct one-phonon process, there are several two-phonon processes, which are more effective at high temperatures, of which the most relevant are named after Orbach and Raman, respectively. The former consists in the absorption of a phonon ω_1, exciting the spin into the $|s\rangle$ level, which lies in the phonon continuum, and the emission of a second phonon ω_2 from $|s\rangle$ to $|n\rangle$, as schematized in Fig. 5.1. In this way the phonons which are responsible for the relaxation have energies Δ_1 and $\Delta_2 \approx \Delta$, much larger than that of the direct process and are therefore much more numerous than those which allow the direct process (Fig. 5.2).

The Raman relaxation must be distinguished in a first- and a second-order process. The former occurs when the dynamic hamiltonian has matrix elements different from zero between the two levels $|m\rangle$ and $|n\rangle$ involved in the relaxation. The transition between the two can occur via absorption of one phonon ω_1 and emission of a phonon ω_2 with the condition that $\hbar\omega_{mn} = \hbar\omega_2 - \hbar\omega_1$. Phonons of energy close to the maximum in the diagram of Fig. 5.2 can contribute in this case, thus making this process much more effective than the direct one at high temperatures. In the second-order Raman process the two states $|m\rangle$ and $|n\rangle$ are both coupled to an excited state and a transition between them can be induced by the absorption and emission of two virtual phonons outside the phonon continuum.

Each of these processes has different temperature dependences. The Orbach process is characterized by:

$$T_1^{-1} \propto \Delta^3/[\exp(\Delta/kT)-1], \tag{5.4}$$

which becomes simply proportional to $\exp(-\Delta/kT)$ if $\exp(\Delta/kT) \gg 1$.

The temperature dependence of the Raman process is different for Kramers and non-Kramers states. For the latter it is found that

$$T_1^{-1} \propto T^7, \tag{5.5}$$

while for the former

$$T_1^{-1} \propto T^9 \tag{5.6}$$

under the condition that $T < \Theta$. Θ is the maximum phonon frequency available in the medium, expressed as temperature. In a lattice this corresponds to the Debye temperature. When $T > \Theta$ both for Kramers and non-Kramers ions in the Raman process, T_1^{-1} becomes proportional to T^2.

Finally, when the ground state is a multiplet, then the $|m\rangle$ and $|n\rangle$ levels in the EPR transition can be coupled to the other states of the multiplet. In this case the relaxation process is named after Blume and Orbach. It leads at low temperatures to a dependence of the type:

$$T_1^{-1} \propto T^5, \tag{5.7}$$

In an exchange-coupled pair, by definition, there are a number of electronic levels with energies comparable to kT. Therefore, these can provide effective relaxation pathways, which make in general T_1^{-1} larger for pairs than for isolated ions, through two phonon relaxation processes. In fact, the modulation processes, which are already present for single spins, may become more efficient due to the presence of new, low lying levels which are generated by the exchange interaction.

The modulation processes of the exchange interaction can be split as usual into isotropic, anisotropic, and antisymmetric components. Modulation of isotropic exchange can yield relaxation only to the second order for pairs, but it can be effective also to the first order for larger clusters. In fact, the selection rules for isotropic exchange require $\Delta S = 0$ and $\Delta M = 0$, but for couples only one total spin state with a given value of S is possible. This condition is no longer met for larger clusters, where more total spin states with the same value of S can exist. Anisotropic and antisymmetric exchange, on the other hand, can couple states with different S and M, therefore, although they are generally much smaller than the isotropic one, they can produce sizeable effects to the first order in pairs.

With this background we may now examine some examples. The experimental studies of spin-lattice relaxation have focused essentially on either simple doped lattices, such as Ir^{4+} pairs in $(NH_4)_2PtCl_6$, or materials of biological interest, among which we choose iron-ferredoxins as interesting examples.

5.3 Iridium(IV) Pairs

The studies of iridium(IV) pairs constitute a classic in relaxation of exchange-coupled systems, and we will work them out in detail, following the original papers by Owens and Yngvesson [5.5, 6]. The temperature dependence of the relaxation rate for iridium (IV) doped $(NH_4)_2PtCl_6$ is shown in Fig. 5.3. The experiments were performed on single crystals, at various doping levels. At low doping levels T_1^{-1} does not depend markedly on concentration, while the dependence is much more dramatic when the concentration increases. This has been taken as evidence of an intrinsic relaxation of pairs to the lattice, not involving cross-relaxation to higher nuclearity intermediates.

The procedure for a theoretical estimation of T_1^{-1} requires knowledge of the ground state and of the excited states which can be admixed into it and of the normal modes of vibration of the pair. Ir^{4+} has a ground $^2T_{2g}$ state in octahedral symmetry, which is split by spin-orbit coupling and low symmetry effects to give a ground doublet and an excited quartet. These two states can be described by effective spins $S = 1/2$ and $S = 3/2$, respectively. The ligand field states are given in Table 5.1. Coupling the two doublets yields a singlet and a triplet, corresponding to the ground manifold of a pair. Excited manifolds correspond to one ion in the $S = \frac{1}{2}$ state and the other in the excited $S = 3/2$ state, yielding two $S = 2$ and two $S = 1$ states. At higher energy we find the states generated by the interaction of the two ions in the excited $S = 3/2$ states, yielding $S = 3, 2, 1$, and 0. In order to determine the relaxation process we must calculate the admixture of the excited states into the low lying singlet and triplet states via the exchange and the Zeeman hamiltonian and then evaluate the matrix elements of the dynamic hamiltonian including the vibrations. Since the energy separation between the

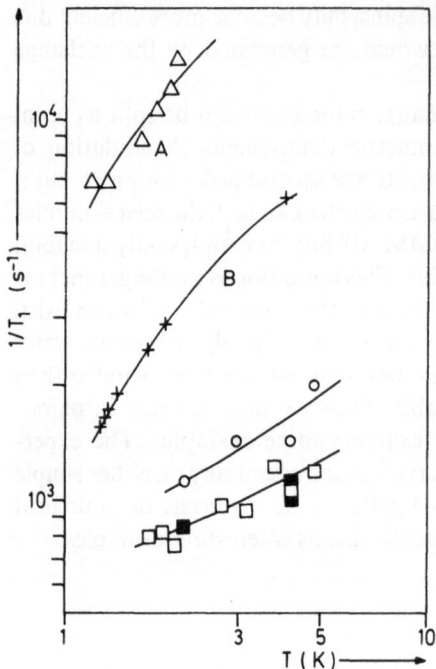

Fig. 5.3. Relaxation data for iridium(IV) pairs in single crystals of ammonium chloroplatinate at 9.5 GHz. The external magnetic field is in the (111) plane. The sets of data refer to different iridium concentrations: □1.2%, ○2.3%; +5.0%; △8.5%. ■ and ● refer to a different crystal orientation. After [5.5]

Table 5.1. Basis functions for the ground manifold of $^2T_{2g}$ of $IrCl_6^{2-}$

$|\frac{1}{2}\frac{1}{2}\rangle = -1/\sqrt{3}[|yz^-\rangle + i|xz^-\rangle + |xy^+\rangle]$

$|\frac{1}{2}-\frac{1}{2}\rangle = 1/\sqrt{3}[-|yz^+\rangle + i|xz^+\rangle + |xy^-\rangle]$

$|3/2\ 3/2\rangle = -1/\sqrt{2}[|yz^+\rangle + i|xz^+\rangle]$

$|3/2\ \frac{1}{2}\rangle = 1/\sqrt{6}[-|yz^-\rangle - i|xz^-\rangle + 2|xy^+\rangle]$

$|3/2-\frac{1}{2}\rangle = 1/\sqrt{6}[|yz^+\rangle - i|xz^+\rangle + 2|xy^-\rangle]$

$|3/2-3/2\rangle = 1/\sqrt{2}[|yz^-\rangle - i|xz^-\rangle]$

---

total spin states is large compared to the anisotropic exchange and Zeeman interaction, the correct functions can be obtained through a perturbation treatment.

The unperturbed ground state and the relevant excited states can be written using standard vector coupling techniques (Chap. 3). By using the isotropic and

126

anisotropic exchange and Zeeman hamiltonians, the singlet and triplet functions of the ground manifold are modified as shown in Table 5.2, where $\frac{3}{2}\Delta$ is the energy difference between the $S = 1/2$ and the $S = 3/2$ manifolds of the individual ions, brought about by ligand field effects. J is the isotropic exchange, D is the axial parameter of anisotropic exchange, defined by the hamiltonian $H = D(3S_{AZ}S_{BZ} - S_A \cdot S_B)$, and B is the external magnetic field, taken parallel to the z molecular axis. The line joining the two iridium ions defines the x axis.

There are only five normal modes of the individual ions which are relevant to the relaxation process (Fig. 5.4), because the totally symmetric one has only diagonal matrix elements which cannot give rise to relaxation. Therefore, there are ten active modes for the pair which can be expressed as symmetric, F_k, and antisymmetric, G_k, combinations of the 21 modes of the individual ions, E_k:

$$F_k = (E_k^1 + E_k^2)/\sqrt{2}; \tag{5.8}$$

$$G_k = (E_k^1 - E_k^2)/\sqrt{2}; \tag{5.9}$$

for $k = 1 - 5$. The magnitudes of the strains associated to the pair modes, f_k and

Table 5.2. Basis functions for the ground manifold of Ir^{4+} pairs[a]

Unperturbed	Perturbed
$\lvert\frac{1}{2}\frac{1}{2}00\rangle$	$\lvert\frac{1}{2}\frac{1}{2}00\rangle - 2/(3\Delta)\,[2/3\,\mu_B B(k+2)\lvert\frac{1}{2}\,3/2\,10\rangle_a +$ $(3J/2 - 3D/8)\lvert\frac{1}{2}\,3/2\,2\,0\rangle_a]$
$\lvert\frac{1}{2}\frac{1}{2}1\,1\rangle$	$\lvert\frac{1}{2}\frac{1}{2}1\,1\rangle - 2/(3\Delta)\{-[J/4 + 3D/16 + \mu_B B(k+2)/3]$ $\lvert\frac{1}{2}\,3/2\,1\,1\rangle_s - [\sqrt{3}J/4 + 3\sqrt{3}D/16 + \mu_B B(k+2)/\sqrt{3}]$ $\lvert\frac{1}{2}\,3/2\,2\,1\rangle_s\}$
$\lvert\frac{1}{2}\frac{1}{2}1\,0\rangle$	$\lvert\frac{1}{2}\frac{1}{2}1\,0\rangle - 2/(3\Delta)\{[J/2 + 9D/16]\,\lvert\frac{1}{2}\,3/2\,1,\,0\rangle_s - 2\mu_B B$ $(k+2)/\sqrt{3}\lvert\frac{1}{2}\,3/2\,2\,0\rangle_s\}$
$\lvert\frac{1}{2}\frac{1}{2}1\,-1\rangle$	$\lvert\frac{1}{2}\frac{1}{2}1\,-1\rangle - 2/(3\Delta)\{-[J/4 + 3D/16 - \mu_B B(k+2)/3]$ $\lvert\frac{1}{2}\,3/2\,1\,-1\rangle_s - [\sqrt{3}J/4 + 3\sqrt{3}D/16 - \mu_B B(k+2)/\sqrt{3}]$ $\lvert\frac{1}{2}\,3/2\,2\,-1\rangle_s\}$

[a]The functions are labeled according to the individual and total spin states: $\lvert S_A S_B SM\rangle$. The subscripts a and s denote antisymmetric and symmetric combinations, respectively, of $\lvert\frac{1}{2}\,3/2\,SM\rangle$ and $\lvert 3/2\,\frac{1}{2}\,SM\rangle$. k is the orbital reduction factor of the Zeeman operator. The contributions of $\lvert 3/2\,3/2\,SM\rangle$ states have been neglected.

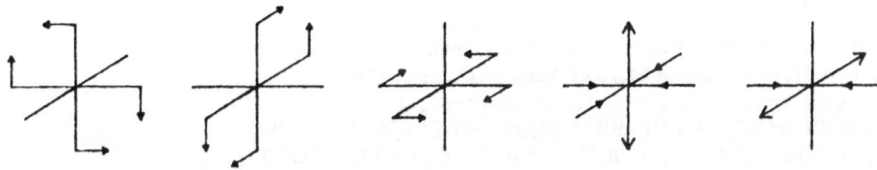

Fig. 5.4. Normal modes of vibration of octahedral $IrCl_6^{2-}$ which are relevant to relaxation processes

g_k, are related to the mean strain, q_o, of a lattice wave by:

$$f_k = c_k q_o; \tag{5.10}$$

$$g_k = b_k q_o. \tag{5.11}$$

The expressions appropriate for b_k and c_k are given in Table 5.3. (l, m, n) are the cosines of the propagation direction of a sound wave whose strain of amplitude q_o is directed along a (p, q, r) direction. For a longitudinal wave (l, m, n) is parallel to (p, q, r), while for a transverse wave (l, m, n) is orthogonal to (p, q, r).

The relaxation of the system is evaluated through a perturbation hamiltonian of the type:

$$H = \Sigma_k (g_k G_k + f_k F_k), \tag{5.12}$$

where G_k and F_k operate only on the electronic coordinates. They depend on the modulation which is responsible for the relaxation and have the same symmetries as the corresponding vibration modes. Symmetric perturbations can give matrix elements between the triplet states, and so even modes give rise to a direct relaxation process. The antisymmetric perturbations couple states of different parity, and so they can induce relaxation between the singlet and triplet states through an Orbach process. For crystal field modulation G_k and F_k can be conveniently expressed as a function of the operators of the individual ions (Table 5.4). The wavelength of low frequency phonons is much larger than the

Table 5.3. b_k and c_k coefficients[a]

k	b_k	c_k
1	$\sqrt{3}\alpha/2\,(1p-mq)(1+m)$	$(1p-mq)$
2	$\alpha/2\,(1p+mq-2nr)(1+m)$	$1/\sqrt{3}(1p+mq-2nr)$
3	$\sqrt{3}\alpha/2\,(mp+lq)(1+m)$	$(mp+lq)$
4	$\sqrt{3}\alpha/2\,(np+lr)(1+m)$	$np+lr$
5	$\sqrt{3}\alpha/2\,(nq+mr)(1+m)$	$(nq+mr)$

[a] $\alpha = (2/3)^{\frac{1}{2}}\pi a/\lambda$, where a is the iridium-iridium distance and λ is the wavelength of the phonon.

Table 5.4. Crystal field hamiltonian operators corresponding to strains active for relaxation in individual Ir^{4+} ions[a]

$\varepsilon_1 = -A/21\,\langle r^2 \rangle (O_2^{+2}+O_2^{-2}) - B/63\,\langle r^4 \rangle (O_4^{+2}+O_4^{-2})$
$\varepsilon_2 = 2A/21\sqrt{3}\,\langle r^2 \rangle O_2^0 - B/84\sqrt{3}\,\langle r^4 \rangle (2O_4^0 - 7O_4^{+4} - 7O_4^{-4})$
$\varepsilon_3 = C/42i\,\langle r^2 \rangle (O_2^{+2}-O_2^{-2}) + D/126i\,\langle r^4 \rangle (O_4^{+2}-O_4^{-2})$
$\varepsilon_4 = C/21\,\langle r^2 \rangle (O_2^{+1}+O_2^{-1}) - B/252\,\langle r^4 \rangle (O_4^{+1}+O_4^{-1}+7O_4^{+3}+7O_4^{-3})$
$\varepsilon_5 = C/21i\,\langle r^2 \rangle (O_2^{+1}-O_2^{-1}) - B/252i\,\langle r^4 \rangle (O_4^{+1}-O_4^{-1}-7O_4^{+3}-7O_4^{-3})$

[a] A, B, C, and D are constants which need not be further specified.

iridium-iridium distance, therefore, the two ions vibrate very nearly in phase, and the $F_k(G_k)$ operators are simply given by the sum(difference) of the operators in Table 5.4.

Let us consider the direct process for the $|1\ 1\rangle \rightarrow |1\ 0\rangle$ transition. In this case only the symmetric modes are relevant, because the two levels have the same parity. Matrix elements of the type:

$$\langle 1\ 1|F_k|1\ 0\rangle\langle \Phi|f_k|\Phi'\rangle \tag{5.13}$$

must be evaluated, where $|\Phi\rangle$ and $|\Phi'\rangle$ are the state vectors of the phonon system before and after the transition. The expression for the direct relaxation rate determined by ligand field modulation is given by:

$$T_{1D}^{-1} = kT\delta^2/(\pi\rho\hbar^4)\Sigma_k[\langle c_k\rangle_l^2/v_l^5 + 2\langle c_k\rangle_t^2/v_t^5]|\langle 11|F_k|10\rangle|^2, \tag{5.14}$$

where δ is the energy separation between the two levels, ρ is the number of phonons of energy in the range $\delta - \delta + d\delta$, v is the velocity of the sound, and the indices l and t refer to longitudinal and transverse waves, respectively.

The orbital integral is different from zero only for $k = 4$ and 5; further, only the excited states admixed by Zeeman and anisotropic exchange are relevant, while isotropic exchange mixing is in this particular case ineffective. Performing the integral on the orbital coordinates and averaging the c_k coefficients on all the possible propagation directions, the relaxation rate can be expressed as:

$$T_{1D}^{-1} = [4\mu_B B(k+2)/(9\Delta) - D/(16\Delta)]^2 2PT, \tag{5.15}$$

where P is a constant. If we compare (5.15) with the analogous expression for the single ion:

$$T_{1D}^{-1} = [4\mu_B B(k+2)/(9\Delta)]^2 4PT, \tag{5.16}$$

we see that neglecting D,

$$(T_{1D}^{-1})_{pair} = \tfrac{1}{2}(B_{pair}/B_{single})^2(T_{1D}^{-1})_{single}. \tag{5.17}$$

Therefore, as far as the direct process is concerned, the relaxation rate of the pair induced by crystal field modulation can be slower than that of the single ion.

Modulation of the anisotropic exchange is also symmetric in the spin coordinates, therefore, it can induce direct relaxation processes between the triplet states. Since this term is not operative for single ions, its inclusion determines a faster relaxation for the pair. The hamiltonian can be expressed as:

$$D'(3S_{AZ}S_{BZ} - S_A \cdot S_B) \tag{5.18}$$

with $D' = r_o\ \delta D/\delta r_{AB}$. The contribution of this term to the direct process is similar to (5.14). The matrix element of the electronic hamiltonian has been estimated numerically to correspond to ≈ 7 cm^{-1}, in such a way that the contribution of the anisotropic exchange to the direct relaxation rate is

$$T_{1D}^{-1} \propto 400 \text{ T s}^{-1}. \tag{5.19}$$

The triplet state functions can relax also by an Orbach process involving the singlet level. In this case it is the antisymmetric modes G_k which are relevant, because they can couple the two spin multiplets of opposite parity. The calculated relaxation rate is of the type:

$$T_1^{-1} = B/[1 - \exp(-J/kT)]. \tag{5.20}$$

The temperature dependence in (5.20) is different from that in (5.4), because in this case the states involved in relaxation are higher in energy than the state $|s\rangle$ of the Orbach process. This is a process which has no counterpart in the single ion. The experimental data of Fig. 5.3 were fitted with this dependence, using the J value 5.2 cm^{-1}. However, the predicted rate is within a factor of 3 of the observed rate, so that it was concluded only that it is probable that (5.20) is a major contribution to the relaxation process.

5.4 Copper Pairs

Another simple system which has been investigated is that of copper pairs in zinc(II) bis(diethyl-dithiocarbamato), $Zn(dtc)_2$, single crystals [5.7]. A typical spectrum of a single crystal containing 5% copper(II) is shown in Fig. 5.5. In addition to the four intense lines in the center of the spectrum, which arise from isolated copper(II) ions, two well-resolved groups of septets, arising from copper pairs, are resolved. The relaxation rate was measured using pulse and spin-echo techniques. At $T < 4$K the experimental data were found to follow the temperature dependence expected for a direct process, with $T_1^{-1} = 40$ T at 28.8 GHz and $T_1^{-1} = 140$ T at 9.0 GHz, while in the range $4 < T < 12$ K the data could be fitted by an Orbach process: $T_1^{-1} = 19 \times 10^3/[\exp(\delta/kT) - 1]$ with $\delta = 13 \pm 1$ cm^{-1}. δ was considered to correspond to the singlet-triplet splitting, the triplet lying lower.

The calculated relaxation rate for the direct process is 2.5 times faster than that for the single ion, however, the experimental result is that relaxation in the dinuclear species is 16 times faster than in the single ion. A possible explanation for this difference lies in the fact that single ions and pairs have different structures.

Another important feature of the relaxation rate of the direct process is its frequency dependence. In fact, (5.14) requires a *direct* linear dependence of T_1^{-1} on $\hbar\omega$, while the experimental data on Cu:$Zn(dtc)_2$ rather suggest an *inverse* linear dependence. This anomaly has been explained on the basis of the dominant role of isotropic exchange modulation in weak fields.

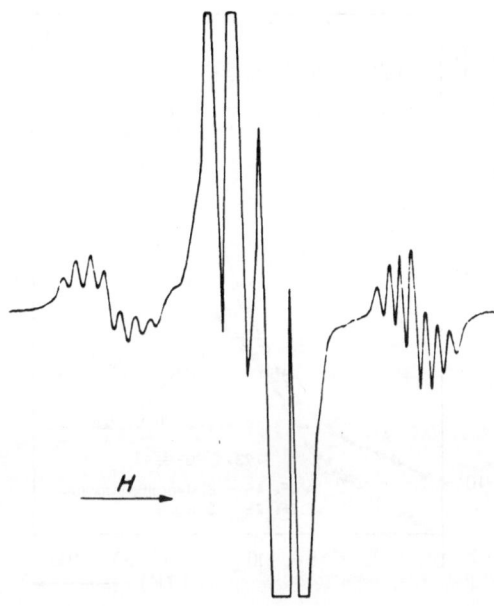

Fig. 5.5. Single crystal spectrum of copper(II) doped $Zn(dtc)_2$ at 4.2 K and 37.0 GHz. After [5.7]

5.5 Two-Iron-Two-Sulfur Ferredoxins

More complicated pairs, which have been studied, however, in some detail, are met in the so-called two-iron ferredoxins. The structure and properties of these metalloproteins are described in more detail in Sect. 9.3.2. Here, it is sufficient to say that in the reduced form they contain pairs with coupled high spin iron(III) and iron(II). The coupling is antiferromagnetic, yielding a ground $S = \frac{1}{2}$ state. Beyond the natural products there are now also several synthetic model compounds which have been studied, mimicking several of the properties of the metalloproteins. We want to mention some relaxation studies here in order to show how they can in principle provide useful information on the exchange in coupled pairs.

The spin-lattice relaxation time of the ferredoxin from *Spirulina maxima* has been studied [5.8-12] with several different experimental techniques in the range 1.8 – 130 K. The data, shown diagrammatically in Fig. 5.6, were fitted by the equation:

$$T_1^{-1} = 0.9T^2 + 3.5 \times 10^{-10}T^9 I_8(60/T) + 7.3 \times 10^{10} \exp(-350/T), \qquad (5.21)$$

where $I_n(\Theta_D/T)$ is the integral:

$$I_n(\Theta_D/T) = \int_0^{\Theta_D} x^n e^x/(e^x - 1)^n \, dx. \qquad (5.22)$$

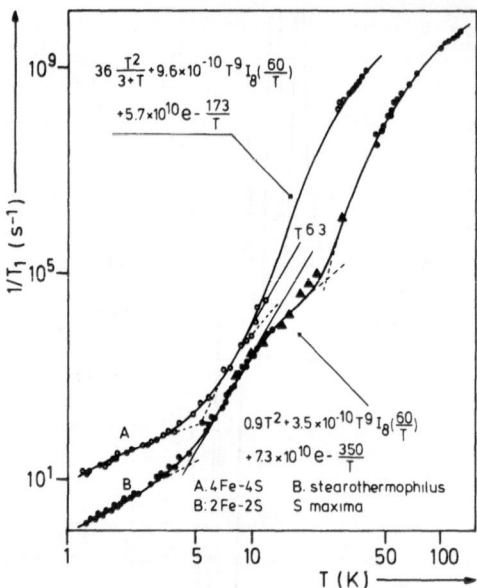

Fig. 5.6. Temperature dependence of the relaxation rate of *A Spirulina maxima* ferredoxin; *B Bacillus stearotermophilus*. After [5.9]

The first term corresponds to a direct bottlenecked process, the second to a Raman, and the third to an Orbach process. The last has been attributed to a transition to an excited $S = 3/2$ level, orginating from the exchange interaction between $S = 5/2$ of iron(III) and $S = 2$ of iron(II). In this way J was estimated to be 166 cm^{-1}. Therefore, the experimental determination of the spin-lattice relaxation rate can provide indirect information on the isotropic exchange, a feature which is particularly important in metalloproteins for which magnetic susceptibility measurements can be rather difficult to perform. However, it must be mentioned that the validity of (5.21) has been questioned [5.11], on the grounds that it assumes a three-dimensional nature of the phonons which are responsible for the relaxation. In fact, the protein should be better described by a fractal model [5.13], which assigns to it nonintegral dimensionality. Following this approach, an alternative fit of the experimental data was found, which differs from (5.21) essentially in the Raman term, which requires a temperature dependence of the type $T^{5.666}$, in agreement with a dimensionality of the protein $d = 1.65$.

5.6 Relaxation in Larger Clusters

The main difference between pairs and larger clusters is that while for the former isotropic exchange modulation cannot provide effective relaxation mechanisms, for the latter this pathway may also become important. In fact, in a triad, or in a larger cluster, there may be several total spin states with the same S, and consequently fast relaxation can occur via modulation of the isotropic exchange. A simple example where this condition is met is that of three $S = \frac{1}{2}$ spins coupled. There are two doublets thus formed (Fig. 5.7). The transitions between the $+\frac{1}{2}$ and $-\frac{1}{2}$ levels of the two doublets can be very fast, thus producing a very effective relaxation. This pathway has been considered to be responsible for the fast relaxation of single ions iridium(IV) in $(NH_4)_2PtCl_6$ due to cross-relaxation effects to triads [5.5]. The relaxation between the doublet spin levels, in an order of magnitude estimate, has been assumed to be of the type $T_1^{-1} = 2.0 \times 10^8 \exp(-5.5/T)$, much faster than the relaxation of the corresponding pair.

Some other indirect evidence for this is available in the literature. For instance, in tetrads of copper ions possessing S_4 symmetry, with antiferromagnetic coupling, the ground state is a singlet, with three excited triplets and a quintet [5.14]. The EPR spectrum of the $S = 2$ state has been clearly resolved at room temperature, while at low temperature the spectrum of the lowest triplet is observed. The spectra of the other two triplets, which are degenerate, are never observed, presumably as a consequence of fast relaxation. In this case it is possible that fast relaxation is induced by transitions within the degenerate levels or also to the lowest multiplet, which then can be observed only because it is still populated at low temperature, when relaxation is sufficiently frozen out.

Another feature which can be determined by fast relaxation is the fact that in several triads of spin $S = \frac{1}{2}$ only one signal is observed at all temperatures [5.15–17]. Since two doublets and a quartet are generated by the exchange interaction, one should anticipate at least three signals, but only one average signal is observed. The effective g values have been found to be temperature-dependent, a fact which has been attributed to the thermal population of the different spin levels, characterized by different g values.

An example of an accurate determination of the spin-lattice relaxation rate is that of four iron-four sulfur ferredoxins from *Bacillus Stearothermophilus* [5.9]. In Fig. 5.6B the temperature dependence of T_1^{-1} is shown, which was fitted according to:

$$T_1^{-1} = 36T^2/(3+T) + 9.6 \times 10^{-10}T^9I_8(60/T) + 5.7 \times 10^{10}\exp(-173/T). \tag{5.23}$$

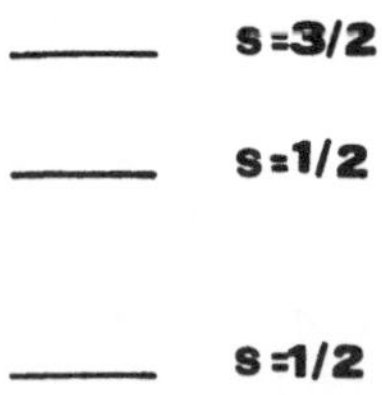

‎——— s=3/2

‎——— s=1/2

‎——— s=1/2

Fig. 5.7. Scheme of the energy levels of three coupled $S = \frac{1}{2}$ spins

This shows that below 4 K the condition is that of a phonon bottleneck, for $4 < T < 10$ K a Raman process is dominant, with a Debye temperature Θ_D $= 60$ K, like in the 2Fe-2S protein, while for $T > 10$ K an Orbach process becomes dominant with an excited level at 173 K. In all the temperature ranges the relaxation rate for the 4 Fe protein is at least an order of magnitude faster than that of the 2Fe protein, except that in the range where the Raman process is dominant, where there is only a factor of 3 between the two.

References

5.1 Bertini I, Luchinat C (1985) NMR of Paramagnetic Molecules in Biological Systems, Benjamin/Cummings: Menlo Park
5.2 Abragam A, Bleaney B (1970) Electron Paramagnetic Resonance of Transition Ions, Clarendon Press, Oxford
5.3 Stevens, KWH (1967) Rep. Progr Phys 30, 189
5.4 Owen J, Harris E A (1972) in Electron Paramagnetic Resonance; Geschwind, S., Ed.; Plenum Press, New York London
5.5 Harris EA, Yngvesson KS (1968) J. Phys. C1, 990
5.6 Harris EA, Yngvesson KS (1968) J. Phys. C1, 1011
5.7 Al'tshuler SA, Kirmse R, Solovev BV (1975) J. Phys. C, 8, 1907
5.8 Gayda JP, Gibson JF Cammack R, Hall DO, Mullinger R (1976) Biochim. Biophys. Acta 434, 154
5.9 Bertrand P, Gayda JP, Rao KK (1982) J. Chem. Phys. 76, 4715
5.10 Gayda JP, Bertrand P, Deville A, More C, Roger G, Gibson JF, Cammack R (1979) Biochim. Biophys. Acta, 581, 15
5.11 Stapleton HJ, Allen JP, Flynn CP, Stinson DG, Kurtz SR (1985) Phys. Rev. Letters 45, 1456
5.12 Bertrand P, Roger G, Gayda JP (1980) J. Magn. Reson. 40, 539
5.13 Tarasov VV (1958) J. Am. Chem. Soc. 80, 5052
5.14 Bencini A, Gatteschi D, Zanchini C, Haasnot JG, Prins R, Reedijk J (1987) J Am. Chem. Soc. 109, 2926
5.15 Benelli C, Gatteschi D, Zanchini C, Latour JM, Rey P (1986) Inorg. Chem. 25, 4242
5.16 Banci L, Bencini A, Dei A, Gatteschi D (1983) Inorg. Chem. 22, 4018
5.17 Banci L, Bencini A, Gatteschi D (1983) Inorg. Chem. 22, 2681

6 Spectra in Extended Lattices

6.1 Exchange and Dipolar Interactions in Solids

When the number of interacting spins becomes very large, we enter the realm of extended interactions. These can occur in one-, two- and three-dimensional spaces, as well as in spaces with fractal dimensions, with an almost infinite number of possible variations. Historically three-dimensional solids have been treated first, while the lower dimensional lattices have become of interest only in the last few years. In the following we will treat first the three-dimensional case, which for some aspects is less complicated than the lower dimensional cases, and we will treat the latter only after the former has been fully accounted for.

It is perhaps instructive at this point to define what is meant by a one-, two-, and three-dimensional magnetic lattice. Let us consider a general spin embedded in the lattice in which we may be interested. In a one-dimensional lattice this is "strongly" coupled to two other spins, with a constant J. It may be coupled to other spins as well, but the relative J' constants must be much smaller than J. In Fig. 6.1a we have depicted a possible realization of a one-dimensional lattice.

In a possible realization of a two-dimensional lattice a spin in general position is bound to four spins with a strong constant J (Fig. 6.1b). All the spins lie approximately in the same plane. The coupling to spins in other planes is determined by J', which is smaller than J. Finally, in a three-dimensional lattice every spin is bound to more than four spins with a strong coupling constant J. It must be recalled here that the factor determining the magnetic dimensionality is the connectivity through exchange-coupling constants, which may or may not correspond to the structural connectivity.

In order to analyze the spectrum of an extended lattice let us consider first a set of spins $S = 1/2$, with isotropic **g** tensors corresponding to the free electron value and no hyperfine interaction, arranged in a three-dimensional lattice. In the absence of any external perturbation other than a static magnetic field, the system is characterized by the Zeeman hamiltonian:

$$H_z = \Sigma_i \mu_B B g S_{zi}, \tag{6.1}$$

and the absorption of microwaves of energy $\hbar\omega_o$ will occur at $\omega_o = g\mu_B B_o/\hbar$. The sum in (6.1) is over all the paramagnetic centers in the lattice. This situation corresponds to the case when all the spins are uncorrelated and in this limit the spectrum consists of one line, characterized by the natural width of the absorption. The eigenstates of (6.1) can be written as an antisymmetrized product

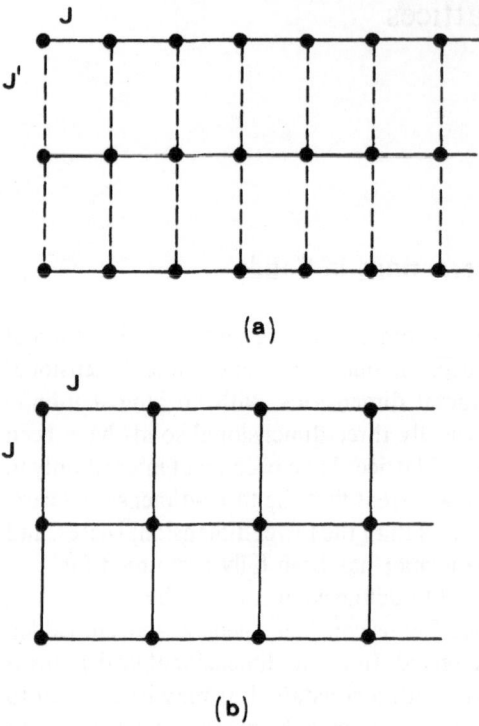

(a)

(b)

Fig. 6.1a, b. Scheme of **a** a one-dimensional; **b** a two-dimensional lattice

of the eigenstates $\Phi_i(m_i)$ of the individual terms of the Zeeman hamiltonian as:

$$\psi = A\{\Phi_1(m_1)\Phi_2(m_2)\ldots\Phi_j(m_j)\ldots\Phi_N(m_N)\}, \tag{6.2}$$

where A is an antisymmetrizer, and the subscripts refer to the sites in the lattice. The function ψ can be characterized by the total spin component $M = \Sigma_i m_i$.

In a real system there are two possible interactions between spins which can be turned on: one is the magnetic dipolar and the other is the exchange interaction. As it was shown in Chapter 2 both of them are generally represented by dyadics which can be decomposed into a scalar, a vector, and a tensor component. In the systems we are considering the vector components are zero for both the interactions, while the scalar is zero for the magnetic dipolar and the tensor is zero for the exchange interaction. The hamiltonian for the latter can be written as:

$$H_{ex} = \Sigma_{j<k} J_{jk} S_j \cdot S_k, \tag{6.3}$$

and if the distance between paramagnetic centers is large for any couple of spins the hamiltonian for the former is given by:

$$H_{dip} = g^2 \mu_B^2 \Sigma_{j<k} \{ S_j \cdot S_k / r_{jk}^3 - 3(S_j \cdot r_{jk})(S_k \cdot r_{jk}) / r_{jk}^5 \}, \tag{6.4}$$

where r_{jk} is the vector connecting the sites of the j and k spins and the sums are extended to the entire lattice.

H_{dip} can be rewritten according to the relation:

$$H_{dip} = \Sigma_{j<k} (g^2 \mu_B^2 / r_{jk}^3) \{ [-S_{jz}S_{kz} + 1/4(S_{j+}S_{k-} + S_{j-}S_{k+})] (3\cos^2 \Theta_{jk} - 1)$$

$$\tag{6.5}$$

$$- 3/2(S_{jz}S_{k+} + S_{j+}S_{kz}) \sin \Theta_{jk} \cos \Theta_{jk} \exp(-i\Phi_{jk}) -$$

$$- 3/2(S_{jz}S_{k-} + S_{j-}S_{kz}) \sin \Theta_{jk} \cos \Theta_{jk} \exp(i\Phi_{jk}) -$$

$$- 3/4[S_{j+}S_{k+} \exp(-2i\Phi_{jk}) + S_{j-}S_{k-} \exp(2i\Phi_{jk})] \sin^2 \Theta_{jk}$$

The first term in the parentheses couples states with $\Delta M = 0$, the second and the third states with $\Delta M = \pm 1$, and the fourth states with $\Delta M = \pm 2$. The effect of H_{dip} therefore is that of modifying the eigenstates of H_z, which are given by (6.2), admixing into them states differing in M by 0, ± 1, and ± 2, respectively. A simple perturbation treatment, which is valid in the assumption that $H_z \gg H_{dip}$, gives the mixing coefficients $c = \mu_B \Sigma_i r_{jk}^{-3}/B$. Transitions between the total spin states can be induced by the transverse magnetization operator, $S_x = \Sigma_i S_{ix}$, connecting states differing in m_i by ± 1. Therefore, the transitions between the perturbed states will be, at the c^2 order of magnitude, between states with $\delta M = 0$, ± 1, ± 2, and ± 3. As a consequence, transitions centered at $\omega = 0$, $2\omega_o$, and $3\omega_o$, with an intensity proportional to c^2, will add to the main absorption at $\omega = \omega_o$.

The second effect of the dipolar interaction is that of broadening the main line centered at $\omega = \omega_o$. The reason for this broadening can become intuitively clear if we build up the infinite lattice by starting from two, three. . . interacting spins. In the limit of weak exchange two spins yield a singlet and a triplet, for which two allowed transitions will be observed between the states described by $M = +1$ and $M = 0$, and $M = -1$ and $M = 0$, respectively. The separation between the two lines, which are symmetrically spaced around ω_o, depends on the direction of the external magnetic field, being at a maximum along the line connecting the two spin sites. In this case the lines are found at $\omega_o \pm 1/2 D_{dip}$. In the case of vanishingly small exchange other transitions will also be observed, but all of them are symmetrically deplaced from ω_o.

For three spins two doublets and a quartet are formed. The first two yield two coinciding lines, centered at ω_o, the quartet yields three lines, one centered at ω_o and two at $\omega_o \pm 2/3 D_{dip}$. The process can be repeated indefinitely, but it is clear that adding one more spin to the cluster yields an increasing number of transitions, only part of which are centered at ω_o, the rest flanking it at different distances from it. The result is that of a homogeneous broadening of the line. The line shape must be Gaussian, resulting from the sum of individual, sharper transitions (Fig. 6.2).

137

Let us now consider the other limiting case, when isotropic exchange alone is present. Although this may not be a realistic possibility, since if the spins are close enough to have reasonable exchange interactions they will also have nonzero dipolar interactions, it is still useful to consider it. The exchange interaction yields a number of levels, increasing with the number of interacting spins, as is shown for chains of two, three, and four spins in Fig. 6.3. As long as the various multiplets are not split in zero field, the resonance will be centered at ω_o, and no shift or broadening will occur. In other words, since H_{ex} commutes with H_z, it will not affect the spectrum of transitions, while the noncommuting H_{dip} hamiltonian may yield broadening.

Now let us move to the general case in which both the isotropic exchange and the anisotropic dipolar interactions are switched on. Let us further assume that

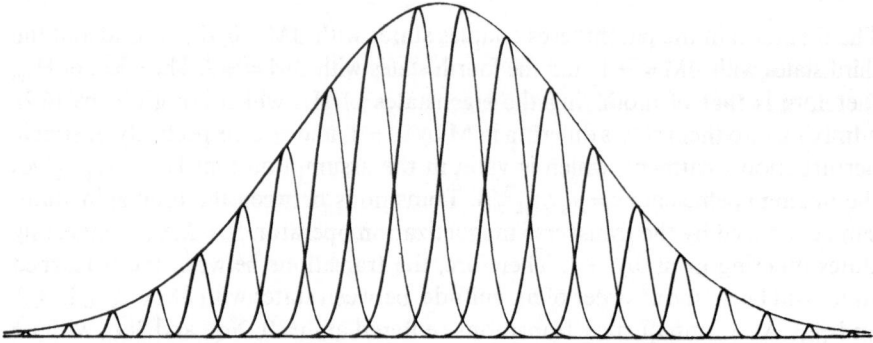

Fig. 6.2. A Gaussian line shape as the result of the sum of individual, sharper transitions

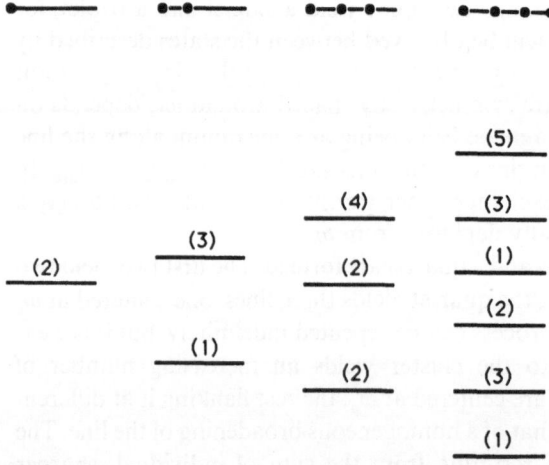

Fig. 6.3. Energy levels and degeneracies for one, two, three, and four $S = \frac{1}{2}$ spins

138

$kT \gg J$. For the tridimensional lattice we are presently considering this is justified by the fact that a paramagnetic phase is possible only if the exchange-coupling constant is smaller than the average thermal energy. For one- and two-dimensional lattices this condition may no longer hold, but we will not consider it here.

Considering that both the exchange and the dipolar interactions are present determines two different limit behaviors, corresponding to the predominance of either the dipolar or the exchange coupling. If $H_{dip} \gg H_{ex}$, the broadening effect dominates, and nothing essentially must be added to the considerations we made above: the line will be Gaussian and the satellite lines will be present. If on the contrary, $H_{dip} \ll H_{ex}$, then a completely new situation occurs, in which the lines are dramatically narrowed, the line widths becoming often comparable to those of the individual spins embedded in the magnetic lattice.

How this can occur even in the presence of nonzero dipolar interaction can be qualitatively understood considering that in the presence of the exchange interaction the spin at a given site sees the spins at the neighboring sites flipping at a rate proportional to J in frequency units. If J is larger than the average energy of the dipolar interaction, the latter will be averaged to zero narrowing the signal, much in the same way as fast motion can average two signals in an experiment in solution. This regime is referred to as exchange narrowing, and it will be treated in the next section.

6.2 Exchange Narrowing

A quantitative approach, which puts the above considerations on a firmer basis was developed initially by van Vleck [6.1], who used the method of moments to approximate the line shape and subsequently by Kubo and Tomita [6.2] and Anderson [6.3]. We will give here a brief outline of the stochastic treatment of the line shape performed by Kubo [6.4], because it is relatively simple, and, although it has been substituted by more sophisticated treatments using memory function formalisms, it is still widely used as a basis for the discussion of the properties in infinite lattices.

In the stochastic theory of the line shape of an exchange-narrowed EPR line, it is assumed that the forces acting on a spin system are random in nature. Being so general the theory applies equally well to exchange-narrowed and broadened lines, to motionally narrowed lines, and also to NMR experiments.

In the linear response theory, the spin precession in an external magnetic field, $B(t)$, given by the sum of a static field B_o and of a time-dependent field $B_1(t)$, which averages to zero, is described by an equation of motion

$$dM_x/dt = i\omega(t)M_x, \tag{6.6}$$

where $\omega(t) = \gamma B(t)$, γ being the gyromagnetic ratio for the considered spin. Equation (6.6) is just the equation of motion of a randomly modulated oscillator.

The random modulation $\omega(t)$ has a definite time average ω_0 given by:

$$\omega_0 = \lim_{T \to \infty} \frac{1}{T} \int_0^T \omega(t)\,dt \equiv \overline{\omega(t)} \qquad (6.7)$$

where the bar on the right-hand side denotes a time average. We can write $\omega(t)$ in the same form as B(t), i.e.,

$$\omega(t) = \omega_0 + \omega_1(t), \qquad (6.8)$$

where $\overline{\omega_1(t)} = 0$. The time-dependent part $\omega_1(t)$ is called a fluctuation. In statistical physics fluctuations are characterized by the probability distribution of their values and by their correlation functions which are defined as the average value of the product of the fluctuations at the times t and $t + \tau$, respectively, the average being over the statistical ensemble, i.e., $\langle \omega_1(t)\omega_1(t+\tau) \rangle$. In the following we will assume that the process represented by $\omega_1(t)$ is ergodic and stationary in time. This means that the probability distribution of the value of $\omega_1(t)$ is Gaussian and that the correlation function does not explicitly depend on t. This last hypothesis allows one to substitute a time average in the ensemble average i.e., $\langle \omega_1(t)\omega_1(t+\tau) \rangle = \overline{\omega_1(t)\omega_1(t+\tau)}$. Under this assumption (6.6) can be rewritten as:

$$M_x(t) = M_x(0)\exp\left\{i \int_0^t \omega(t')\,dt'\right\} =$$

$$= M_x(0)\exp\left\{i\omega_0 t + i \int_0^t \omega_1(t')\,dt'\right\}, \qquad (6.9)$$

decomposing $\omega(t)$ into its time-independent, ω_0, and its time-dependent, $\omega_1(t)$, parts.

The assembly of spins in the lattice can be considered to form a statistical ensemble, so that the expectation value of $M_x(t)$ can be evaluated by performing an ensemble average of the random process $\omega(t)$:

$$\langle M_x(t) \rangle = M_x(0)e^{i\omega_0 t} \left\langle \exp\left\{i \int_0^t \omega_1(t')\,dt'\right\} \right\rangle, \qquad (6.10)$$

where $M_x(0)$ is the initial value of M_x at $t = 0$. $\left\langle \exp\left\{i \int_0^t \omega_1(t')\,dt'\right\} \right\rangle$ defines a function $\Phi'(t)$, which is called the relaxation function of the oscillator. It describes the decay of the transverse magnetization in a frame rotating at the resonance angular frequency ω_0. Equation (6.10) can be rewritten as:

$$\langle M_x(t)M_x(0)^* \rangle = |M_x(0)|^2 e^{i\omega_0 t}\,\Phi'(t), \qquad (6.11)$$

where * indicates complex conjugation.

The fluctuation-dissipation theorem for a linear response system shows that the resonance absorption spectrum at the angular frequency ω is given by the Fourier transform of $\Phi'(t)$:

$$I(\omega - \omega_0) = 1/2\pi \int_{-\infty}^{+\infty} e^{-i(\omega - \omega_0)t} \Phi'(t) \, dt, \tag{6.12}$$

where $I(\omega - \omega_0)$ is normalized to unit. The absorption line is centered at ω_0 and is broadened by $\omega_1(t)$. In other words, the resonance at ω_0 is modulated by the time-dependent perturbation $\omega_1(t)$.

In order to characterize the modulation process two parameters are used, namely its amplitude and its correlation time. The former, which measures the magnitude of the modulation, is defined by:

$$\Delta^2 = \langle \omega_1{}^2 \rangle. \tag{6.13}$$

$\langle \omega_1{}^2 \rangle$ is the average of the square of the modulation process. If we refer to Fig. 6.2, $\langle \omega_1{}^2 \rangle$ is the average of the square of the width of each of the individual resonances which are contained under the envelope of the line. In the hypothesis of Gaussian modulation of $\omega_1(t)$ this quantity is equivalent to the second moment of the resonance, which is defined by:

$$M_2 = \Delta^2 = \int_{-\infty}^{\infty} (\omega - \omega_0)^2 I(\omega - \omega_0) d\omega. \tag{6.14}$$

Clearly, the larger M_2, the broader is the overall resonance.

The correlation time of the modulation is defined by:

$$\tau_c = \int_0^\infty \psi(t) \, dt / M_2, \tag{6.15}$$

where $\psi(t)$, the correlation function of the modulation:

$$\psi(\tau) = \langle \omega_1(t) \omega_1(t + \tau) \rangle. \tag{6.16}$$

$\psi(\tau)$ is normalized to M_2 for $\tau = 0$. $\psi(\tau)$ shows how much the modulation at time $t + \tau$ is influenced, i.e., correlated, by the value of the modulation at time t. In fact, if $\psi(\tau)$ is close to 1, it means that $\omega_1(t + \tau) \approx \omega_1(t) \approx \omega_1(0)$, while if $\psi(\tau)$ is close to 0, it means that there is no correlation between the two values. In Fig. 6.4 a typical behavior of the correlation function is plotted. It is seen to decrease rapidly at the beginning, and then more slowly for the long times. The characteristic time τ_c corresponds to the time after which the correlation function is reduced to less than one-half the initial value. τ_c is a measure of the speed of the modulation and it tells us how fast the modulation vanishes. The exchange

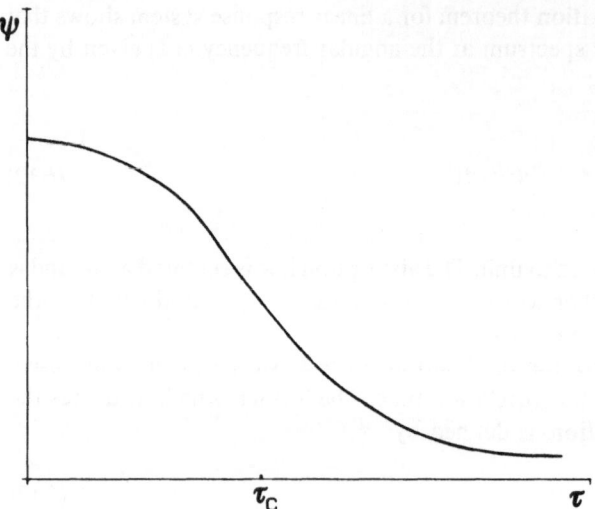

Fig. 6.4. Time dependence of the spin correlation function $\psi(\tau)$

interaction determines a time variation of the spin states of the system which is inversely proportional to the value of the coupling constant in frequency units. Therefore, τ_c is often taken as:

$$\tau_c \approx \hbar/J. \tag{6.17}$$

In order to characterize the modulation process we must compare Δ and τ_c. In fact, for a modulation ω_1 of frequencies to be really effective in broadening a line, it must have a lifetime, τ_c, long enough to allow it to complete at least one cycle of 2π of phase before the frequency changes to another state ω'_1. This condition is fulfilled if the correlation frequency τ_c^{-1}, which is a measure of how fast the modulation shifts from a frequency to another, is much smaller than the modulation amplitude, Δ, i.e., if $\tau_c \Delta \gg 1$. This regime is referred to as slow modulation. If the reverse relation applies, i.e., $\Delta\tau_c \ll 1$, the modulation is in the fast regime. In this case the lifetime τ_c is so short that the modulation has no time to complete a full cycle, and is therefore ineffective in broadening the line.

For a Gaussian modulation process in the lattice, the relaxation function $\Phi'(t)$ is related to the correlation function $\psi(\tau)$ by:

$$\Phi'(t) = \exp\left[- \int_0^t (t-\tau)\psi(\tau)\,d\tau \right]. \tag{6.18}$$

In order to study the behavior of $\Phi'(t)$, it is useful to consider two regions, one in which $t \ll \tau_c$, and the other in which $t \gg \tau_c$. When $t \ll \tau_c$, then $\psi(t) \approx \psi(0) = M_2$, the integration of (6.18) yields

$$\Phi'(t) = \exp[-M_2 t^2/2], \tag{6.19}$$

142

which means that the relaxation function has a Gaussian time dependence.

On the other hand, when $t \gg \tau_c$, $\psi(\tau) \approx 0$, so that the upper limit of the integral in the exponent of $\Phi'(t)$ in (6.18) can be replaced by ∞. It is then easy to see from (6.18) that the relaxation function has an exponential dependence on t:

$$\Phi'(t) = \exp(-M_2 \tau_c t). \tag{6.20}$$

The two approximations of the relaxation function are plotted in Fig. 6.5. The two curves, Gaussian and exponential, coincide for $t = 2\tau_c$. For a given modulation amplitude, Δ, in the limit of slow modulation ($\Delta\tau_c \gg 1$), the Gaussian approximation of $\Phi(t)$ is valid for most of the t range, except for very large t, where the exponential behavior sets in. The Fourier transform of a Gaussian is also Gaussian, therefore:

$$I(\omega - \omega_o) = 1/(\sqrt{2\pi}\Delta) \exp\{-(\omega - \omega_o)^2/(2\Delta^2)\}. \tag{6.21}$$

Therefore, the line shape in this approximation is Gaussian with a half-width at half-height $\sqrt{2\ln 2}\,\Delta = 1.177\Delta$ (Fig. 6.6).

In the fast modulation limit ($\Delta\tau_c \ll 1$) the condition $t \gg \tau_c$ holds for most of the t range, except that close to the origin, and $\Phi'(t)$ will be approximated by the

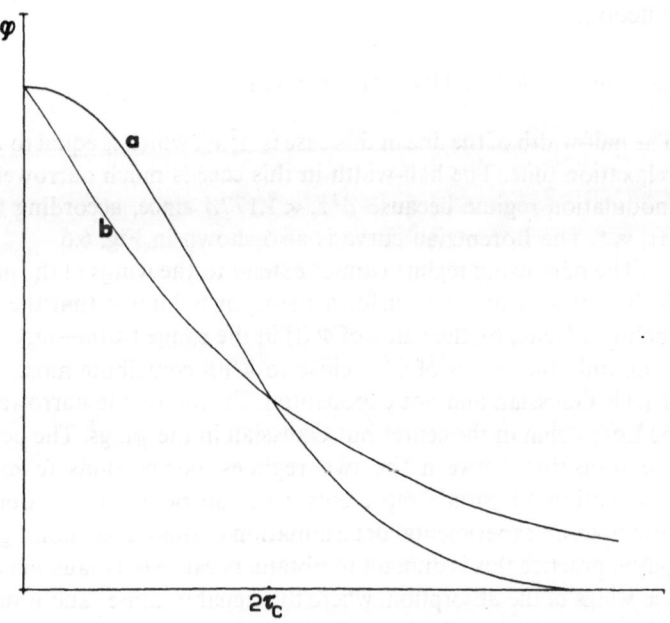

Fig. 6.5. Time dependence of the spin relaxation function $\Phi'(t)$: *a* Gaussian approximation; *b* exponential approximation

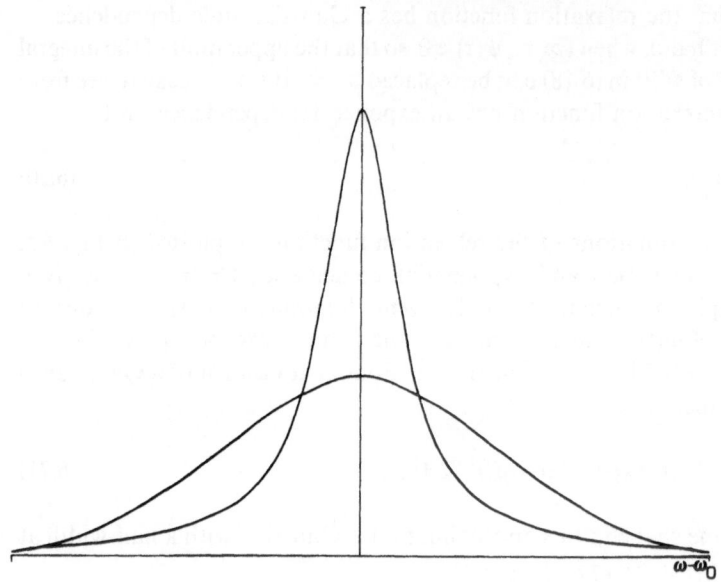

Fig. 6.6. Gaussian and Lorentzian line shapes

exponential function. The Fourier transform of this function is a Lorentzian function:

$$I(\omega-\omega_o)=1/\pi \Delta^2\tau_c/[(\omega-\omega_o)^2+\Delta^4\tau_c^2].$$ (6.22)

The half-width of the line in this case is $\Delta^2\tau_c$, which is equal to τ_r^{-1}, where τ_r is the relaxation time. The half-width in this case is much narrower than in the slow modulation regime because $\Delta^2\tau_c \ll 1.177\Delta$ since, according to the hypothesis, $\Delta\tau_c \ll 1$. The Lorentzian curve is also shown in Fig. 6.6.

The narrowing regime cannot extend to the wings of the line shape function. In fact, from Fourier transform theory it is known that the $I(\omega-\omega_o)$ value is mainly affected by the value of $\Phi'(t)$ in the range $t \approx (\omega-\omega_o)^{-1}$: for very large $\omega-\omega_o$ only the values of $\Phi'(t)$ close to $\Phi'(0)$ contribute most, and in this region $\Phi'(t)$ is Gaussian and not exponential. Therefore, the narrowed line shape must be Lorentzian in the center but Gaussian in the wings. The point, where there is the transition between the two regimes, corresponds to $\omega-\omega_o=\Delta$, i.e., the modulation amplitude represents a cut-off point for the Lorentzian curve. In principle, an experimental determination of this cut-off point gives the value of Δ but in practice this is difficult to obtain, because the Gaussian behavior sets in at the wings of the absorption, where the signal-to-noise ratio is unfavorable. On the other hand, the broadening regime cannot apply at $\omega \approx \omega_o$, because in this case it is the exponential part of $\Phi'(t)$ which determines the line shape, and the curve must be Lorentzian. However, in slow modulation this condition is limited to a

144

very narrow range around the center, so that the Lorentzian section is totally undetectable.

The above treatment is very general in nature, and can be applied to any magnetic resonance experiment, ESR, NMR..., with either motional or exchange modulation.

In the EPR spectra of a magnetic solid the slow dipolar interaction broadens the lines yielding a Gaussian shape, while the exchange interactions can be fast enough to bring the system in the exchange-narrowing regime. The speed of the modulation can be estimated from the frequency of the interaction: in Chap. 2 we have seen that the dipolar interactions for paramagnetic centers, separated by at least 300 pm, are typically of the order of $10^{-1}-10^{-2} \, cm^{-1}$, which on a frequency scale corresponds to 10^8-10^9 Hz, while exchange interactions can be at least one order of magnitude larger than the upper limit of the former. Therefore, the narrowing condition is fulfilled in many cases, and, since it is exchange-determined, it is called exchange narrowing. In the following we will focus on this scheme, and will try to express the $\psi(\tau)$ and $\Phi'(t)$ functions using the hamiltonians appropriate to the system, relating the calculated absorption curves to parameters which depend on the nature of the interaction between the paramagnetic centers.

The exchange-narrowing condition is fulfilled if $H_{ex} \gg H_{dip}$. In this case we can split the total hamiltonian, H, into an unperturbed and a perturbed component:

$$H = H_o + H', \tag{6.23}$$

where H_o comprises the Zeeman and exchange hamiltonians, and H' includes the dipolar term. Since the two hamiltonians describe largely different frequencies, H_o being much faster than H', it is convenient to redefine $\Phi'(t)$ in such a way to remove the fast Zeeman and exchange frequencies and focus on the slow perturbing frequencies. We will not develop the treatment in detail, but rather will touch upon the relevant points, trying to underline what is relevant to the physics of the problem, without entering the mathematics, which is rather involved.

The relaxation function of the system defined by (6.10) can be rewritten using time-dependent perturbation theory as:

$$\Phi(t) = \langle \tilde{M}_+(t)M_-(0)\rangle / \langle M_+M_-\rangle, \tag{6.24}$$

where $\tilde{M}_+(t)$ is in the interaction representation with respect to the unperturbed hamiltonian:

$$\tilde{M}_+(t) = \exp(-iH_o t/\hbar)M_+(t)\exp(iH_o t/\hbar), \tag{6.25}$$

and $M_+(t)$ is in the Heisenberg representation with respect to the total hamiltonian, i.e., $M_+(t) = \exp(iHt/\hbar)M_+\exp(-iHt/\hbar)$. Heisenberg and representation interaction are the two different schemes for expressing the time dependence of operators and wave-functions. It is easy to verify that according to (6.25) the time evolution of $\tilde{M}_+(t)$ is only determined by the slow, dipolar perturbation. $\Phi(t)$ in (6.24) is related to $\Phi'(t)$ by the relation $\Phi(t) = e^{i\omega_o t}\Phi'(t)$.

In order to evaluate $\Phi(t)$ two thermal averages must be performed according to (6.24). The thermal average of the transverse magnetization $\langle M_+ M_- \rangle$ relates it to the intensity of the EPR signal or the static susceptibility via:

$$\langle M_+ M_- \rangle = kT\chi(T) = 2/3\, S(S+1)\mu_B^2 g^2 \chi(T)/\chi_c, \tag{6.26}$$

where χ_c is the static magnetic susceptibility for a system with the same individual spin and the same g tensor of the system under investigation, but following the Curie law:

$$\chi_c = \mu_B^2 g^2 S(S+1)/3kT. \tag{6.27}$$

$\chi(T)/\chi_c$ is a measure of the spin correlations at any temperature: if the ratio is equal to 1 the susceptibility of the system is that of a Curie paramagnet, and the spins are totally uncorrelated, while deviations from this value express larger spin-spin interactions.

The thermal average of the numerator of (6.26) is more difficult to evaluate directly, therefore, in order to obtain $\Phi(t)$ it is necessary to use (6.18), and the correlation function $\psi(\tau)$ defined by (6.16) must be explicitly rewritten expressing the time dependence of the transverse magnetization through commutators required by time-dependent perturbation theory as:

$$\psi(\tau) = (1/\hbar^2)\langle [H'(\tau), M_+(0)][M_-(0), H'(0)]\rangle / \langle M_+ M_- \rangle \tag{6.28}$$

where $H'(\tau)$ is in the interaction representation:

$$H'(\tau) = \exp[-iH_o\tau/\hbar]H'\exp(iH_o\tau/\hbar)]. \tag{6.29}$$

In the evaluation of $\psi(\tau)$ the perturbing dipolar hamiltonian $H'(\tau)$ can be reexpressed in an equivalent way, by considering the fact that H_{ex} commutes with the Zeeman hamiltonian. The time dependence relative to the latter, for a pair (ij) in the lattice, is given by:

$$\exp(ig\mu_B BS_z\tau/\hbar)S_i^{(m_i)}S_j^{(m_j)}\exp(-ig\mu_B BS_z\tau/\hbar) =$$
$$= S_i^{(m_i)}S_j^{(m_j)}\exp[i(m_i+m_j)\omega_o\tau], \tag{6.30}$$

with $m_i = -1, 1$ or 0 for S_i^-, S_i^+, and S_i^z, respectively. If we define $M = m_i + m_j$, we see that M can be $0, \pm 1, \pm 2$ and the $H'(\tau)$ hamiltonian can be rewritten as:

$$H'(\tau) = \Sigma_M \exp[iM\omega_o\tau]G_M(\tau), \tag{6.31}$$

where $G_M(\tau)$ shows time dependence on the exchange hamiltonian:

$$G_M(\tau) = \exp[-iH_{ex}\tau/\hbar]\,G_M(0)\exp[iH_{ex}\tau/\hbar], \tag{6.32}$$

$G_M(0)$ is that part of the dipolar hamiltonian which induces a change M in the total spin. Using $(6.29-32)$ $\psi(\tau)$ becomes:

$$\psi(\tau) = (1/\hbar^2)\Sigma\langle g_M(\tau)g_M(0)\rangle\cos(M\omega_o t), \tag{6.33}$$

where $g_M(\tau)$ is a time correlation function defined as:

$$g_M(\tau) = [G_M(\tau),\ M_+(0)], \tag{6.34}$$

which is related to that part of the perturbing hamiltonian which induces a change M in the total Zeeman quantum number. The sum extends from $M=0$ to $M=2$, due to the fact that only bilinear spin-spin terms are taken into account. Historically the terms with $M=0$ are called secular, while the others are collectively called nonsecular. The total time dependence associated with the Zeeman hamiltonian is contained in $\cos(M\omega_o\tau)$, while the time variation associated with the exchange interaction is contained in $\langle g_M(\tau)g_M(0)\rangle$.

In the case of dipolar interactions the functions $\langle g_M(\tau)g_M(0)\rangle$ are of the type:

$$\langle g_M(\tau)g_M(0)\rangle = \Sigma_{ijkl}F_{ij}^{(M)}F_{kl}^{(-M)}S_{ijkl}^{(M)}(\tau), \tag{6.35}$$

where $F_{ij}^{(M)}$ is a dipolar factor as shown in Table 6.1 and $S_{ijkl}^{(M)}(\tau)$ is a time correlation function involving four spin operators which are obtained by combining the spin operators of Table 6.1 for the (ij) and (kl) pairs.

The physical meaning of a spin correlation function $\langle S_{i\alpha}(\tau)S_{j\beta}(\tau)S_{k\gamma}S_{l\delta}\rangle$ is that its value is a statement of a probability that at time τ a spin deviation α is on site i, while at the same time on sites j, k, and l the spin values are β, γ, and δ, respectively. In the Gaussian modulation limit these can be decoupled into

Table 6.1. Dipolar factors $F_{ij}^{(M)}$ and relative spin operators[a]

M	$F_{ij}^{(M)}$	$S_i^{mi}S_j^{mj}$
0	$1/2(3\cos^2\Theta_{ij}-1)$	$-S_{iz}S_{jz}+1/4(S_{i+}S_{j-}+S_{i-}S_{j+})$
1	$-3/4\sin\Theta_{ij}\cos\Theta_{ij}\exp(-i\Phi_{ij})$	$S_{iz}S_{j+}+S_{i+}S_{jz}$
-1	$-3/4\sin\Theta_{ij}\cos\Theta_{ij}\exp(i\Phi_{ij})$	$S_{iz}S_{j-}+S_{i-}S_{jz}$
2	$-3/8\sin^2\Theta_{ij}\exp(-2i\Phi_{ij})$	$S_{i+}S_{j+}$
-2	$-3/8\sin^2\Theta_{ij}\exp(2i\Phi_{ij})$	$S_{i-}S_{j-}$

[a] The dipolar factors in units $g^2\mu_B^2r_{ij}^3$.

products of two spin correlation functions, allowing for relatively simple calculations. The decoupling leads to:

$$\langle S_{i\alpha}(\tau)S_{j\beta}(\tau)S_{k\gamma}S_{l\delta}\rangle = \langle S_{i\alpha}(\tau)S_{k\gamma}\rangle\langle S_{j\beta}(\tau)S_{l\delta}\rangle + \langle S_{i\alpha}(\tau)S_{j\beta}(\tau)\rangle\langle S_{k\gamma}S_{l\delta}\rangle$$
$$+ \langle S_{i\alpha}(\tau)S_{l\delta}\rangle\langle S_{j\beta}(\tau)S_{k\gamma}\rangle, \tag{6.36}$$

and the four-spin correlation functions are evaluated by summing over all the possible pairs in the lattice. In three dimensions the disturbances move relatively freely in the lattice, so that the nearest neighbor contributions are dominant and on increasing the distance between the spins, the correlation goes rapidly to zero. Matters are different for one-dimensional lattices where any disturbance should eventually affect any spin in the chain, so that the decay of the disturbance is much slower. Two-dimensional lattices are intermediate between these two limits. This has paramount effects on the spin dynamics in lower dimensional systems, as will be discussed in Sect. 6.4.

In three dimensions $\psi(\tau)$ decays in a sufficiently rapid way so that τ_c is small and the exchange-narrowing condition is fulfilled. Therefore, it is not necessary to evaluate $\psi(\tau)$ in detail, and only the region close to zero where its value is given by the second moment is relevant. Equation (6.22) is valid, and the line width can be easily evaluated calculating the dipolar second moment. In order to do this (6.28) reduces to:

$$\psi(0) = M_2 = \langle [H', M_+][M_-, H']\rangle/\langle M_+M_-\rangle. \tag{6.37}$$

The average of an operator is given by:

$$\langle O \rangle = \text{Tr}[e^{-H/kT}O]/\text{Tr}[e^{-H/kT}], \tag{6.38}$$

where H is the hamiltonian appropriate to the system and Tr denotes the trace. In the high temperature limit $e^{-H/kT} \approx 1$ and the average is simply calculated as $\langle O \rangle = \text{Tr}O/\text{Tr}1$ where 1 is the unit operator.

If $H_z \gg H_{ex}$, then it is possible to evaluate the second moment including only the secular terms because the other ones couple states differing in M, which, in the above hypothesis are widely spaced. In other words, the $M \neq 0$ terms in (6.33) determine a very fast decay of the $\psi(\tau)$ function in this limit. Within this approximation, for a single spin characterized by the quantum number S, we find

$$\langle M_+M_-\rangle = \langle S_+S_-\rangle = \Sigma_M(S+M)(S-M+1),$$

where we have used the definition of shift operators. This can be rewritten as:

$$\langle M_+M_-\rangle = \Sigma_M(S^2+S-M^2+M).$$

The indicated sum yields:

$$\Sigma_M(S^2+S) = S(S+1)(2S+1); \quad \Sigma_M M = 0;$$
$$\Sigma_M M^2 = 1/3\, S(S+1)(2S+1).$$

Therefore, for a single spin we find

$$\langle M_+ M_- \rangle = 2/3\, S(S+1)(2S+1).$$

Finally passing to N spins:

$$\langle M_+ M_- \rangle = 2/3\, N\, S(S+1)(2S+1)^N.$$

In the same way for the numerator of (6.37) we find:

$$\langle [H', M_+][M_-, H'] \rangle = N/2\, S^2(S+1)^2(2S+1)^N \hbar^{-2} \mu_B^4 g^4 \Sigma_j (3\cos^2\Theta_{jk}-1)^2/r_{jk}^6.$$

Therefore, (6.37) becomes:

$$M_2(B) = 3/4\, S(S+1)\, \mu_B^4\, g^4 \Sigma_j (3\cos^2\Theta_{jk}-1)^2/r_{jk}^6, \qquad (6.39)$$

where the sum is performed by choosing one paramagnetic center in the lattice and including all the other centers in the sum. Due to the r_{jk}^{-6} dependence the number of sites to be included in the sum is relatively small. In (6.39) we used the relation between the second moments in a field, $M_2(B)$, and in a frequency, $M_2(\omega)$, swept experiment:

$$M_2(\omega) = (1/\hbar^2) g^2 \mu_B^2 M_2(B). \qquad (6.40)$$

If M_2 must be expressed in Gauss and r_{jk} in Å, then (6.39) can be written as:

$$M_2 = 6.45 \times 10^7\, S(S+1)\, g^2 \Sigma_j\, (3\cos^2\Theta_{jk}-1)^2/r_{jk}^6. \qquad (6.41)$$

If, on the other hand, the condition $H_z \gg H_{ex}$ is relaxed, then the nonsecular terms must also be included in the calculation of the second moment. The result is that M_2 becomes:

$$M_2 = 3/4\, S(S+1)\, \mu_B^2 g^2 \Sigma_j r_{jk}^{-6} \{(3\cos^2\Theta_{jk}-1)^2 +$$
$$+ \sin^4\Theta_{jk} \exp[-1/2(2\omega_o/\omega_e)^2] +$$
$$+ 10 \cos^2\Theta_{jk} \sin^2\Theta_{jk} \exp[-1/2(\omega_o/\omega_e)^2]\}, \qquad (6.42)$$

where ω_o and ω_e are the Zeeman and exchange frequency, respectively.

If we take the powder average of (6.39) and (6.42) we find that the second moment calculated including the nonsecular terms is 10/3 larger than the second

Table 6.2. Ratios of experimental line widths at various frequencies and calculated J values in two manganese salts

	9 GHz	14 GHz	25 GHz	35 GHz	$J\,(cm^{-1})$
$Mn(CH_3COO)_2 \cdot 4H_2O$	1	0.91	0.78	0.69	0.12
$Mn(HCOO)_2 \cdot 2H_2O$	1	0.89	0.82	0.68	0.14

moment calculated not including them. This means that nonsecular terms contribute 7/3 of the broadening, the rest coming from secular terms. This larger broadening effect in the limit of exchange which is much larger than the Zeeman frequency has been often referred to as the "10/3 effect".

In the intermediate case, the line width will be frequency-dependent. In fact, at high frequencies $H_z \gg H_{ex}$ may hold, yielding lines 10/3 narrower than at low frequencies, where $H_{ex} \gg H_z$.

This behavior has been observed, for instance, in a number of manganese compounds [6.5], such as $Mn(CH_3COO)_2 \cdot 4H_2O$ and $Mn(HCOO)_2 \cdot 2H_2O$. Line widths were measured in polycrystalline powders at 9, 14, 25, and 36 GHz. The ratios of the experimental line widths at the various experimental frequencies are given in Table 6.2. It is apparent that the line width decreases with increasing frequency, indicating that as ω_0 increases the ratio ω_0/ω_e decreases, and the nonsecular terms become less important. The line width at a given frequency ω_0 is:

$$\Delta B_{pp} \propto 1 + 5/3 \, \exp[-\tfrac{1}{2}(\omega_0/\omega_e)^2] + 2/3 \, \exp[-2(\omega_0/\omega_e)]. \tag{6.43}$$

Therefore, measuring it at several frequencies affords the determination of J. The values calculated in this way are given in Table 6.2.

6.3 Additional Broadening Mechanisms

Now it is time to introduce more complications in the above treatment, which for most readers may already be too involved. In fact, in order to have a tool available for the analysis of the spectra of real systems it is necessary to take into account additional broadening mechanisms, such as hyperfine coupling, anisotropic and antisymmetric exchange contributions, g anisotropy, and crystal field effects.

Starting from g anisotropy, we may consider the frequent case of a lattice with two magnetically unequivalent sites, which can, for instance, occur in a monoclinic crystal. The two sites are identical, except for the spatial orientation of the spins. Let us suppose that the two paramagnetic centers are characterized by anisotropic g tensors. If the z axis of one of them makes an angle α with the C_2 crystal axis, the z axis of the other one must make an angle $-\alpha$ (Fig. 6.7).

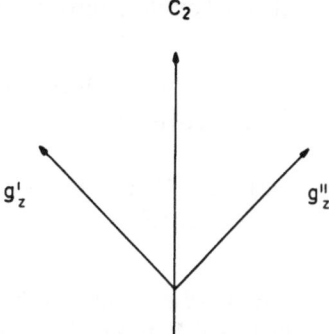

Fig. 6.7. Relative orientations of the z axes of the g tensors of two sites in a monoclinic lattice

Indicating the anisotropic **g** tensors for the two different sites with a prime and a double prime, respectively, in the absence of exchange interactions we should observe two lines at $\hbar\omega' = g'\mu_B B$ and $\hbar\omega'' = g''\mu_B B$, respectively. If the exchange interaction is strong enough, i.e., if $J > |g' - g''|\mu_B B/2$, we will observe one narrowed line at $g = 1/2(g' + g'')$. By referring to a coordinate system [6.6] in which the z axis is in the direction defined by $\frac{1}{2}(\mathbf{g}' + \mathbf{g}'') \cdot \mathbf{B}$ and the x axis is in the plane of $\mathbf{g}' \cdot \mathbf{B}$ and $\mathbf{g}'' \cdot \mathbf{B}$, the half-width of the resulting Lorentzian line can be estimated to be:

$$\Delta B_{pp} = (\hbar\mu_B/4Jg)\sqrt{3\pi}/[2S(S+1)z] \{[(\mathbf{g}' \cdot \mathbf{B})_z - (\mathbf{g}'' \cdot \mathbf{B})_z]^2 + \\ + 2(\mathbf{g}' \cdot \mathbf{B})_x^2 \exp[-3\hbar^2\omega_0^2/(8S(S+1)\,zJ^2)]\}, \tag{6.44}$$

where z is the number of nearest neighbors. Equation (6.44) has been evaluated assuming that the anisotropy effects for the magnetic moment associated with the oscillating field applied parallel to x can be neglected. The first term in square brackets is the secular one, and gives a contribution which increases with the Zeeman frequency ω_0 because it depends on the external field. The broadening process is not an indefinite one, however, because when the difference in the square brackets becomes too large two lines will be observed. This case will be briefly treated below. The second term is the nonsecular one, which gives broader lines at low frequencies. Often this term is small, and it can be usually neglected. As a result the line width is generally estimated to be simply given by:

$$\Delta B_{pp} \approx (\mu_B/8gJ)[(g' - g'')B]^2. \tag{6.45}$$

As expected (6.45) does not give any broadening effect when the two **g** tensors are magnetically equivalent.

In the opposite limit of the strong field two lines are observed which show broadening and shift effects when the exchange frequency becomes comparable to the difference in the Zeeman frequencies of the two lines. The calculation using the Kubo-Tomita approach provides the shifts of the two sets of lines:

$$\delta g' = zJ^2 S(S+1)/(12\mu_B^2 B_o(g'-g''))\{2(1''/1'+m''/m')(1+\cos\Theta)\cos\Theta - -(1+\cos\Theta)^2\};$$
(6.46)

$$\delta g'' = zJ^2 S(S+1)/(12\mu_B^2 B_o(g'-g''))\{2(1'/1''+m'/m'')(1+\cos\Theta)\cos\Theta - -(1+\cos\Theta)^2\}.$$
(6.47)

The coordinate axes in this case are chosen in such a way that z' and z'' are parallel to $\mathbf{g'\cdot B}$ and $\mathbf{g''\cdot B}$, respectively, and x' and x'' are parallel to each other and perpendicular to the $z'-z''$ plane. With this choice Θ is the angle between z' and z'' and the l's, m's, and n's are defined by:

$$M = l'S'_+ + m'S'_+ + m'S'_- + n'S'_z + 1''S''_+ + m''S''_- + n''S''_z,$$
(6.48)

where M is the operating dipole moment.

In the case of two spins $S = \frac{1}{2}$, the derivative absorption shape can be expressed as:

$$Y(B) = N\{[W_2 + (B-B_o)J](W_1^2 + W_2^2) - 4[(B-B_o)W_2 - (\Gamma_o - J)W_1][(B-B_o)W_1 + (\Gamma - J/2)W_2]\}/(W_1^2 + W_2^2),$$
(6.49)

$$W_1 = (B-B_a)(B-B_b) - (\Gamma_a - J/2)(\Gamma_b - J/2) + J^2/4;$$
(6.50)

$$W_2 = (B-B_a)(\Gamma_b - J/2) + (B-B_b)(\Gamma_a - J/2),$$
(6.51)

Γ_i are the half-widths of the lines, related to the experimentally determined peak-to-peak line-width by $\Gamma_i = \sqrt{3}\Delta B_{pp}^i/2$. N is a normalization factor, B_o and Γ_o are the averages of the resonance fields, B_i, and of the half-widths Γ_i [6.7].

In the above treatment the dipolar broadening effect has been ignored. In the case of weak exchange, when two peaks are resolved, the broadening of each peak can be assumed to come from the anisotropy effects, and the dipolar broadening can be neglected. In the strong exchange limit, in the approximation which allowed us to obtain (6.44) and (6.45), the dipolar contribution is simply additive to that of the **g** anisotropy.

The estimation of the dipolar contribution in the presence of an anisotropic **g** tensor becomes more complicated, especially in the case of **g** and $\mathbf{D_{dip}}$ tensors with nonparallel axes. Given the large number of approximations present in the treatment we have developed above, it has been customary to approximate the dipolar tensor using the average, $\mathbf{g} = (\mathbf{g_x} + \mathbf{g_y} + \mathbf{g_z})/3$, instead of the anisotropic tensor, in all the cases in which the **g** anisotropy is not too large, as it is found for instance in copper(II) compounds.

Another relevant broadening mechanism is provided by hyperfine splitting, mainly metal hyperfine. In the usual strong field approximation for hyperfine $(H_z \gg HI_s)$, the HI_s hamiltonian:

$$HI_s = \Sigma_k \, I_k \cdot A_k \cdot S, \qquad (6.52)$$

where the sum is over all the nuclei with spin different from zero, can be added in the perturbation hamiltonian (6.23). The hyperfine contribution to the second moment has been evaluated for an axial **A** tensor with principal axes parallel to those of the **g** tensor:

$$M_2^{hyp} = I(I+1)/3 \; K^2(\Theta)/g^2(\Theta) + I(I+1)/6[A_{||}^2 A_{\perp}^2/g^2(\Theta) + A_{||}^2 + \\ + E^2(\Theta)], \qquad (6.53)$$

where

$$K^2(\Theta) = A_{||}^2 + g_{||}^2 \cos^2\Theta + A_{\perp}^2 g_{\perp}^2 \sin^2\Theta; \qquad (6.54)$$

$$g^2(\Theta) = g_{||}^2 \cos^2\Theta + g_{||}^2 \sin^2\Theta; \qquad (6.55)$$

$$E^2(\Theta) = (A_{\perp}^2 - A_{||}^2)g_{||}g_{\perp} \sin\Theta \cos\Theta/[K^2(\Theta)g^2(\Theta)]. \qquad (6.56)$$

Another broadening source can be the anisotropic exchange contribution which is similar in nature to the dipolar contribution (see Sect. 2.3). As for isotropic exchange it is customary to consider only the interactions between nearest neighbors. D_{dip} and D_{ex} may have different principal axes, therefore, two different sets of angles will be required to define their contribution, (Fig. 6.8). In several cases it is assumed that the principal axes of D_{ex} are parallel to the principal axes of **g**. The contribution to the second moment from the dipolar and exchange tensors of the sites n and $n+1$ which are the nearest neighbors along

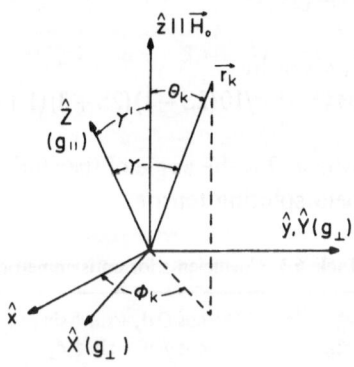

Fig. 6.8. Coordinate systems for dipolar (x, y, z) and exchange (X, Y, Z) anisotropic tensors

the chain to the reference spin in the limit of quasi-isotropic g and axial $\mathbf{D}_{ex}$ and $J \gg g\mu_B B_0$ has been calculated as:

$$
\begin{aligned}
M_2^{n,n+1} = 3/2\ S(S+1) \{ &D_{ex}^2/3(1+\cos^2\gamma') + D_d^2(1+\cos^2\Theta) - \\
&- 1/3\ D_{ex}D_d(3\cos^2\gamma'-1)(3\cos^2\Theta-1) - \\
&- 1/3\ D_{ex}D_d \sin^2\gamma' \sin^2\Theta \cos^2\Phi - \\
&- 10/3\ D_{ex}D_d \sin\gamma' \cos\gamma' \sin\Theta \cos\Theta \cos\Phi \},
\end{aligned}
\tag{6.57}
$$

where $D_d = g^2\mu_B^2 r_1^{-3}$. The contributions of the spins, which are not nearest neighbors along the chain, or which belong to different chains, will not have the exchange contribution, and can be calculated with the same formula as (6.42).

Finally, we may consider antisymmetric exchange. Since in the evaluation of $\psi(\tau)$, according to (6.28), no cross-terms involving the anisotropic and antisymmetric exchange contributions occur, the antisymmetric exchange contribution is simply additive. For the computation of the latter a useful formula is:

$$
M_2^A = (M_{2,0}^A + M_{2,1}^A) F_A \chi_c / \chi(T),
\tag{6.58}
$$

where $M_{2,0}^A$ and $M_{2,1}^A$ are the secular and nonsecular parts, respectively, and F_A is the static spin correlation which depends on the number of nearest neighbors. The secular and nonsecular contributions can be calculated as:

$$
M_{2,0}^A = 2/3\ S(S+1) \Sigma_j (\Omega_{ixjy}^A)^2;
\tag{6.59}
$$

$$
M_{2,1}^A = 1/3\ S(S+1) \Sigma_j [(\Omega_{ixjz}^A)^2 + (\Omega_{iyjz}^A)^2].
\tag{6.60}
$$

The Ω's are given in Table 6.3. The polar angles Θ and Φ are defined in such a way that the polar axis is parallel to the static magnetic field.

For individual spins $S_i > \frac{1}{2}$ another possible broadening mechanism is determined by single ion zero field splitting effects. The second moment for an axial zero field splitting

$$
H = D\Sigma_i S_{iz}^2
\tag{6.61}
$$

is given by
$$
M_2^{zfs} = D^2/10(2S-1)(2S+3)(1+\cos^2\Theta),
\tag{6.62}
$$

where Θ is the angle of the static magnetic field with the unique axis of the zero field splitting tensor.

Table 6.3. Coefficients for antisymmetric exchange

Ω_{ixjy}^A	$\cos\Theta\, d_z + \sin\Phi \sin\Theta\, d_y + \cos\Phi \sin\Theta\, d_x$
Ω_{ixjz}^A	$\cos\Phi\, d_y - \sin\Phi\, d_x$
Ω_{iyjz}^A	$-\sin\Theta\, d_z + \sin\Phi \cos\Theta\, d_y + \cos\Phi \cos\Theta\, d_x$

6.4 Exchange Narrowing in Lower Dimensional Systems

The reason why the structural and magnetic dimensionality can affect the exchange-narrowing mechanism is easily understood by recalling that the spin correlation function $\langle S_{iz}(\tau)S_{iz}\rangle$ is a measure of the probability that a disturbance (a spin deviation) is localized on site i for a time τ. This probability is bound to the mechanism according to which the disturbance can move in the lattice, and it is clear that the movement will be much more difficult in one-dimensional than in three-dimensional exchange-coupled systems, since the number of alternative paths for a disturbance to move from one site to another is much more limited in the former. We have already seen that in three dimensions the characteristic time for the decay of the correlation functions, τ_c, can be estimated to be inversely proportional to J, the exchange frequency. In one dimension, however, this time is too short, and we must explore the behavior of $\psi(\tau)$ also beyond this limit. $\psi(\tau)$ for a one-dimensional system is shown in Fig. 6.9. For low τ, $\psi(\tau)$ is approximated by a Gaussian function, like in the three-dimensional case, but beyond τ_1 the decay of $\psi(\tau)$ is best approximated by $\tau^{-1/2}$ [6.8]. In a three-dimensional system the corresponding curve follows a $\tau^{-3/2}$ dependence, and in general for a d-dimensional lattice $\psi(\tau) \propto \tau^{-d/2}$ beyond τ_1. The origin of the d dependence lies in the spin diffusion mechanism which is responsible for the decay. An important consequence of this behavior is that for one- and two-dimensional materials the integral (6.15) shows divergence for $t \to \infty$, a fact which does not hold for three dimensions where the decay as $\tau^{-3/2}$ is fast enough to ensure convergence. The

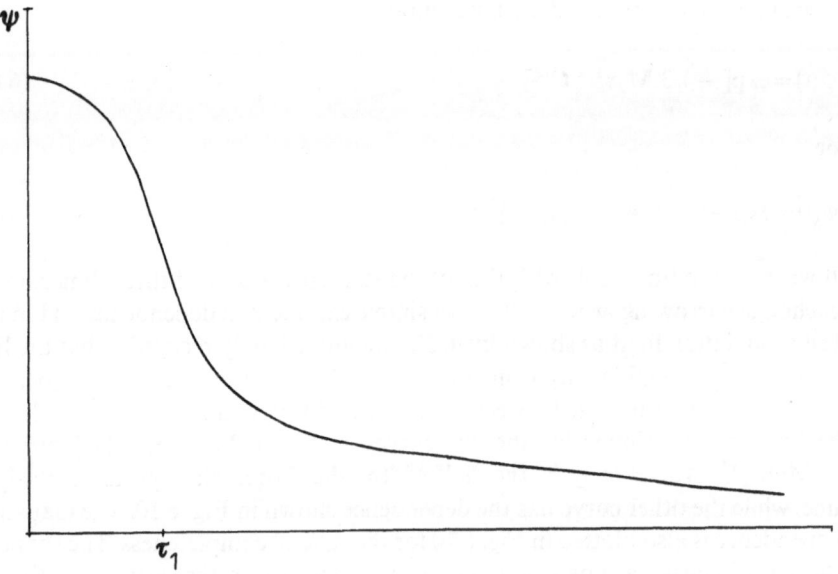

Fig. 6.9. Time dependence of the spin correlation function $\psi(\tau)$ in a one-dimensional system

155

divergence is more pronounced for one dimension, where it has a $t^{1/2}$ dependence, than for two dimensions, where it has a dependence ln (t). This means that the presence of the diffusive tail makes (6.15) and (6.18) invalid for low dimensional systems.

The importance of the diffusive tail in the lower dimensional compared to the three-dimensional case is different for secular and nonsecular terms. This point is well understood by considering the definition (6.33) of $\psi(\tau)$ in which the Zeeman time dependence is explicitly factored out. In fact, the rapid modulation provided by $\cos(M\omega_0\tau)$ for $M \neq 0$ is sufficient to destroy the effects of the long-time divergence, a case which of course cannot apply to the secular part which, having $M = 0$, is time-independent in the rotating frame associated with the Zeeman interaction.

In order to overcome the above difficulties it is therefore necessary to refine the theory. We will consider with more detail the one-dimensional case, which is relatively simple and which has been experimentally observed in a number of well-behaved examples. We will leave the two-dimensional case to Chap. 10.

Since the standard definition of τ_c in this case diverges, an alternative one can be obtained, giving to $\psi(\tau)$ the required $\tau^{-\frac{1}{2}}$ dependence, by:

$$\psi(\tau) = \psi(0)\,(\tau'_c/\tau)^{1/2} \tag{6.63}$$

for $\tau \gg \tau'_c$. Equivalently

$$\psi(\tau) = M_2(\tau'_c/\tau)^{1/2}. \tag{6.64}$$

Recalling the definition of $\Phi(t)$, we obtain

$$\Phi(t) = \exp[-4/3\ M_2\tau_c^{1/2}\ t^{3/2}] \tag{6.65}$$

or

$$\Phi(t) = \exp-[1.2114\ M_2^{2/3}\ \tau_c^{1/3}t]^{3/2}. \tag{6.66}$$

If we compare this result with that of the standard theory of three-dimensional exchange narrowing, where $\Phi(t)$ has a simple exponential dependence on t in the fast modulation limit as shown by (6.20), we immediately recognize that the line shape departs significantly from Lorentzian. In fact, the line shape associated with (6.66) is intermediate between Gaussian and Lorentzian. Experimentally the best way for discriminating the two curves is to plot the inverse I(ω) curve vs $(\omega/\Delta\omega_{1/2})^2$, where $\omega_{1/2}$ is the half-width: the Lorentzian yields a straight line, while the other curve has the dependence shown in Fig. 6.10. The Gaussian dependence is also plotted in Fig. 6.10 for the sake of completeness. The shape of the one-dimensional line is very close to Lorentzian in the center, but then it decreases more rapidly. In order to perform the same type of analysis for a

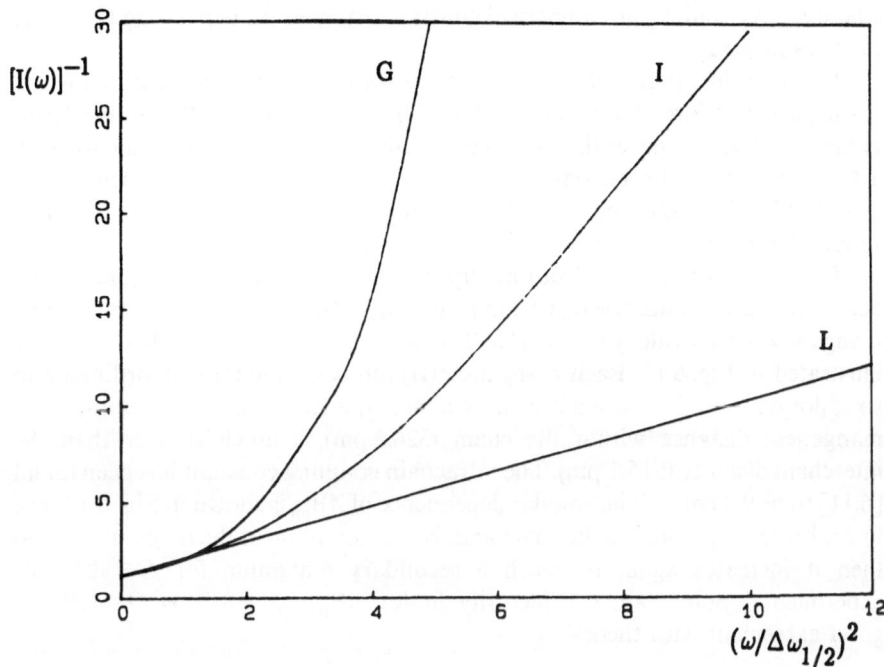

Fig. 6.10. Plot of the inverse $I(\omega)$ versus $(\omega/\Delta\omega_{1/2})^2$ curve for a Lorentzian and a Gaussian, and the line shape appropriate to a one-dimensional system

normal field swept experiment with derivative line shapes, it is convenient to plot $[(B-B_o)/Y(B)]^{\frac{1}{3}}$ vs $(B-B_o)^2$.

Since a Lorentzian line shape is characteristic of fast exchange and a Gaussian of slow exchange, we can state that the slow decay of the correlation in one dimension does not allow the system to reach the fast exchange regime, but leaves it in an intermediate situation. As a consequence lines will be broader in one-dimensional compounds compared to three-dimensional systems. In fact, for the latter $\Delta B_{pp} = M_2/J$, while for the former $\Delta B_{pp} = M_2^{2/3}/J^{1/3}$.

If we use (6.39) for the second moment in (6.65) and (6.66) and if we assume that only dipolar contributions along the chain are relevant, we see that the angular dependence of the secular term in the linear chain case reduces to $(3\cos^2\Theta - 1)^{4/3}$. Θ is the angle between the chain direction and the external magnetic field.

Since the fast Zeeman modulation washes out the anomalous contributions from the long-time diffusive tail for the nonsecular terms, the corresponding component of $\Phi(t)$ has the usual simple exponential dependence on t. Therefore, the explicit form of $\Phi(t)$, taking into account the angular dependence and the frequency dependence, will be:

$$\Phi(t) = \exp\{-[A_1(3\cos^2\Theta - 1)^{4/3}t]^{3/2} - A_2\sin^4\Theta t - A_3\cos^2\Theta\sin^2\Theta t\}, \qquad (6.67)$$

157

where A_1, A_2, and A_3 are numerical constants. A_2 and A_3 depend on the ω_o/ω_e ratio, as in (6.42).

The long-time part of $\Phi(t)$ mainly affects the center of the absorption, where $\omega - \omega_o \approx 0$, therefore, this part of the line will be dominated by the secular terms at most angles. Whenever this occurs the absorption line is intermediate between a Gaussian and a Lorentzian curve. The main exception to this behavior is observed for the magic angle, $\Theta = 54.74°$, when the secular term goes to zero, and Lorentzian behavior is anticipated.

The best practical realization, up to the moment, of one-dimensional behavior, has been observed [6.9] in $[N(CH_3)_4]MnCl_3$, tetramethylammonium manganese trichloride (or TMMC). The crystal structure of TMMC [6.10] is illustrated in Fig. 6.11. Each manganese(II) ion is octahedrally coordinated to six chloride ions. The octahedra share a face to form a chain. The manganese-manganese distance within the chain (324.5 pm) is much shorter than the interchain distance (915.1 pm). The intrachain coupling constant has been found [6.11] to be 9.3 cm^{-1}. The angular dependence of ΔB_{pp} is shown in Fig. 6.12. The line is broadest parallel to the chain axis, is at a minimum at the magic angle, and then it increases again to reach a secondary maximum for $\Theta = 90°$. The experimental points are satisfactorily fit for $\Delta B_{pp} = A + B(3\cos^2\Theta - 1)^{4/3}$, in good agreement with theory.

6.4.1 The Effect of Weak Interchain Coupling

In all the above treatments we have neglected the exchange interactions between chains. In fact, however, neglecting this may induce serious errors and it is now time to consider its effect. We consider it in the linear chain, but the results can be extended to the two-dimensional case as well.

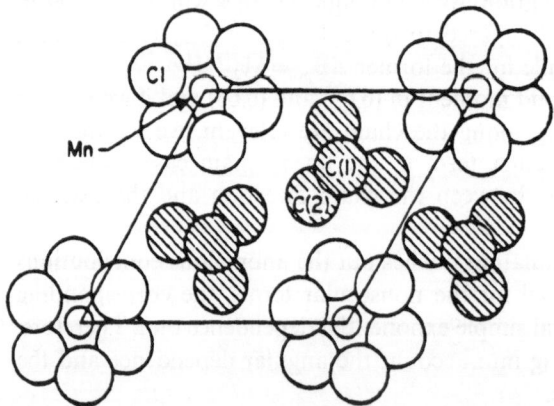

Fig. 6.11. Structure of TMMC as viewed along the magnetic chains. After [6.9]

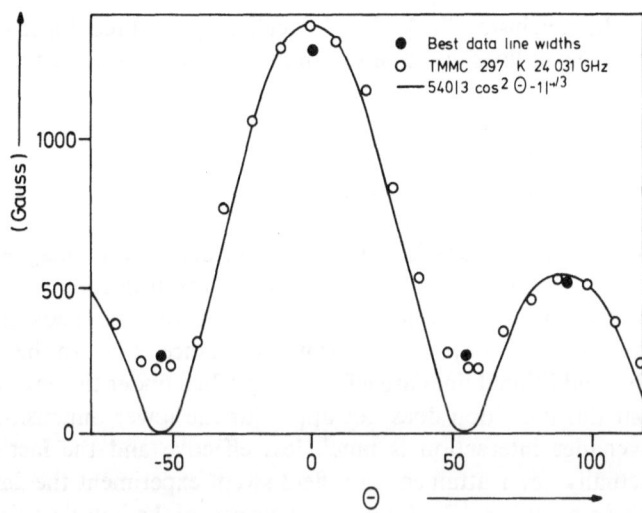

Fig. 6.12. Angular dependence of the line width of TMMC at K-band and 297 K. Θ is the angle between the linear chain and the applied field. After [6.9]

In a qualitative way it is easy to understand that interchain coupling, J', affects the line shape because it affects the spin autocorrelation function. Indeed, in the absence of J' we have seen that the autocorrelation function has a Gaussian behavior for short times, followed by a diffusive regime after a time $\tau_1 \propto J^{-1}$. The inclusion of $J' < J$ determines now an additional time $\tau_2 > \tau_1$ after which a rapid decrease of $\psi(\tau)$ will be observed. Therefore, including J' offers in principle the possibility to the $\psi(\tau)$ function to go to zero, avoiding the problems associated with the long-time diffusive behavior outlined above. In other terms interchain coupling moves the spin system toward the three-dimensional behavior. How far this can go depends on the J'/J ratio: for $J' \ll J$, the one-dimensional limit will be closely approached, but for $J' \approx J$ we go to the three-dimensional case.

Putting this on a quantitative basis is by no means trivial, because in the diffusive region the Kubo-Tomita approach cannot be used any longer. However, some fruitful attempts have been made to go beyond this obstacle. The main result which was obtained [6.12, 13] has been that to estimate τ_2, the characteristic time after which the spin diffusive regime is destroyed by interchain coupling:

$$\tau_2 = \hbar[J'(J'/J)^{1/3}]^{-1}. \tag{6.68}$$

The factor $(J'/J)^{1/3}$, and not (J'/J), which would be expected on the basis of an analysis of moments, occurs because of the slow decay of the spin correlations by intrachain diffusion. Since $J'/J \ll 1$, the 1/3 exponent has a large effect, making the effective exchange frequency, τ_2^{-1}, much larger than in the (J'/J) limit. This means that the long-time divergence is effectively cut off, producing a Lorentzian line, like in the three-dimensional case, also for rather small J'/J ratios.

159

In conclusion, including J' yields a more three-dimensional behavior, with Lorentzian lines and a less pronounced dominance of the secular terms.

6.4.2 Half-Field Transitions

Another characteristic feature of lower dimensional magnets is that a half-field transition can be observed. The origin of this transition is quite simple. Indeed, in Sect. 6.1. we showed that the dipolar perturbation yields satellite lines at $\omega = 0$, $2\omega_0$, and $3\omega_0$, beyond the normal resonance at ω_0. In the fast exchange regime these additional lines are effectively pushed under the envelope of the main line, but this condition does not apply for the lower dimensional cases, where the exchange interaction is much less effective and the fast exchange regime is actually never attained. In a field swept experiment the $2\omega_0$ resonance will be observed at $B = \frac{1}{2}B_0$. Beautiful examples of the half-field line were observed for TMMC and they showed that the intensity of the satellite line is strongly dependent on the crystal orientation, while its line width is practically identical to that of the normal EPR transition. Finally, the intensity of the satellite line is strongly frequency-dependent, becoming roughly B_0^{-2} as expected.

The angular dependence of the intensity for the usual transversal experimental setup, i.e., for spectra recorded with the oscillating microwave field orthogonal to the static magnetic field, in the case of a one-dimensional magnet is of the type [6.14]:

$$I \propto \sin^2 \Theta \cos^2 \Theta / [\Gamma^+(\Theta)]^{3/2}, \tag{6.69}$$

where Θ is as usual the angle between the magnetic axis and the external magnetic field and $\Gamma^+(\Theta)$ is the angular-dependent width of the main line. The overall angular dependence of the intensity of the half-field line in a transverse field experiment is shown in Fig. 6.13. In practice the intensity shows a sharp maximum at the magic angle, due to the $3\cos^2 \Theta - 1$ dependence of the denominator, and goes to zero at both $\Theta = 0°$ and $90°$, due to the $\sin^2 \Theta \cos^2 \Theta$ dependence of the numerator.

Another important feature of the half-field transition is that its intensity can be greatly enhanced compared to the intensity of the main EPR line if a longitudinal field experiment is performed, i.e., if the oscillating microwave field is parallel to the static magnetic field. In Fig. 6.14 such an effect is clearly shown: indeed, the intensity of the half-field line is larger than that of the main line. In this case the angular dependence of the line intensity is of the type:

$$I \propto \sin^4 \Theta, \tag{6.70}$$

and so the maximum intensity must be expected orthogonal to the chain axis, as experimentally observed.

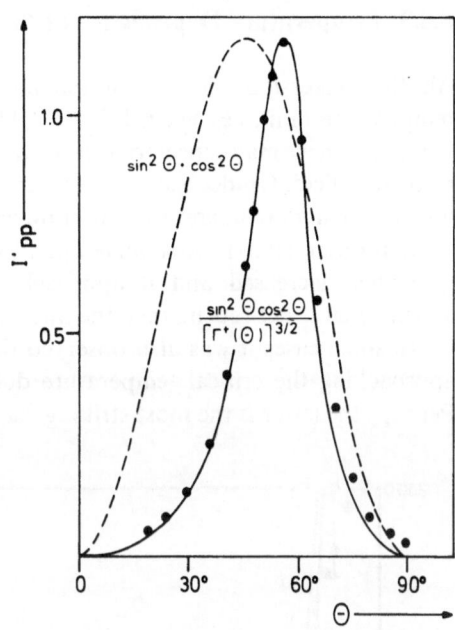

Fig. 6.13. Angular dependence of the intensity of the half-field line in TMMC at Q-band frequency. —— calculated with (6.69), – – – $\sin^2\Theta\cos^2\Theta$. The *dots* are experimental points. After [6.14]

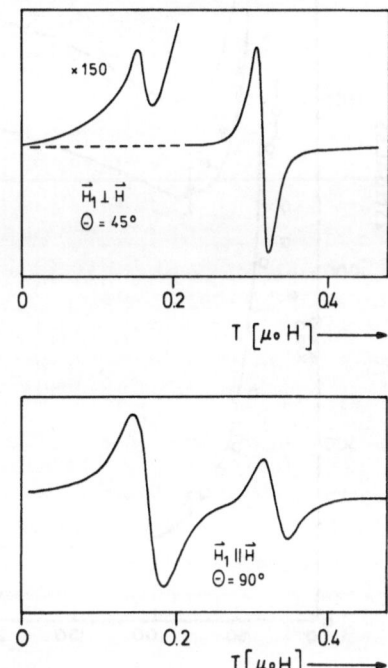

Fig. 6.14. Transverse and longitudinal EPR spectra of TMMC at X-band; upper, transverse, lower, longitudinal. After [6.14]

6.4.3 Temperature Dependence of the Spectra

All the expressions derived in the previous sections are valid in the high temperature limit, i.e., when $kT \gg H_z, H_{ex}\ldots$ When this condition no longer applies, then we may expect to observe variations in the line width and also in the resonance field. On decreasing the temperature from room temperature, but as long as the high temperature condition can still be considered to apply, the line width is observed to have small temperature dependence. When the temperature is further decreased and it approaches the critical point at which three-dimensional order sets in, then the line width increases quite rapidly (Fig. 6.15).

In some cases it was also observed that the resonance frequency, shifts on approaching the critical temperature due to short-range order effects [6.15]. Perhaps the latter is the most striking feature which was observed [6.16, 17] for

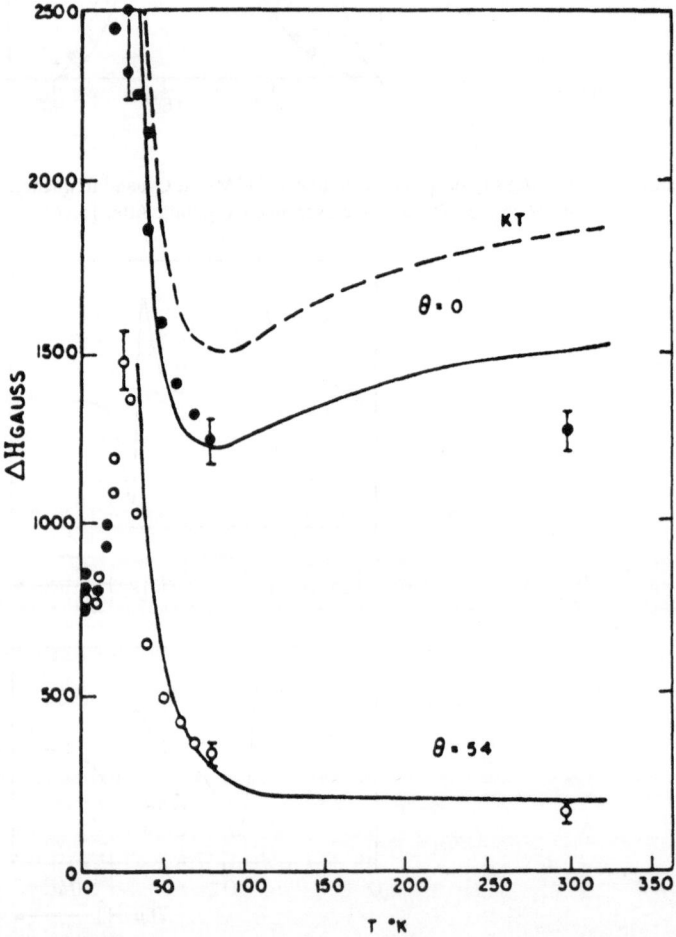

Fig. 6.15. Temperature dependence of the EPR line in TMMC at X-band. After [6.12]

the first time for the linear chain antiferromagnets $CsMnCl_3 \cdot 2H_2O$ and TMMC. In Fig. 6.16 the temperature dependence of the resonance field of single crystals of $CsMnCl_3 \cdot 2H_2O$ at Q-band frequency is shown. The signal observed parallel to the chain axis is seen to shift downfield up to 50 mT, while the signals orthogonal to it shift upfield 50% of this amount at the same temperature. This shift has been attributed to the magnetic dipole interactions among shortrange ordered spins in the chains. Indeed, lower dimensional systems cannot undergo a one- or two-dimensional order at temperatures $T > 0$ K, but, at sufficiently low temperatures it is conceivable that small clusters of, e.g., five to ten spins are present which are strongly correlated to each other. In this way moderate fields are built in the lattice which add or subtract from the external magnetic field, determining a shift of the resonance line. In other terms the effect observed here is similar to that observed in an antiferromagnetic resonance experiment in the ordered three-dimensional state, the main difference lying in the spacial extention of the correlated spins.

A quantitative estimation of the resonance shift was performed using classical spins. The result is:

$$h\omega_{\parallel} = g_{\parallel}\mu_B B + 12\,\alpha g_{\perp}\mu_B B\{(2+ux)/(1-u^2)-2/3x\}/10x; \tag{6.71}$$

$$h\omega_{\perp} = g_{\perp}\mu_B B - 6\,\alpha g_{\perp}\mu_B B\{(2+ux)/(1-u^2)-2/3x\}/10x; \tag{6.72}$$

where $\parallel$ and $\perp$ denote parallel and perpendicular to the chain direction, respectively,

$$u = \coth(K) - 1/K; \tag{6.73}$$

$$K = J S(S+1)/kT = -1/x; \tag{6.74}$$

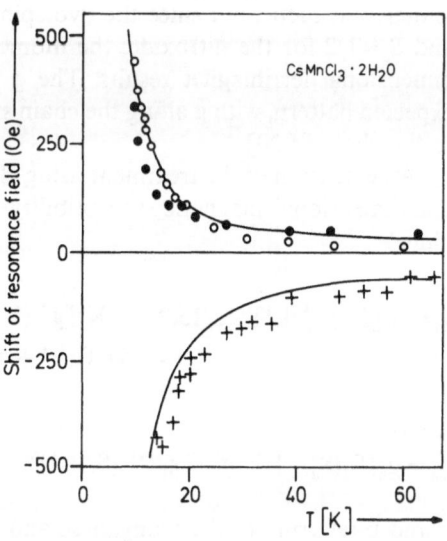

Fig. 6.16. Temperature dependence of the resonance field for $CsMnCl_3 \cdot 2H_2O$ at 34.4 GHz. The field is parallel to a (+), the chain direction, b (○), and c (●). The curves are calculated. After [6.16]

and

$$\alpha = g^2 \mu_B^2/(Jr^3), \tag{6.75}$$

where r is the nearest-neighbor spin-spin distance. Equations (6.71–75) have been obtained for one-dimensional manganese salts, and indeed the use of classical spins seems to be more appropriate for $S = 5/2$ than for $S = 1/2$. However, they have been used also for ferromagnetic copper salts, as we will see in Chap. 10.

The relations (6.71–75) can be generalized in the form [6.18]:

$$B_r^i = \sqrt{\chi_j \chi_k/\chi_i}\, \hbar\omega/g_i \mu_B, \tag{6.76}$$

where B_r^i is the resonance field along the i direction, χ_i, χ_j, and χ_k are the one-dimensional susceptibilities parallel to the i, j, and k directions, and g_i is the corresponding g value.

Equation (6.76) allows us to predict the sign of the g shifts for ideal one-dimensional ferro-, antiferro-, and ferrimagnets, respectively. In fact, if we assume that i is the chain direction for both a ferro- and an antiferromagnet, we must expect that $\chi_i > \chi_j, \chi_k$, and consequently that Δg_i is positive; for a ferrimagnet $\chi_i < \chi_j, \chi_k$, and Δg_i is negative. This is easily understood by considering Fig. 6.17 which shows the preferred spin orientations for one-dimensional magnetic materials.

Beyond the ideal antiferromagnetic case, represented by TMMC, a typical behavior for a one-dimensional ferrimagnet was observed for a compound in which manganese(II) ions alternate regularly in a chain with stable nitronyl-nitroxide radicals [6.19]. The coupling between the metal ions and the radicals is antiferromagnetic, but since the two spins are different, $S = 5/2$ for manganese and $S = 1/2$ for the nitroxide, the moments are not compensated and a one-dimensional ferrimagnet results. The g shifts at low temperature follow the expected pattern, with g along the chain smaller than 2, and larger orthogonally to the chain [6.20].

An extension of the treatment using classical spins yields the formula for the one-dimensional magnetic susceptibility in the assumption of dipolar broadening:

$$\chi_i = g_a' g_b'/r^3 \mu_B^4/k^2 T^2 \{2/15(2-u/K)[g^2/(1-u^2) - \delta^2/(1+u^2)] + \\ + 4/45\, K[g^2(1+u)/(1-u) - \delta^2(1-u)/(1+u)]\}, \tag{6.77}$$

where
$$g_a' = g_a[S_a(S_a+1)]; \quad g_b' = g_b[S_b(S_b+1)]; \tag{6.78}$$

a and b referring to the manganese and the nitroxide, respectively.

$$g = 1/2(g_a + g_b); \quad \delta = 1/2(g_a - g_b). \tag{6.79}$$

164

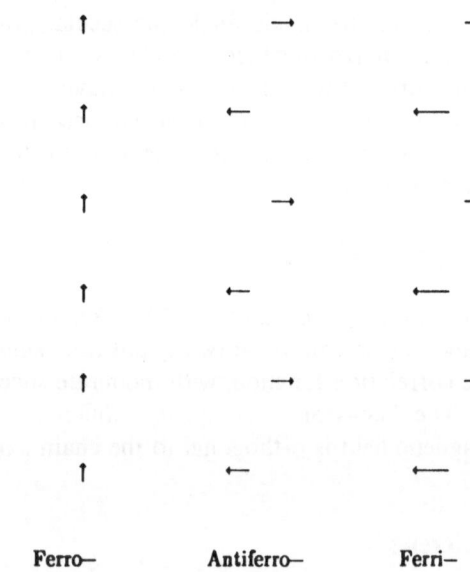

| Ferro– | Antiferro– | Ferri– |

Fig. 6.17. Preferred spin orientation in ferro-, antiferro-, and ferrimagnetic one-dimensional materials

In the range of temperature at which a huge shift of the resonance is observed, also a large increase in the line-width is observed, which presumably has the same origin. However, at the moment no really satisfactory quantitative agreement has been obtained, due to the complexity of the required treatment.

The temperature dependence of the line width for $T \gg T_N$ (T_N being the temperature of three-dimensional ordering) has also been the subject of long discussions in the literature [6.15], and it has not been well settled as yet. In particular much interest has focused on the observed linear relationship between ΔB_{pp} and T and several explanations have been produced. The mechanisms which have been suspected to be responsible for this have been within the framework of spin correlations or of the dependence of the various exchange interactions on temperature, such as phonon modulation of antisymmetric exchange and the temperature dependence of the isotropic exchange. Since at the moment there seems to be no definitive agreement in the literature, we will briefly mention this problem in Chap. 10 where some examples of spectra of low dimensional systems will be reported.

6.4.4 Frequency Dependence of the Line Widths

The frequency dependence of the line widths has been studied for linear chain compounds, at some particular angular setting. Of great interest is the behavior at the magic angle [6.21], where it has been found that:

$$\Delta B_{pp} \propto \nu^{-1/2}. \tag{6.80}$$

In fact, at the magic angle the secular components vanish, and the simple motional narrowing theory can be used to calculate the line width. The difference with normal three-dimensional behavior lies essentially in the intensity of the effect, which is much more pronounced in the low dimensional systems. The experimental frequency dependence of the line widths of TMMC was fitted with the expression:

$$\Delta B_{pp} = A + S \nu^{-1/2}, \tag{6.81}$$

with $A = 28 \pm 3$ G, and $S = 406 \pm 8$ G GHz$^{1/2}$ at 295 K. Attempts were also made to calculate these two quantities using various decoupling techniques of the correlation function, with moderate success [6.22–24].

The behavior is completely different for $\Theta = 90°$, i.e., when the static magnetic field is orthogonal to the chain axis.

References

6.1   van Vleck JH (1948) Phys. Rev. 74:1168
6.2   Anderson PW, Weiss PR (1953) Rev. Mod. Phys. 25:269
6.3   Kubo R, Tomita K (1954) J. Phys. Soc. Jpn. 9:888
6.4   Kubo R (1961) In ter Haar D (ed) Fluctuation, relaxation and resonance in magnetic systems. Oliver and Boyd, Edinburgh p 23
6.5   Pleau E, Kokoszka G (1973) J. Chem. Soc. Faraday Trans. II 69:355
6.6   Anderson PWJ (1954) J. Phys. Soc. Jpn 9:316
6.7   Hoffmann SK (1983) Chem. Phys. Letters 98:329
6.8   Richards PM (1975) In: Local properties at phase transitions. Editrice Compositori Bologna, p 539
6.9   Dietz RE, Merritt FR, Dingle R, Hone D, Silbernagel BG, Richards PM (1971) Phys. Rev. Letters 26:1186
6.10 Morosin B, Graeber (1967) J. Acta Cryst. 23:766
6.11 Hone DW, Richards PM (1974) Annu. Rev. Meter. Sci. 4:337
6.12 Cheung TTP, Soos ZG, Dietz RE, Merritt FR (1978) Phys. Rev. B 17: 1266
6.13 Hennessy MJ, McElwee CD, Richards PM (1975) Phys. Rev. B 7:930
6.14 Benner H, Weise J (1979) Physica 96B:216
6.15 Drumheller (1982) J. Magnetic Res. Rev. 7:123
6.16 Nagata K, Tazuke Y (1972) J. Phys. Soc. Jpn. 32:337
6.17 Tazuke Y, Nagata KJ (1971) J Phys. Soc. Jpn. 30:285
6.18 Karasudani T, Okamoto H (1977) J. Phys. Soc. Jpn. 43:1131
6.19 Caneschi A, Gatteschi D, Rey P, Sessoli R. (1988) Inorg. Chem. 27:1756
6.20 Caneschi A, Gatteschi D, Renard, J-P, Rey P, Sessoli R (1989) Inorg. Chem. 28:1976
6.21 Siegel E, Lagendijk A (1979) Solid State Commun. 32:561
6.22 Lagendijk A (1976) Physica 83B:283
6.23 Lagendijk A, Siegel E (1976) Solid State Commun. 20:709
6.24 Siegel E, Mosebach H, Pauli N, Lagendijk A (1982) Phys. Lett. 90A:309

7 Selected Examples of Spectra of Pairs

7.1 Early Transition Metal Ion Pairs

Very detailed EPR spectra have been recorded for manganese(II) pairs in MgO and CaO. Manganese(II) enters substitutionally the host crystals: at concentrations of about 1% appreciable numbers of isolated pairs are found which were characterized at three frequencies (15.5, 24.5, and 35.5 GHz) over the temperature range 1.2 to 300 K [7.1].

A typical spectrum of Mn^{2+} : CaO at 35.5 GHz is shown in Fig. 7.1. Signals are observed from all the five total spin states: it is apparent that on lowering temperature the low spin states increase the intensity relative to the high spin states. The signals shown in Fig. 7.1 are only part of the spectrum, the scheme of which is given in Fig. 7.2. The temperature dependence of the signal intensity of Mn:MgO was found to conform to the curves of Fig. 3.3, and the order of the levels is shown in Fig. 7.3. The levels deviate considerably from a Landé interval rule (Table 7.1), a result which was explained by including biquadratic terms. The levels on the left in Fig. 7.3 are calculated only with bilinear exchange, while those on the right, which correspond to a good approximation of the experimental data, are obtained including a biquadratic exchange constant j, such that j/J = 0.05. The effect of biquadratic exchange on the energy separations is quite dramatic, according to the relation:

$$E_s - E_{s-1} = JS - jS[S^2 - S_1(S_1 + 1) - S_2(S_2 + 1)]. \tag{7.1}$$

The biquadratic term may have its origin from both an intrinsic higher order exchange and by exchange striction as discussed in Section 2.4. Beyond the departure from Lande interval rule biquadratic exchange has been considered to be responsible also for the departure of the anisotropic exchange parameters from the simple behavior expected using Eq. (3.21). The experimental D and E parameters for the various spin states are given in Table 7.1. According to Table 3.3, the D_i parameters should be given by:

$$D_1 = 37/10 \, D_{MnMn} - 32/5 \, D_{Mn}$$

$$D_2 = 41/42 \, D_{MnMn} - 20/21 \, D_{Mn}$$

$$D_3 = 47/90 \, D_{MnMn} - 2/45 \, D_{Mn} \tag{7.2}$$

$$D_4 = 5/14 \, D_{MnMn} + 2/7 \, D_{Mn}$$

$$D_5 = 5/18 \, D_{MnMn} + 4/9 \, D_{Mn}$$

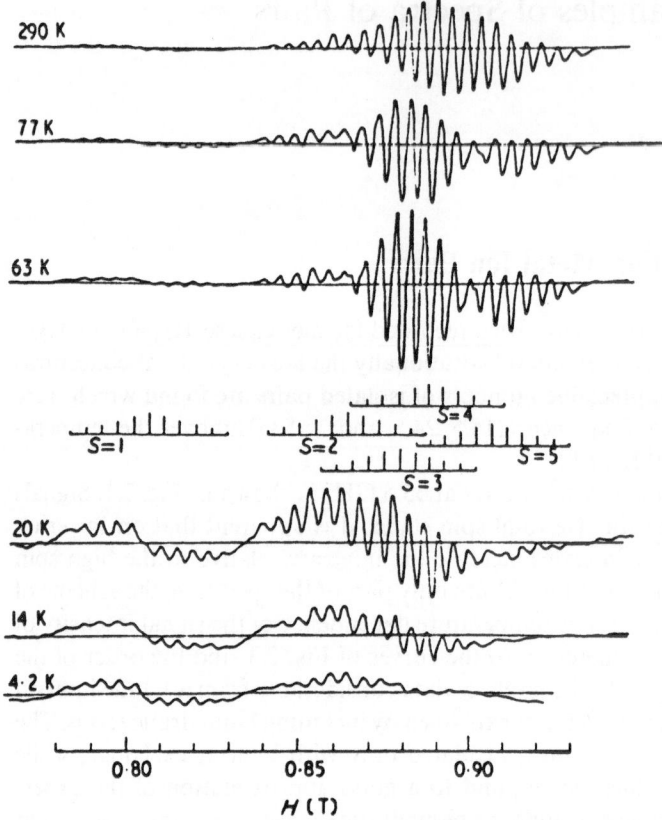

Fig. 7.1. A part of the spectrum of manganese(II) pairs in CaO recorded at Q-band frequency with the magnetic field in the ⟨110⟩ direction at various temperatures. After [7.1]

where D_{MnMn} is the anisotropic spin-spin and D_{Mn} the single ion zero field splitting tensor. It is easy to verify that there is not a single set of parameters D_{MnMn} and D_{Mn} which satisfy (7.2). Attempts were made to rationalize the experimental data using an exchange striction model, but the system becomes in this way so heavily parametrized, that it seems safer to stop here and to consider that experimentation shows how the simplest possible model can provide only a first approximation of the real system.

$CsMgCl_3$, $CsMgBr_3$, and $CsCdBr_3$ are host lattices which have been extensively used for recording EPR spectra of paramagnetic impurities [7.2, 3]. They all have the so-called hexagonal $CsNiCl_3$ structure (Fig. 7.4), which can be described as an array of infinite parallel linear chains composed of $[MX_6]^{4-}$ octahedra sharing opposite faces. When trivalent cations, such as chromium(III) and molybdenum(III) are doped into these lattices, an extraordinary tendency of the ions to cluster into pairs is observed, even at concentrations of 10^{-3} M. This behavior has been attributed to the necessity of charge compensation: two

168

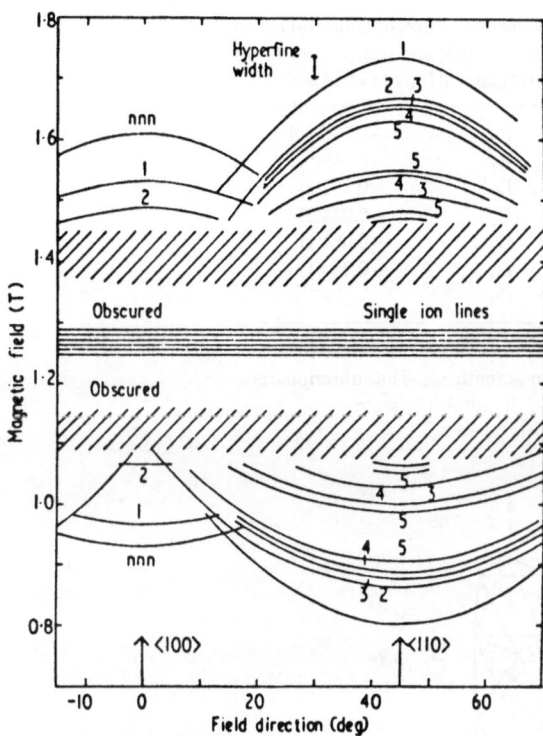

Fig. 7.2. Angular dependence of the resonance fields of manganese(II) pairs in CaO. After [7.1]

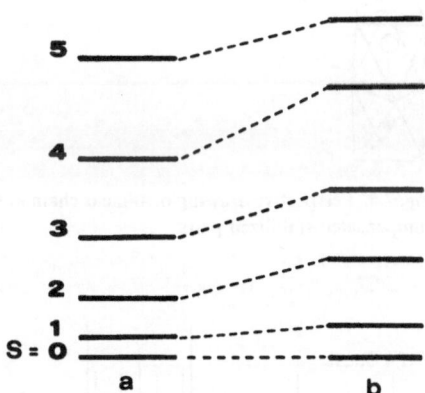

Fig. 7.3. The energy levels of manganese(II) pairs in MgO: *a* without biquadratic exchange terms; *b* including biquadratic exchange terms

trivalent ions occupy two sites of two bivalent ions only leaving a vacancy in between them (Fig. 7.4).

Representative spectra of chromium(III) pairs in $CsMgCl_3$ are shown in Fig. 7.5. The external magnetic field is parallel to the trigonal axis of the chain. The

Table 7.1. Experimental values of the zero field splitting parameters of manganese(II) pairs in MgO[a]

D_1	$-0.776(10)$
E_1	$-0.149(5)$
D_2	$-0.182(2)$
E_2	$-0.024(3)$
D_3	$-0.084(1)$
D_4	$-0.050(1)$
D_5	——

[a]Values in cm^{-1}, estimated errors in parentheses. The subscripts refer to the total spin state.

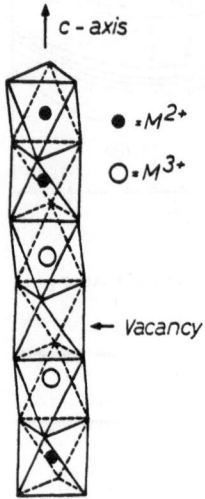

Fig. 7.4. Perspective drawing of a linear chain in the hexagonal CsNiCl$_3$ structure, showing charge-compensated stabilized pairs

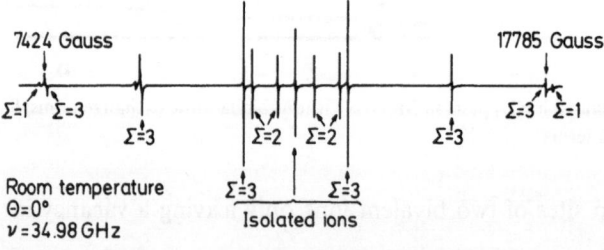

Fig. 7.5. Q-band EPR spectrum of chromium(III) pairs in a single crystal of CsMgCl$_3$. The external magnetic field is parallel to the hexagonal axis. Σ indicates the total spin state. After [7.3]

signals have been attributed to transitions within the total spin states $S = 3, 2$ and 1 which originate from the interaction of two $S = 3/2$ ions. The analysis of the spectra of these multiplets can be performed by considering separately the various total spin multiplets. In this case the analysis requires the inclusion of second-order terms in the spin hamiltonian for $S = 2$ and of fourth- and sixth-order terms for $S = 3$. The usual analysis (Table 3.3) suggests that the following relations must hold for the zero field splitting parameters:

$$D_1 = 17/10\, D_{CrCr} - 12/5\, D_{Cr};$$

$$D_2 = 1/2\, D_{CrCr}; \tag{7.3}$$

$$D_3 = 3/10\, D_{CrCr} + 2/5\, D_{Cr}.$$

Again in this case the three equations cannot be solved exactly, but average values of D_{CrCr} and D_{Cr} can be calculated as: $D_{CrCr} = -0.0140\ \mathrm{cm}^{-1}$, $D_{Cr} = -0.224\ \mathrm{cm}^{-1}$.

A more rigorous procedure can also be used by diagonalizing the 16×16 matrix obtained using the direct product basis of the two $S = 3/2$ states. In this case a more accurate reproduction of the transition fields of the $S = 2$ state could be obtained, so that it was found at 297 K, $J = 0.80\ \mathrm{cm}^{-1}$, $D_{CrCr} = -0.0079\ \mathrm{cm}^{-1}$, $D_{Cr} = -0.221\ \mathrm{cm}^{-1}$, while at 77 K, $J = 0.96\ \mathrm{cm}^{-1}$, $D_{CrCr} = -0.0081\ \mathrm{cm}^{-1}$, $D_{Cr} = -0.222\ \mathrm{cm}^{-1}$. Similar results were obtained also for molybdenum(III) pairs and for mixed chromium(III)−molybdenum(III) pairs [7.4]. A typical spectrum is shown in Fig. 7.6. The isotropic exchange of the Cr–Mo pairs in $CsMgCl_3$ at 77 K was determined to be $1.61\ \mathrm{cm}^{-1}$, an intermediate value between $0.96\ \mathrm{cm}^{-1}$ for Cr–Cr and $2.75\ \mathrm{cm}^{-1}$ for Mo–Mo pairs.

Quite often fluid solution spectra of pairs comprising two $S = 1/2$ transition metal ions cannot be recorded, because the zero field splitting of the triplet is too large. One distinct exception to this rule is provided by oxovanadium(IV) *dl*-tartrate whose room temperature solution spectra are shown in Fig. 7.7. The 15 lines are determined by the hyperfine interaction with two equivalent $^{51}V, I = 7/2$

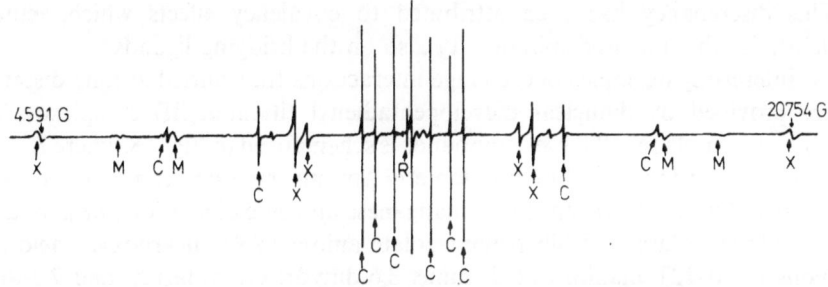

Fig. 7.6. Q-band EPR spectrum of a $CsMgCl_3$ crystal doped with Cr^{3+} and Mo^{3+} at 77 K. The external magnetic field is parallel to the c crystal axis. C indicates chromium(III) pairs; M molybdenum pairs; and X chromium-molybdenum pairs. R is DPPH. After [7.4]

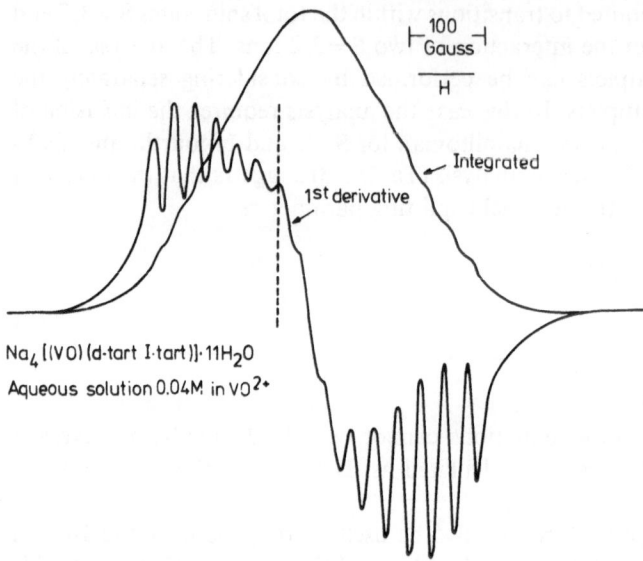

Fig. 7.7. X-band EPR spectrum of sodium oxovanadium(IV) *dl*-tartrate at room temperature. After [7.5]

nuclei, and the splitting is nearly one-half that expected for a mononuclear oxovanadium(IV) species. Frozen solution spectra of the same compound yielded $D = 0.0334(2)$ cm^{-1} [7.5].

Vanadium(II) ions (d^3 configuration, $S = 3/2$) can be doped into KMgF$_3$, and a measurable amount of pairs is formed, which can be investigated by EPR spectroscopy [7.6]. The ions are coupled antiferromagnetically to give $S = 3, 2, 1$ and 0, with $J = 4.5 \pm 1.0$ cm^{-1}. Surprisingly, only the signals of the $S = 2$ total spin manifold could be detected. This has been explained by the fact that crystal field effects on the $S = 2$ state are zero, as shown by (7.3), while they are different from zero for the other total spin multiplets. The observed zero field splitting parameter, $D = -0.0339$ cm^{-1} is 35% larger than the classical dipolar value. This discrepancy has been attributed to covalency effects which actually delocalize the unpaired spin density also on the bridging ligands.

Interesting examples of exchange interactions transmitted at long distances are provided by dinuclear dicyclopentadienyl titanium(III) complexes. The general formula for all these compounds is schematized in Fig. 7.8, where B is just an indication for some bridging moiety. Among these one can mention cyanurato and uracilato groups [7.7], manganese and zinc dichloride [7.8], pyrazolate, biimidazolate, and bibenzimidazolate anions [7.9], dicarboxylic acid dianions [7.10–12], dianions of thymine, 3,6-dihydroxypyridazine, and 2,3-dihydroxyquinoxaline [7.13]. A typical spectrum is shown in Fig. 7.9, with six transitions corresponding to a rhombic hamiltonian. In this case the bridge is the oxalato anion. The observed zero field splitting parameter, $D = 0.0116$ cm^{-1},

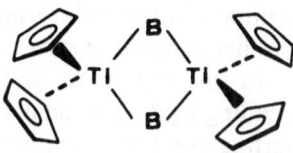

Fig. 7.8. Scheme of dinuclear dicyclopentadienyl titanium(III) complexes

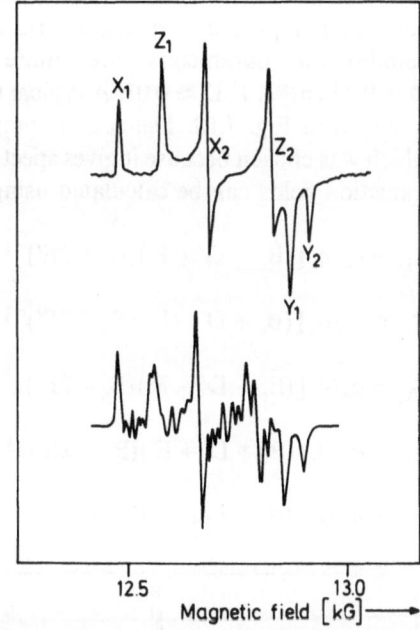

Fig. 7.9. Q-band spectrum of oxalate bridged dicyclopentadienyl titanium(III) dimers. *Upper*, experimental; *lower*, simulated. The noise in the simulated spectrum is a computational artifact. After [7.12]

corresponds to a metal-metal distance of 597 pm according to the assumption of dominant dipolar interaction. The calculated value compares well with 585 pm expected for two titanium ions bridged by an oxaloto ion. The fact that the zero field splitting tensor is essentially dipolar for these complexes is not surprising, given the small effect of the spin-orbit coupling on titanium (III), as evidenced by the g values close to 2. With this procedure metal-metal distances as long as 1080 pm were estimated.

7.2 Copper Pairs

Copper is by far the metal ion which has been most often investigated through EPR, and the pairs formed by it are no exception. We have already given several examples, in particular using them in order to obtain correlations between the spin hamiltonian parameters and the electronic structure of pairs.

In this section we will provide more examples which can be helpful in shedding light on some particular feature of the EPR spectra of pairs. A useful starting point may be given by complexes of the copper acetate hydrate type, which was studied at the very beginning of the applications of EPR spectroscopy. Indeed, by interpreting the spectra Bleaney [7.14] correctly suggested the dimeric nature of the compound, which was subsequently confirmed by X-ray crystal structure [7.15]. After that hundreds of derivatives with similar structures have been reported and characterized by EPR. In most of these the spin hamiltonian parameters are quite similar, with $g_{\parallel} \approx 2.3$, $g_{\perp} \approx 2.08$, $D \approx 0.34 \text{ cm}^{-1}$, $E/D \approx 0.01$. A typical spectrum recorded at Q-band frequency is shown in Fig. 7.10. The actual compound is copper chloroacetate hydrate, which was chosen because it gives spectra with many well-resolved features. The transition fields can be calculated using the following equations:

$$B_{x_1} = g_e/g_x [(B_o - D' + E')(B_o + 2E')]^{\frac{1}{2}}; \tag{7.4}$$

$$B_{x_2} = g_e/g_x [(B_o + D' - E')(B_o - 2E')]^{\frac{1}{2}}; \tag{7.5}$$

$$B_{y_1} = g_e/g_y [(B_o - D' - E')(B_o - 2E')]^{\frac{1}{2}}; \tag{7.6}$$

$$B_{y_2} = g_e/g_y [(B_o + D' + E')(B_o + 2E')]^{\frac{1}{2}}; \tag{7.7}$$

$$B_{z_1} = g_e/g_z [(B_o - D')^2 - E'^2]^{\frac{1}{2}}; \tag{7.8}$$

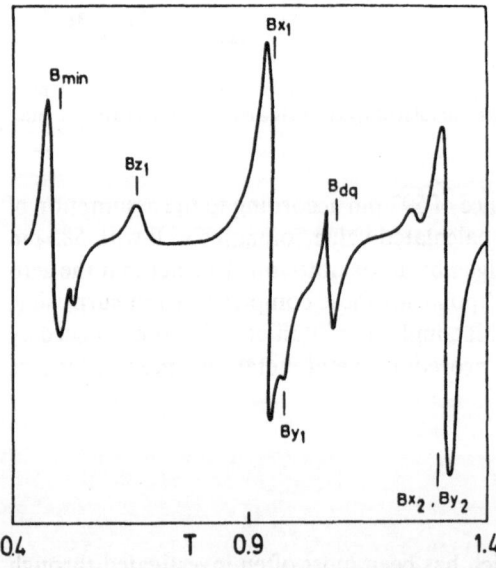

Fig. 7.10. Polycrystalline powder EPR spectrum of copper(II) chloroacetate hydrate

$$B_{z_2} = g_e/g_z[(B_o + D')^2 - E'^2]^{\frac{1}{2}}; \tag{7.9}$$

$$B_{min} = g_e/g_{min}[B_o^2/4 - D'^2/3 - E'^2]^{\frac{1}{2}}; \tag{7.10}$$

$$B_{dq} = g_e/g_{av}[B_o^2 - D'^2/3 - E'^2]^{\frac{1}{2}}; \tag{7.11}$$

where $g_{min} = [g_\perp^2 \sin^2\alpha + g_z^2 \cos^2\alpha]^{\frac{1}{2}}$, $g_\perp = [g_x g_y]^{\frac{1}{2}}$, $g_{av} = [2/3g_\perp^2 + 1/3g_z^2]^{\frac{1}{2}}$,

$B_0 = (h\nu/\mu_B g_e)$, $D' = D/(\mu_B g_e)$, $E' = E/(\mu_B g_e)$, and

$$\alpha^2 = \cos^{-1}\{[9 - 4(D/h\nu)^2]/[27 - 36(D/h\nu)^2]\} \tag{7.12}$$

Beyond the six allowed $\Delta M = \pm 1$ transitions, also two forbidden $\Delta M = \pm 2$ transitions are clearly resolved. One corresponds to the absorption of one photon which induces a transition between the two $M = +1$ and $M = -1$ levels, and the other, indicated as double quantum transition, corresponds to the absorption of two quanta. This transition can be observed at relatively high microwave power, when the separations between the $M = -1$ and $M = 0$ and the $M = 0$ and $M = 1$ levels are about equal.

In the spectra of copper(II) pairs quite often one observes, beyond the transitions within the triplet, also features corresponding to simple $S = 1/2$ species. These spectra are often referred to as "mononuclear" impurities, and may originate from defective sites in the lattice. This phenomenon has not received much attention, although it is potentially very important in order to characterize the defects in molecular solids. Indeed, single crystals of molecular materials can host widely different species, as has been shown through EPR spectroscopy in the following example.

[(dien)Cu(ox)Cu(tmen) (H₂O)₂] (ClO₄)₂ is an asymmetric dinuclear copper(II) species (dien = diethylenetriamine; ox = oxalate; tmen = N, N, N', N'-tetramethylethylenediamine) with the structure illustrated in Fig. 7.11, and a coupling constant $J = 75.5$ cm^{-1} [7.16]. The EPR spectra of this compound at high temperature are broad and ill-characterized, but at 4.2 K, when the magnetic susceptibility data clearly indicate that the dinuclear species is in the ground singlet state, a beautiful triplet spectrum is resolved [7.17]. In Fig. 7.12 a typical spectrum of a single crystal is reported. The two septets are attributed to the $-1 \rightarrow 0$ and $0 \rightarrow 1$ transitions, split by the interaction with two *equivalent* copper ions. The complete analysis of the spectra yielded $|D| = 0.0163$ cm^{-1}, $E/D \approx 1/3$, with the principal directions which cannot be reconciled with any structural feature of the [(dien)Cu(ox)Cu(tmen)(H₂O)₂]²⁺ cation. The explanation of this behaviour is that in this lattice, during the synthesis of the compound, beyond the asymmetric species [(dien)Cu(ox)Cu(tmen)(H₂O)₂]²⁺, also a symmetric one is formed. Since [(dien)Cu(ox)Cu(dien)]²⁺, shown in Fig. 7.11 is independently known to exist and to have a very weak coupling between the two copper ions, it seems extremely feasible that this is indeed the species which is formed in the lattice and which appears at low temperature when the host compound becomes diamagnetic.

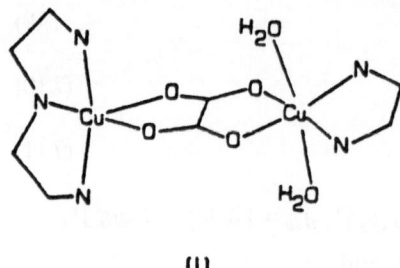

(I)

(II)

Fig. 7.11. Schematic representation of the structures of $[(dien)Cu(ox)Cu(tmen)(H_2O)_2]^{2+}$ and $[(dien)Cu(ox)Cu(dien)]^{2+}$

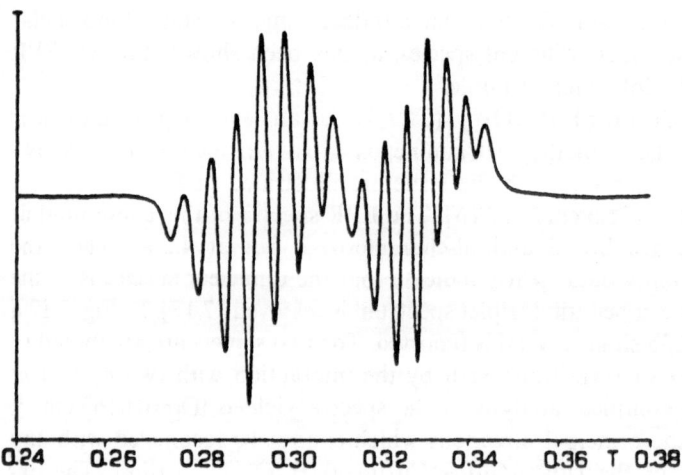

Fig. 7.12. Single crystal EPR spectrum of $[(dien)Cu(ox)Cu(tmen)(H_2O)_2](ClO_4)_2$ at X-band frequency and 4.2 K. After [7.17]

An interesting example of the use of EPR in determining weak exchange interactions has been reported for the compound $[Cu_2(tren)_2(OCN)_2](BPh_4)_2$, where tren is 2,2′,2″-triamino triethylamine. The structure of this compound is shown in Fig. 7.13 [7.18]. The EPR spectra are characteristic of magnetically

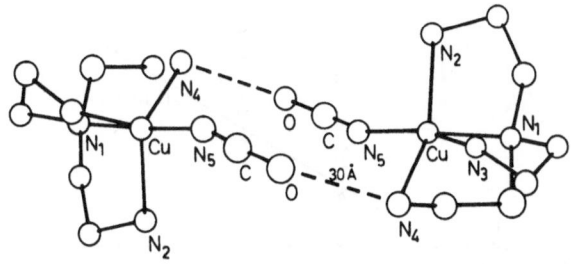

Fig. 7.13. Molecular structure of the dimer cation [Cu(tren)(OCN)]₂. After [7.18]

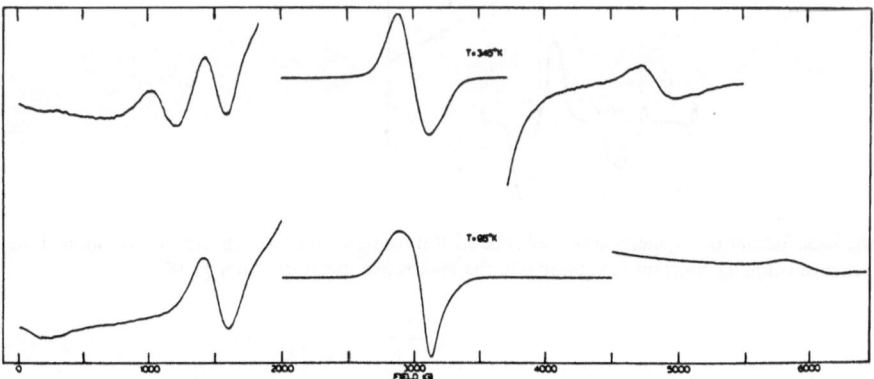

Fig. 7.14. Temperature dependence of the X-band EPR spectrum of [Cu(tren)(OCN)](BPh₄). After [7.18]

nondilute copper(II) complexes, but, beyond the usual absorption in the $g = 2$ region, they show additional features both at high and low field as shown in Fig. 7.14. These have been assigned to transitions involving singlet and triplet levels, as indicated by the energy level scheme of Fig. 7.15. The two transitions are expected to occur at $B = B_0 \pm J/g\mu_B$, where B_0 is the resonance field of the normal $\Delta M = 1$ transition. Therefore, measuring B allows the direct determination of J. In the present case it has been found to depend on temperature, ranging from 0.09 to 0.16 cm^{-1}. A more complete discussion of the conditions under which these "forbidden" transitions can be observed is reported in Sec. 7.4.

The single crystal EPR spectra of [Cu₂(Me₅dien)₂(N₃)₂](BPh₄)₂ are instructive for illustrating the role of second-order zero field splitting effects determining the hyperfine splitting [7.19]. The compound contains dinuclear units [7.20], with the structure shown in Fig. 7.16, with a coupling constant of 13 cm^{-1}. The crystal is monoclinic, with the dinuclear units in general position, therefore, four $\Delta M = \pm 1$ transitions are expected for a general orientation of the static magnetic field (two fine structure transitions for either magnetically nonequivalent site). The single crystal spectrum corresponding to the static

177

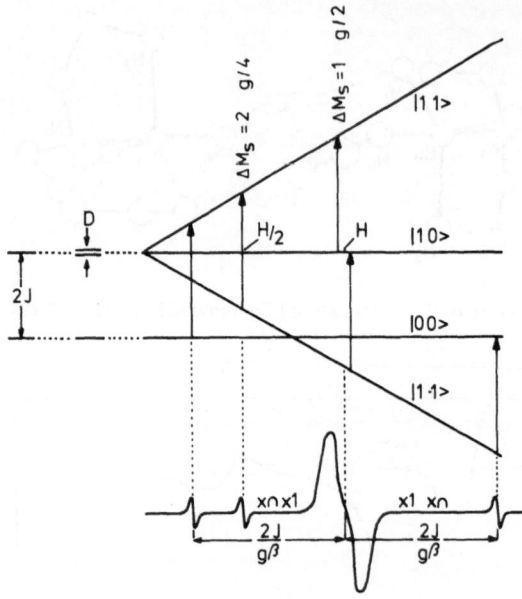

Fig. 7.15. Scheme of the energy levels, allowed EPR transitions, and EPR spectrum in the limit of the exchange-coupling constant comparable to the microwave quantum. After [7.18]

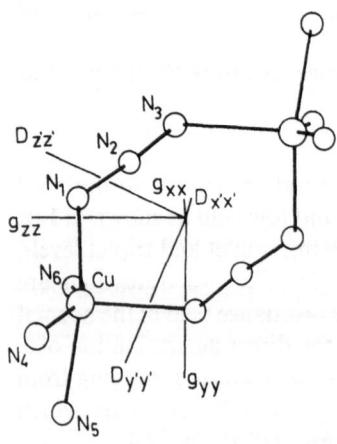

Fig. 7.16. Scheme of the coordination environments of the copper ions in $[Cu_2(Me_5dien)_2(N_3)_2](BPh_4)_2$ and orientation of the **g** and **D** tensors in the molecular frame. After [7.19]

magnetic field in the (100) plane is shown in Fig. 7.17. Apparently there are four transitions, three of which are split into seven components, the splitting corresponding to 37, 85, and 53 G, respectively, on passing from low to high field. The third transition from low field does not show any measurable splitting. It

178

must also be noted that the spacings of the hyperfine components are not equal within one transition and the values given above are just average values. The observed features must be assigned to the fine structure components of the two sites present in the lattice. Under these conditions it is difficult to recognize the two pairs of fine structure transitions. However, a complete analysis of the spectra showed that the two outermost signals belong to the same transition, as do the internal ones. The marked difference in the hyperfine splitting between the two components of the fine structure depends on the second-order mixing of the $M = \pm 1$ states within the $M = 0$ level, due to second-order zero field splitting effects [7.21]. This effect, in turn, depends on the ratio $D/h\nu$ as confirmed by the Q-band spectra which show a more regular behaviour, with much smaller differences in the hyperfine splitting of the two fine structure components.

Similar effects of large differences in the hyperfine splitting have been observed also in polycrystalline powder spectra [7.22]. For instance, in the spectra of Fig. 7.18 the parallel high-field feature has markedly different splitting compared to the low-field one. This behaviour is due to the misalignment of the A and D tensors.

An interesting example of the resolution of weak interdimer interactions via EPR has been reported for copper(II) maleonitriledithiolate complexes, $[Cu(mnt)_2]^{2-}$. When the methylene blue cation, MB^+ is used as a counterion, a compound of formula $(MB)_2Cu(mnt)_2$. acetone is obtained [7.23]. The anions are paired and the pairs are stacked along the c crystal axis. The intrapair Cu–Cu distance is 711.5 pm, and the shortest interpair distance is 1074.3 pm. The coupling within the dimer is antiferromagnetic, $J = 5.2\ cm^{-1}$. A typical EPR spectrum of a single crystal is shown in Fig. 7.19. The 14 internal lines are attributed to two fine structure components of the $S = 1$ state, each of which is

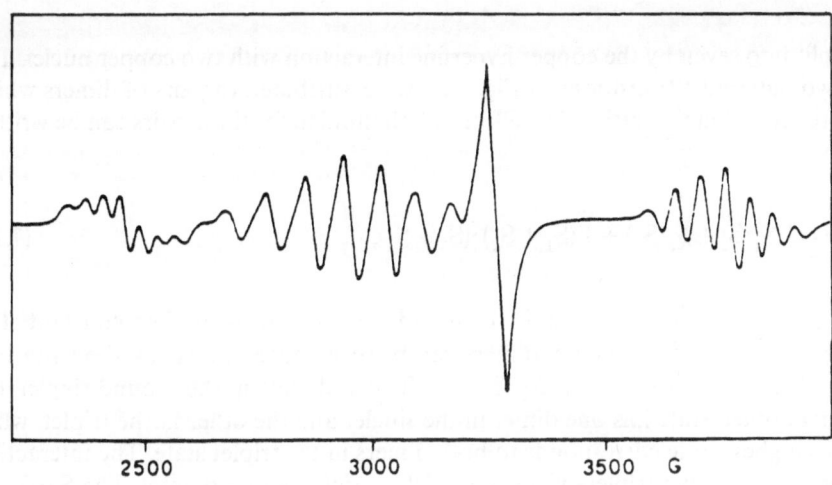

Fig. 7.17. Single crystal X-band EPR spectrum of $[Cu_2(Me_5dien)_2(N_3)_2(BPh_4)_2$. After [7.19]

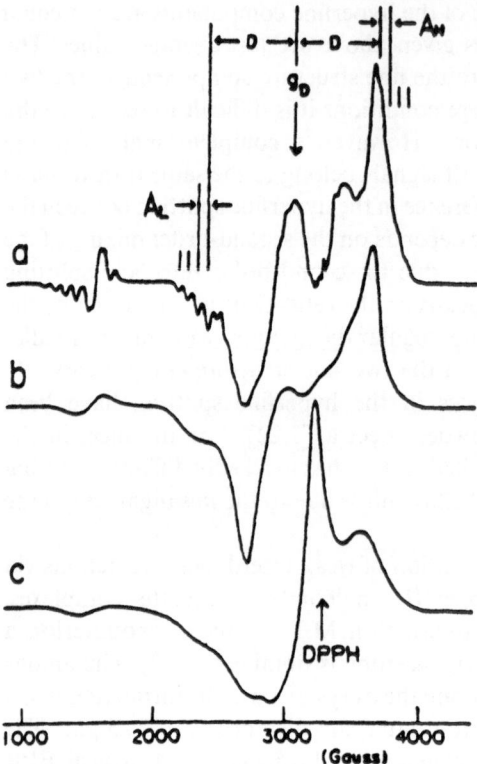

Fig. 7.18. X-band polycrystalline powder EPR spectra of bis (N-methyl salicylideneiminato) copper(II): *a* in toluene at 77 K; *b* and *c* polycrystalline powders at room temperature and 77 K, respectively. After [7.22]

split into seven by the copper hyperfine interaction with two copper nuclei. The two outermost transitions in Fig. 7.19, were attributed to pairs of dimers which are present in the lattice. The effective hamiltonian for these pairs can be written as:

$$H = J(S_1 \cdot S_2 + S_3 \cdot S_4) + J'(S_1 + S_2) \cdot (S_3 + S_4), \tag{7.13}$$

where J is the intra- and J' is the interdimer exchange-coupling constant. The energy levels of the pair of dimers can be schematized [7.24] as shown in Fig. 7.20. The lowest state corresponds to the two dimers in the ground singlet, the first excited state has one dimer in the singlet and the other in the triplet, while the highest state corresponds to both dimers in the triplet state. The interaction between the two triplets yields a singlet, a triplet, and a quintet. The $S = 0$ and $S = 2$ states can be admixed by either interdimer zero field splitting and hyperfine

180

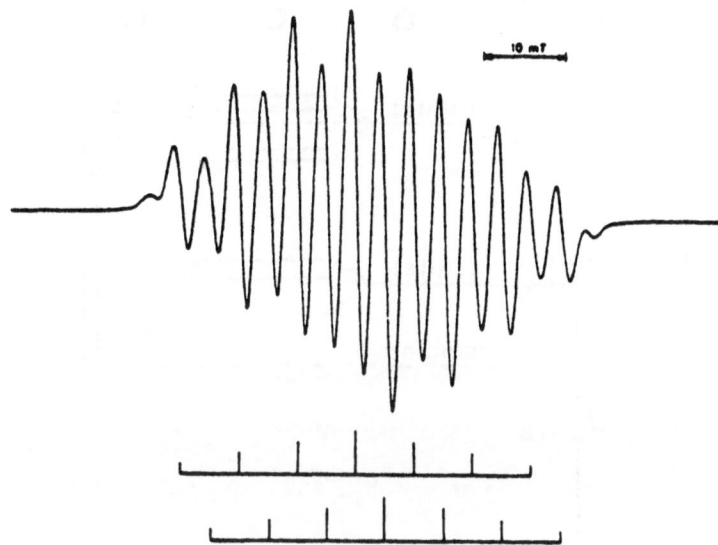

Fig. 7.19. EPR spectrum of (MB)[Cu(mnt)$_2$] with magnetic field in an arbitrary orientation. After [7.23]

coupling, while the triplet remains as a pure state. Therefore, the interdimer transitions can be labeled as triplet and mixed singlet-quintet, respectively.

The probability of observing n neighboring dimers in the triplet state is given by:

$$P_n = 2^{2-n} x^n (1-x)^{2-n}, \tag{7.14}$$

where n = 1, 2 and

$$x = 3 \exp(-J/kT)/[1 + 3 \exp(-J/kT)]. \tag{7.15}$$

The ratio P_2/P_1 is a very sensitive function of temperature as shown by the experimental spectra in the range 2–4.2 K (Fig. 7.21). The spectra at 2 K only show the isolated triplets. The splittings of the bands are due to the two different copper isotopes and they are washed out at higher temperatures due to the broadening of the bands. Eight transitions are observed because the zero field splitting in this orientation is equal to the hyperfine splitting. When the temperature is increased, additional absorptions are neatly resolved, which were attributed to the pairs of dimers. The complete analysis of these spectra showed that they are given by pairs of dimers clustering along the a axis. The isotropic interdimer coupling was found to be negative (ferromagnetic) and < 5 × 10^{-4} cm^{-1} in absolute value. The interdimer zero field splitting was assumed to be given by the point-dipolar approximation.

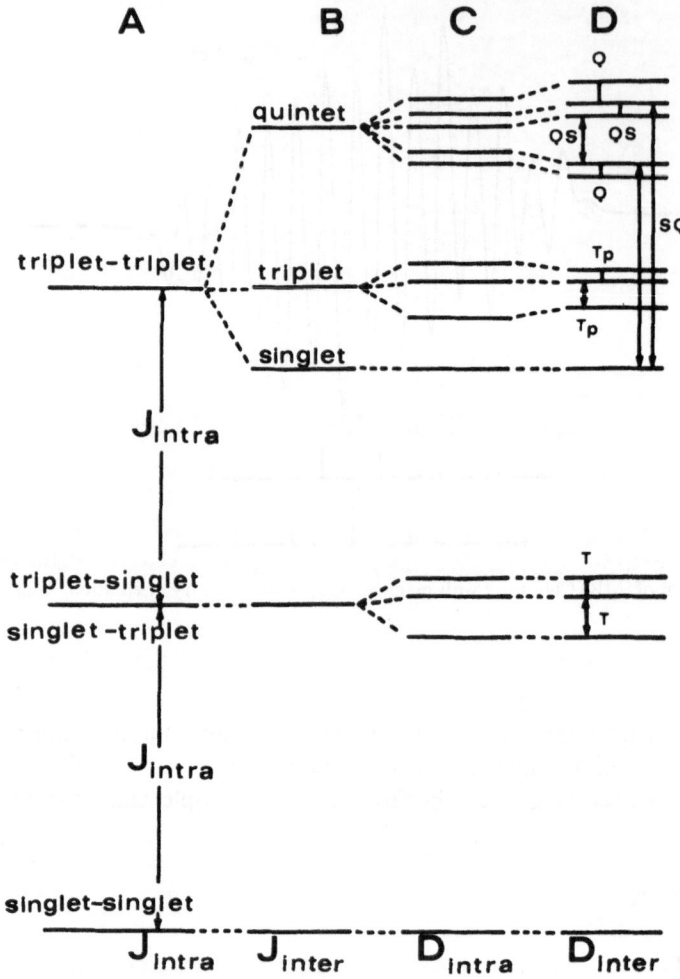

Fig. 7.20. Scheme of the energy levels of pairs of dimers present in $(MB)[Cu(mnt)_2]$. After [7.24]

7.3 Heterometallic Pairs

In this section we focus on pairs containing different metal ions. Studies on such systems are much less numerous than those on homonuclear pairs, although in the last few years they have attracted increasing interest [7.25–27]. We have already reported some examples in Sect. 7.1, and will cover here others, without any attempt to be completely exhaustive.

A relatively numerous class of heterometallic pairs includes high spin nickel(II) and copper(II) ions. In Table 7.2 are shown the g values for some pairs, and the g values calculated for the nickel(II) centers using the coefficients of Table 3.3 and the g values of the copper centers. Since the experimental spectra

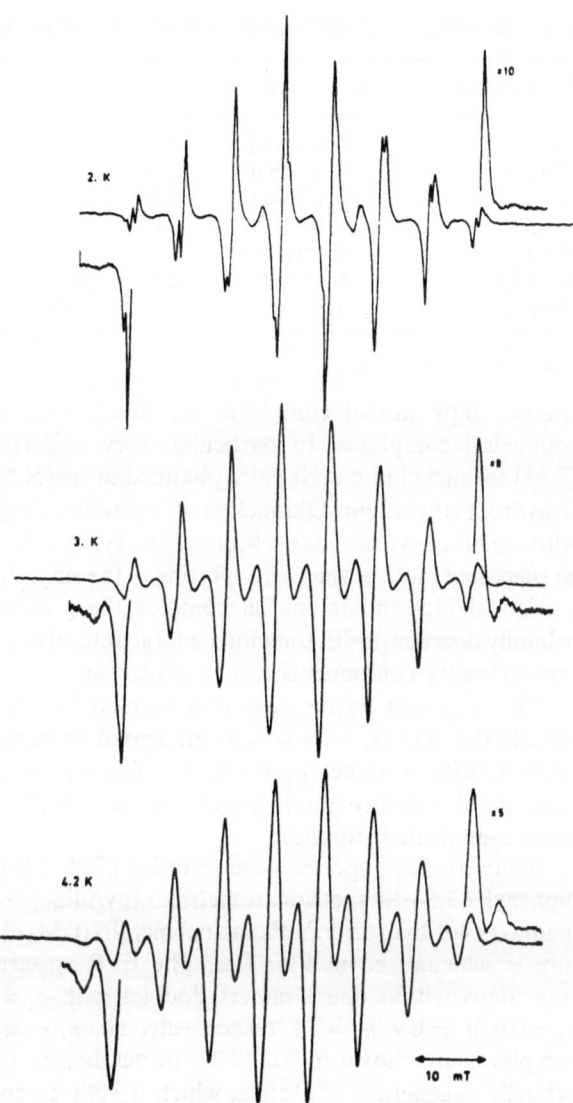

Fig. 7.21. Experimental single crystal EPR spectra of $(MB)[Cu(mnt)_2]$ in the range 2–4.2 K, with the static magnetic field parallel to a. After [7.24]

correspond to the $S = 1/2$ total spin state, the g values of the pair are given by $g = 4/3\, g_{Ni} - 1/3\, g_{Cu}$. In Sect. 3.6 we showed, however, how the single ion zero field splitting can induce variations in this scheme. An indirect check of the goodness of the simple formula is provided by the calculated g values for the nickel center: for instance, for an octahedral complex the **g** tensor is expected to be quasi-isotropic, and close to 2.2, and indeed the values in Table 7.2 appear to conform to this prediction. The calculated **g** tensors of square pyramidal and

Table 7.2. g values for nickel(II) complexes calculated from the spectra of Cu-Ni pairs

Chromophore	Coord. Geom.	g_1	g_2	g_3	Ref.
NiO_4N_2	Octahedral	2.18	2.20	2.21	[7.28]
NiN_4O_2	Octahedral	2.16	2.17	2.21	[7.29]
NiO_4O_2	Octahedral	2.20	2.24	2.24	[7.30]
NiO_4O_2	Octahedral	2.26	2.26	2.26	[7.31]
NiN_2O_2Cl	Sq. pyram.	2.12	2.32	2.38	[7.32]
NiO_2Cl_2O	Sq. pyram.	2.17	2.36	2.48	[7.33]
NiN_3O_2	Trig. bip.	2.06	2.09	2.61	[7.34]

trigonal bipyramidal complexes are much more anisotropic than those of octahedral complexes. In particular, very anisotropic values are calculated [7.34] for nickel in Cu-Ni pairs obtained in bis(N,N-bis(2-diethylamino)ethyl) (2-hydroxyethylamino-O)dinickel(II) diperchlorate [7.35], $Ni_2(bdhe)_2(ClO_4)_2$. Although these values may reflect the low symmetry of the complexes, it can also be suspected that the zero field splitting of the single ions can play some role. For a trigonal bipyramidal nickel complex the ground state (C_{3v} symmetry) is orbitally degenerate ^3E, therefore, a large interplay of the Jahn-Teller effect and low symmetry components can be suspected.

That this can be the case is confirmed by the spectra of the analogous cobalt(II)-nickel(II) pairs which correspond to transitions within one Kramers doublet, with g values $g_1 = 3.4$, $g_2 = 0.8$, $g_3 = 0.6$. These values cannot be reproduced with any simple formula of the type (3.34), and must reflect a much more complicated situation.

More tractable spectra were recorded [7.28, 29] for Co–Ni pairs in bis(1, 5-diphenyl-1,3,5-pentanetrionato) tetrakis (pyridine) dimetal(II), M_2trik(py)$_4$, and diaquo (1,4-dihydrazionphthalazine) metal(II), $M_2(dhph)(H_2O)_4$, whose structure is schematized in Fig. 7.22. The EPR spectra recorded at 4.2 K show transitions within one Kramers doublet with $g_z = 2.1$, $g_x = 1.2$, $g_y = 0.3$ and $g_z = 0.6$, $g_x = 0.9$, $g_y = 2.1$, respectively. The x, y, and z directions for the two complexes are shown in Fig. 7.23. In octahedral symmetry cobalt(II) has an orbitally degenerate $^4T_{1g}$ state, which is split by spin-orbit coupling and low symmetry components to yield a ground Kramers doublet. The simplest approach to rationalize the observed g values is to consider the ground Kramers doublet as an effective $S = 1/2$ spin and couple it with $S = 1$ for nickel. In this frame the g values of the pair might be associated with the total spin $S = 1/2$. However, use of the coefficients of Table 3.3 fails to give reasonable values, when the g values of the nickel and cobalt center, obtained from the spectra of the Cu-Ni and Co-Zn pairs, respectively, are used. A more sophisticated approach takes into account the orbital degeneration of the cobalt(II) ion, through a hamiltonian

$$H = H_{Co} + H_{Ni} + H_{CoNi}, \tag{7.16}$$

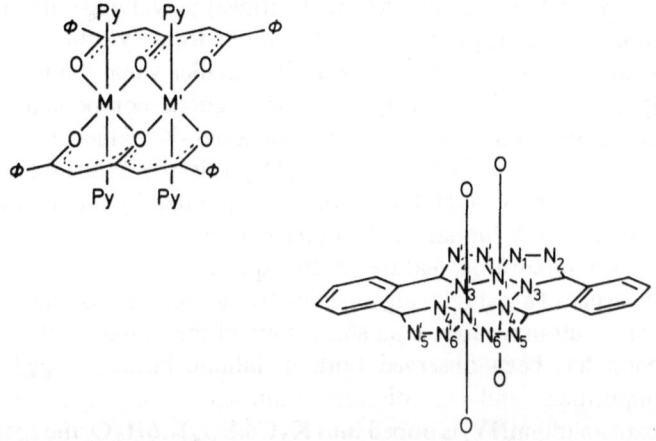

Fig. 7.22. Scheme of the structure of $M_2(trik)(py)_4$ and $M_2(dhph)_4(H_2O)_4$

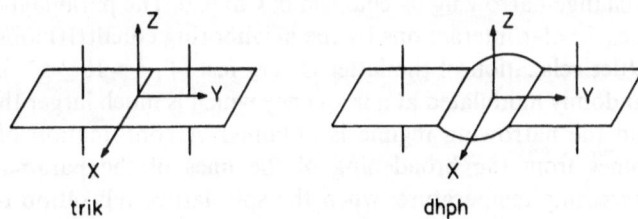

Fig. 7.23. Reference frames for $M_2(trik)(py)_4$ and $M_2(dhph)_4(H_2O)_4$

where H_{Co} and H_{Ni} are the hamiltonians appropriate to the single ion cobalt(II) and nickel(II) centers, including spin-orbit coupling and low symmetry effects. H_{CoNi} is the interaction hamiltonian, which, in the assumption of C_{2v} symmetry for the pair, can be written in the form [7.36] of Eq. (2.61):

$$H_{CoNi} = \Sigma_\Gamma J_\Gamma S_{Co} \cdot S_{Ni}, \qquad (7.17)$$

where the sum is over the irreducible representations Γ of the symmetry group of the pair, spanned by the direct product $A_{2g} \otimes T_{1g}$ (A_{2g} is the orbital symmetry of the ground state of octahedral nickel(II)). By diagonalizing the hamiltonian matrix of (7.16), including also the Zeeman terms, it is possible to fit the experimental g values. In order to reduce the number of parameters to a minimum, those relative to H_{Co}, H_{Ni}, and to the Zeeman hamiltonian were obtained independently from the analysis of the spectra of the Cu–Ni and Co–Zn pairs. Therefore, only the J_Γ parameters remain to be fitted. They were found to be: $J_{A1} = 4(4)$ cm^{-1}, $J_{A2} = 45(5)$ cm^{-1}, $J_{B2} = 45(5)$ cm^{-1} for CoNi(trik) (py)$_4$ and $J_{A1} = J_{A2} = J_{B2} = 30(10)$ cm^{-1} for CoNi(dhph) (H$_2$O)$_4$. The J_Γ parameters

185

can be decomposed into a sum of pathways involving different magnetic orbitals: in particular, J_{A1} contains contributions from xy orbitals on cobalt interacting with $x^2 - y^2$ orbitals on nickel. The smaller value found for J_{A1} in CoNi(trik) $(py)_4$ may be justified by the ferromagnetic component associated with this exchange mechanism (see the Goodenough-Kanamori rules of Sect. 1.2). On the other hand, in CoNi(dhph) $(H_2O)_4$ this mechanism becomes less efficient, because the two metal ions are now separated by a polyatomic bridge, and J_{A1} becomes very similar to the other two.

An interesting feature of the spectra of heteronuclear pairs is that the presence of a fast relaxing ion close to another one, to which it is weakly coupled, can result in a substantial sharpening of the signals of the latter. This phenomenon has been observed both in infinite lattices doped with paramagnetic impurities, and in discrete dinuclear complexes. For instance, when oxovanadium(IV) is doped into $K_2 Co(SO_4)_2.6H_2O$, the spectrum shown in Fig. 7.24 is observed at room temperature [7.37]: the lines are as narrow as they would be expected to be in a diamagnetic lattice, and the ^{51}V hyperfine is clearly resolved. The interpretation of this interesting phenomenon is bound to the exchange-narrowing mechanism of Chap. 6. The paramagnetic impurity undergoes dipolar interactions by the neighboring cobalt(II) ions, but, since the spin lattice relaxation of the latter is very fast ($T_1^{-1} > 10^{11}$ s^{-1}), the dipolar field is randomly modulated at a frequency which is much larger than the perturbation and the narrowing regime is obtained. A confirmation of this interpretation comes from the broadening of the lines of the paramagnetic impurity on decreasing temperature: when the spin lattice relaxation of the cobalt(II) ion becomes slower, the narrowing regime is no longer obtained.

An interesting example of a triplet spectrum in a heterodinuclear complex is provided by $CuVO(fsa)_2 en.CH_3OH$, where $(fsa)_2 en^{4-}$ is N,N',-(2-hydroxy-3-carboxybenzilidene)-1,2-diamino ethane [7.38]. This compound has already been mentioned in Chapter 1 as an example of a moderate ferromagnetic coupling. Its powder EPR spectrum has been interpreted with $|D| = 0.24$ cm^{-1}

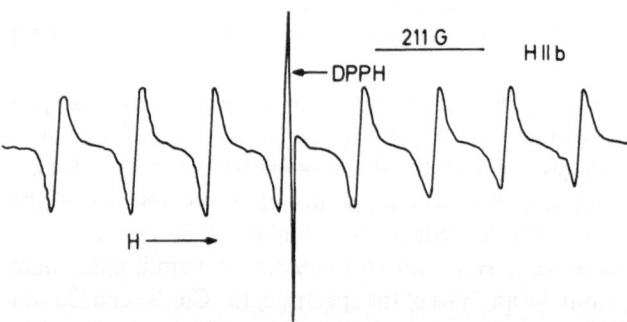

Fig. 7.24. Room temperature single crystal X-band EPR spectrum of oxovanadium(IV) doped $K_2Co(SO_4)_2 \cdot 6H_2O$. After [7.37]

and $E/D \approx 0.17$. Since the copper-vanadium distance seen in the crystal structure is 298.9 pm, the dipolar value of D is not expected to exceed 0.1 cm^{-1}, therefore, even in this case exchange contributes significantly.

7.4 Organic Biradicals

Biradicals are formed by the linkage of two molecular fragments, each containing one unpaired electron. Several different types of such molecules have been synthesized and characterized, but perhaps the most numerous class is that of bis-nitroxides, and we will refer mainly to these in the following. An excellent review of their properties has been given by Luckhurst [7.38], and we will often refer to this review.

In the study of nitroxide biradicals it is mainly the fluid solution spectra which are relevant, due to the possible use of these molecules as spin probes, although data are available also for spectra in oriented matrices. The scalar hamiltonian can be written as:

$$H = g_1 \mu_B B S_{1z} + g_2 \mu_B B S_{2z} + A(S_{1z}I_{1z} + S_{2z}I_{2z}) + JS_1 \cdot S_2, \tag{7.18}$$

In (7.18) it was assumed that the hyperfine interactions are small compared to the electron Zeeman splitting and nonsecular hyperfine terms were neglected.

The electronic states in (7.18) can be grouped as singlet and triplet. If the two g_i values are identical, and the hyperfine can be neglected, the total spin S is a good quantum number, and only transitions within the triplet state can be detected. However, if at least one of the above conditions does not apply, then the two total spin levels with $M = 0$ can be admixed, and the number of observed transitions increases. Assuming that the g_i values are identical, the resonance fields are given by:

$$B_r = \tfrac{1}{2} A m \pm |A \, \delta m| [(1 - kL)/(1 + kL)], \tag{7.19}$$

where A is in gauss, $m = m_1 + m_2$, and $\delta m = m_1 - m_2$, m_1 and m_2 are the nuclear spin components,

$$L = R/\{|\delta m| + [|\delta m|^2 + 1]^{\frac{1}{2}}\}, \tag{7.20}$$

$R = J/A$ and k can be either ± 1. The upper sign applies to the transitions involving the $M = 0$ state which has a more triplet character, and the lower sign to those involving a more singlet character. The relative intensities of the transitions are given by:

$$I = 1 + [2kL/(1 + L^2)]. \tag{7.21}$$

For nitroxides the largest hyperfine interaction is with ^{15}N, $I = 1$. A stick plot showing the transition fields and the relative intensities for various values of R is given in Fig. 7.25. When R = 0 three transitions, corresponding to the interaction of the unpaired electrons with each of the three m_i components, are observed. For $0 < R \leqslant 4/3$, 15 transitions can be observed, 9 of the "triplet" and 6 of the "singlet". The singlet transitions are separated from the others by roughly the exchange energy, so that for $R > 4/3$ all the singlet resonances lie outside the main spectrum. On increasing R the intensities of the singlet transitions decrease until eventually only the triplet transitions can be observed. For $R \to \infty$ the spectrum shows five lines with relative intensities $1:2:3:2:1$, as expected in the strong exchange limit. Two representative examples of spectra are shown in Fig. 7.26. The spectrum of Fig. 7.26a shows the resolved lines, from which the exchange integral can be calculated [7.39].

It must be stressed here that the above ratios of intensities are only valid if J is time-independent, while matters can be much different in the case when J is

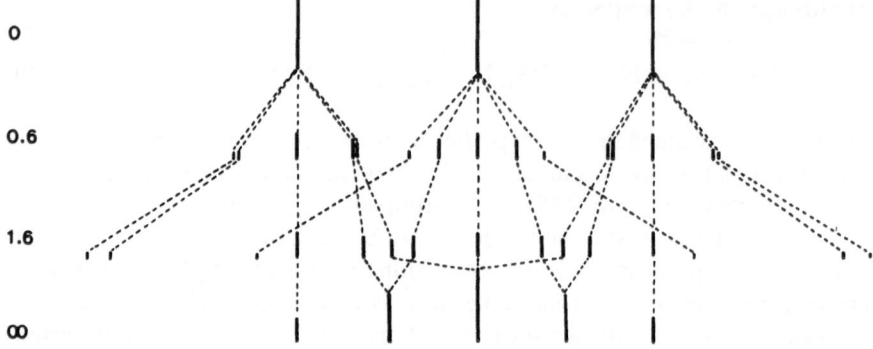

Fig. 7.25. Stick plot diagram of the transition fields of dinitroxide radicals as a function of $R = J/A$. The heights of the sticks are proportional to the relative intensities of the transitions

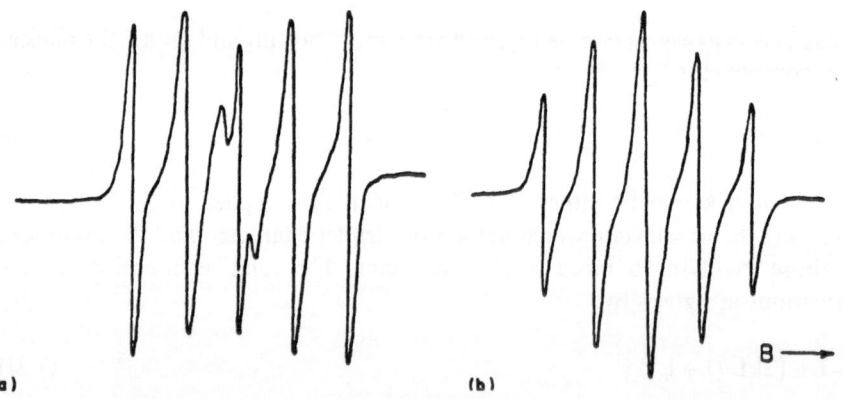

Fig. 7.26. EPR spectra of two different biradicals dissolved in tetrahydrofuran. After [7.39]

modulated by some time-dependent perturbation [7.40]. This can occur, for instance, for a flexible biradical such as the one shown in Fig. 7.27. The intramolecular motion of the biradical yields nitroxide-nitroxide distances which are time-dependent, and consequently time-dependent J. According to (7.21), the transitions which are most affected by the R ratio, i.e., by the time modulation of

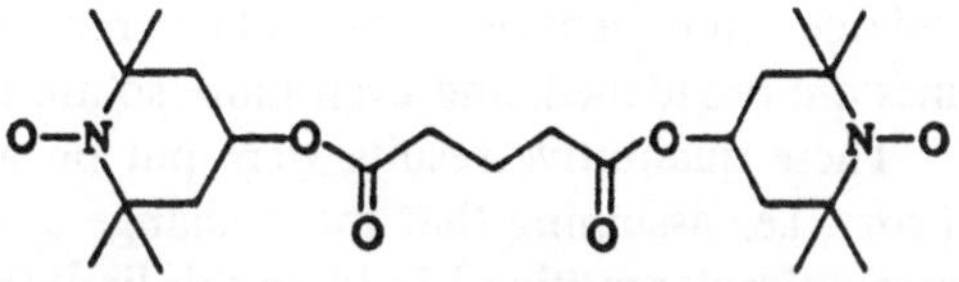

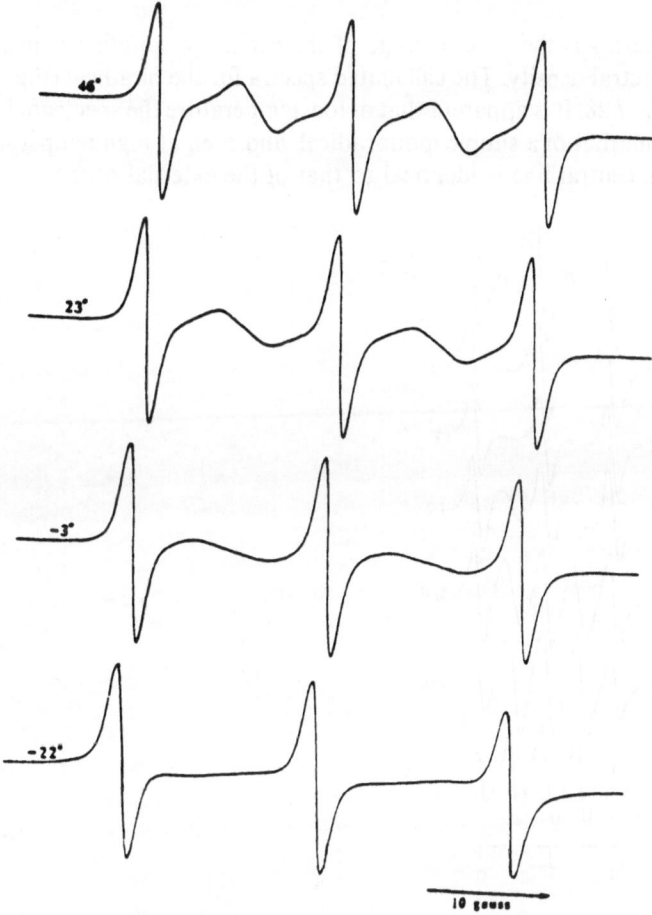

Fig. 7.27. Scheme of the structure of a flexible dinitroxide radical

Fig. 7.28. Calculated spectra of the radical of Fig. 7.27 as a function of temperature. After [7.41]

J, are, in order of increasing effect, the ones for which $\delta m = \pm 1$ and those for which $\delta m = \pm 2$. On the other hand, the transitions with $\delta m = 0$ are independent of R. Therefore, in a qualitative way we may predict that the line width of the $\delta m = 0$ transitions are unaffected by the modulation of J, while the $\delta m = \pm 1$ and $\delta m \pm 2$ will be broadened, the latter much more than the former. As a consequence, line width alternation of the five lines in the strong exchange limit can be expected for flexible biradicals. In fact, the lowest and highest field lines and one component of the central line remain unaffected, while the intermediate lines are broadened, and even more so are two components of the central line.

These qualitative results were put on a quantitative basis using Redfield theory, i.e., assuming that the exchange coupling is modulated rapidly by the intramolecular motion [7.41]. In this limit the line width is expected to be given by:

$$T_2^{-1}(\delta m) = [A^2 (\delta m)^2 / 4 J^2] j(J), \tag{7.22}$$

where J is the time average of the exchange-coupling constant and j(J) is the spectral density. The calculated spectra for the biradical (Fig. 7.27) are shown in Fig. 7.28. It is apparent that at low temperature the spectrum is indistinguishable from that of a simple monoradical, and even at high temperatures the height of the central line is identical to that of the external ones.

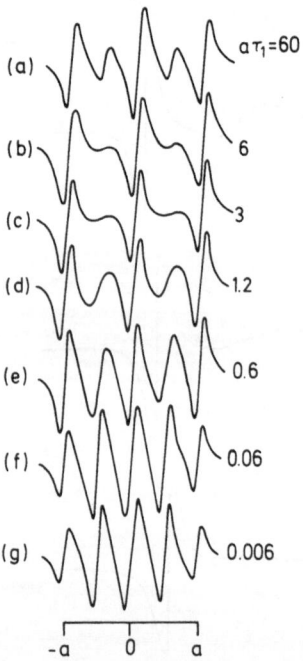

(a) $a\tau_1 = 60$

(b)   6

(c)   3

(d)   1.2

(e)   0.6

(f)   0.06

(g)   0.006

    -a   0   a

Fig. 7.29. Calculated spectra of a dinitroxide radical existing in two configurations. τ_1 is the lifetime in each configuration, a is the hyperfine splitting parameter. After [7.42]

The model has been extended also to the case of slow motion. In Fig. 7.29 the results of calculations within this generalized model are shown [7.42]. The nitroxide is supposed to inter-convert between two conformations with equal lifetimes: in one configuration the exchange integral is zero and in the other it is 30 times the hyperfine coupling constant. The calculated spectra are obtained for a lifetime parameter $a\tau_1$.

Calculations are available also for the weak exchange limit, where the main time-dependent perturbation is the modulation of anisotropic exchange.

Another type of interesting biradicals is provided by radical anions bound to a diamagnetic metal ion [7.43]. Examples of such radical anions are obtained by alkaline earth reduction of 1,1,4,4-tetramethyltetralin-2,3-dione: bis-complexes of the type shown in Fig. 7.30 are obtained, which give triplet EPR spectra. Data collected in frozen solutions show D values in the range 40–250 G, with small E/D ratios. The observed zero field splitting was justified within a simple model which considers the effect of delocalization of the unpaired electrons in the π orbitals of the ligands. The conformation of the complexes giving the best fit between the experimental and the calculated D and E values was chosen.

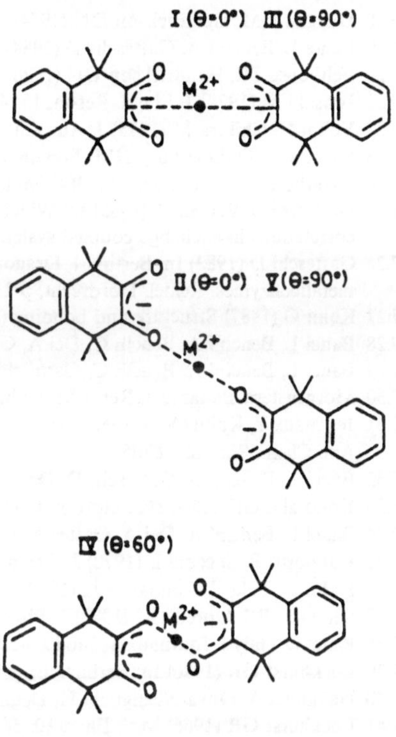

Fig. 7.30. Conformation of alkaline earth complexes giving the best fit between experimental and calculated D and E values. Θ is the angle of rotation of one radical relative to the other around the dashed line. After [7.44]

191

References

7.1   Harris EA (1972) J. Phys. C 5: 332
7.2   McPherson GL, Nodine MH, Devaney KO (1978) Phys. Rev. 18: 6011
7.3   McPherson GL, Heung W, Barraza JJ (1978) J. Am. Chem. Soc. 100: 469
7.4   McPherson GL, Varga JA, Nodine MH (1970) Inorg. Chcm. 18: 2189
7.5   Belford RL, Chasteen ND, So H, Tapscott RE (1969) J. Am. Chem. Soc. 91: 4675
7.6   Smith SRP, Owen J (1971) J. Phys. C 4: 1399
7.7   Fieselmann BM, McPherson AM, McPherson GL (1978) Inorg. Chem. 17: 1841
7.8   Setukowski D, Jungst R, Stucky GD (1978) Inorg. Chem. 17: 1848
7.9   Fieselmann BF, Hendrickson DN, Stucky GD (1978) Inorg. Chem. 17: 2078
7.10 Kramer LS, Clauss AW, Francesconi LC, Corbin DR, Hendrickson DN, Stucky GD (1981) Inorg. Chem. 20: 2070
7.11 Francesconi LC, Corbin DR, Clauss AW, Hendrickson DN, Stucky GD (1981) Inorg. Chem. 20: 2078
7.12 Francesconi LC, Corbin DR, Clauss AW, Hendrickson DN, Stucky GD (1981) Inorg. Chem. 20: 2059
7.13 Corbin DR, Francesconi LC, Hendrickson DN, Stucky GD (1981) Inorg. Chem. 20: 2084
7.14 Bleaney B, Bowers KD (1952) Proc. R. Soc. A214: 451
7.15 van Niekerk, Schoening FRL (1953) Acta Cryst. 6: 227
7.16 Julve M, Verdaguer M, Gleizes A, Philoche-Levisalles M, Kahn O (1984) Inorg. Chem. 23: 3808
7.17 Bencini A, Gatteschi D, Zanchini C, Kahn O, Verdaguer M, Julve M (1986) Inorg. Chem. 25: 3181
7.18 Duggan DM, Hendrickson DN (1974) Inorg. Chem. 13: 2929
7.19 Banci L, Bencini A, Gatteschi D (1984) Inorg. Chem. 23: 2138
7.20 Felthouse TR, Hendrickson DN (1978) Inorg. Chem. 17: 444
7.21 Iwasaki M (1974) J. Magn. Reson. 16: 417
7.22 Yokoi H, Chikira M (1975) J. Am. Chem. Soc. 97: 3975
7.23 Snaathorst D, Doesburg HM, Perenboom JA, Keijzers CP (1981) Inorg. Chem. 20: 2526
7.24 Snaathorst D, Keijzers C P (1984) Mol. Phys. S1: 509
7.25 Gatteschi D, Bencini A (1985) In: Willett RD, Gatteschi D, Kahn O (eds) Magneto-Structural correlations in exchange coupled systems. Willett RD, Reidel, Dordrecht, p 241
7.26 Gatteschi D (1983) In: Bertini, I, Drago RS, Luchinat C (Eds) The coordination chemistry of metalloenzymes. Reidel, Dordrecht, p 215
7.27 Kahn O (1987) Structure and bonding (Berlin) 68: 89
7.28 Banci L, Bencini A, Benelli C, Dei A, Gatteschi D (1981) Inorg. Chem. 20: 1399
7.29 Banci L, Bencini A, Benelli C, Gatteschi D (1982) Inorg. Chem. 21: 3868
7.30 Morgenstern-Badarau I, Rerat M, Kahn O, Jaud K, Galy J (1982) J. Inorg. Chem. 21: 3050
7.31 Journaux Y, Kahn O, Morgenstern-Badarau I, Galy J, Jaud J, Bencini A, Gatteschi D (1985) J. Am. Chem. Soc. 107: 6305
7.32 Banci L, Bencini A, Gatteschi D, Dei A (1979) Inorg. Chim. Acta 36, L419
7.33 Kokoszka GF, Allen HC, Gordon G (1967) J. Chem. Phys. 46: 3020
7.34 Banci L, Bencini A, Dei A, Gatteschi D (1981) Inorg. Chem. 20: 393
7.35 Dapporto P, Sacconi L (1970) J. Chem. Soc. A 681
7.36 Kahn O, Tola P, Coudanne H (1979) Chem. Phys. 42: 355
7.37 Saraswat RS, Upreti GC (1978) J. Phys. Soc. Jpn. 44: 1142
7.38 Kahn O, Galy J, Journaux Y, Jaud J, Morgenstern-Badarau (1982) J. Am. Chem. Soc. 104: 2165
7.39 Luckhurst GR (1976) In: Berliner LJ (ed) Spin labeling. Academic, New York, p 133
7.40 Nakajima A, Ohya-Nishiguchi H, Deguchi Y (1973) Bull. Chem. Soc. Jpn. 45: 713
7.41 Luckhurst GR (1966) Mol. Phys. 10: 543
7.42 Luckhurst GR, Pedulli GF (1970) J. Am. Chem. Soc. 92: 4738
7.43 Parmon VN, Zhidomirov GM (1974) Mol. Phys. 27: 367
7.44 Brustolon M, Pasimeni L, Corvaja C (1981) JCS Dalton 604

8 Coupled Transition-Metal Ions-Organic Radicals

8.1 Introduction

Systems in which a transition metal ion is directly bound to a stable organic radical are still relatively rare, but the number of examples reported in the literature has increased in the last few years. There are several reasons for this increased interest, the first being of course a theoretical one, because bringing into close contact two atoms formally carrying unpaired electrons can allow one to study direct exchange interactions rather than indirect superexchange. Presumably the direct interactions can be fairly strong, and rather peculiar magnetic behaviors can be expected to arise.

The stimulating influence of biological studies cannot be ignored also for this class of compounds: in fact, a number of systems are known to contain organic radicals and transition metal ions actually or potentially interacting. Just as examples it is possible to refer to the oxygen-evolving photosynthetic system or to enzymes such as ribonucleotide reductase, as discussed in Chap. 9. Another possibility is given by the interaction of the widely used spin probes and spin labels with paramagnetic metal ions [8.1].

Although in principle any organic radical can be induced to interact with a metal ion, the most widely studied systems at the moment are the nitroxides and the semiquinones. Both classes of organic radicals are relatively stable, and can interact with the metal ions via the oxygen atoms, which can act as donors. In this case if the overlap between the magnetic orbitals is large a normal covalent bond will be formed and spins are paired. If, on the other hand, the overlap is small or zero then direct exchange can be operative, yielding either antiferro- or ferromagnetic coupling, as will be shown in the following sections. However, quite numerous are also the examples of weakly coupled systems in which the atom(s) bearing the unpaired electrons on the radical are not directly bound to the metal ion, but nevertheless the spins are interacting, either through superexchange or by magnetic dipolar coupling.

8.2 Nitroxides Directly Bound to Metal Ions

Nitroxides, which we have already mentioned several times, are well-known stable organic radicals in which the NO group bearing the unpaired electron is protected by bulky methyl groups (Fig. 8.1). In the free ligand the unpaired electron is in a π^* orbital and spends approximately 50% of its time on the

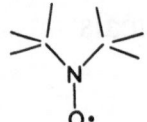

Fig. 8.1. Skeleton of a nitroxide

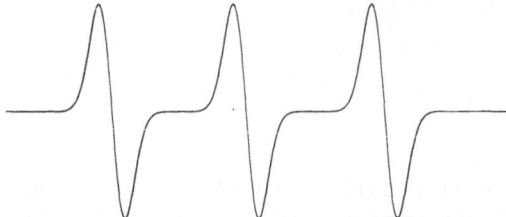

Fig. 8.2. Typical fluid solution spectrum of a nitroxide

Table 8.1. Principal values of the **g** and **A** tensors for some typical nitroxides[a]

Structure	g_{xx}	g_{yy}	g_{zz}	A_{xx}	A_{yy}	A_{zz}
	2.0088	2.0062	2.0027	7.6	6.0	31.8
	2.0103	2.0069	2.0030	—	—	—
	2.0095	2.0064	2.0027	—	—	—
	2.0104	2.0074	2.0026	5.2	5.2	31
	2.0088	2.0058	2.0032	—	—	31
	2.0088	2.0058	2.0022	5.9	5.4	32.9
	2.0088	2.0061	2.0027	6.3	5.8	33.6

[a] The x axis is parallel to the NO direction, z is orthogonal to the plane C–NO–C. The hyperfine components in gauss. After [8.4].

nitrogen atom [8.2], as shown by the fluid solution EPR spectra, which consist of a triplet, due to the interaction with the ^{14}N nucleus (Fig. 8.2). The **g** tensor is only slightly anisotropic, as expected, but the hyperfine tensor is highly anisotropic, the largest component being observed orthogonal to the plane defined by the NO group and the neighboring carbon atoms. In Table 8.1 the principal g and A values for some representative nitroxides are reported.

The oxygen atom of the nitroxide can bind to a metal ion, and quite a few different complexes have been reported up to now [8.4]. Perhaps the most numerous are the copper(II) complexes, which therefore give the best opportunity to rationalize the structural dependence of the exchange interaction. In fact, several square pyramidal copper(II) complexes with nitroxides have been

194

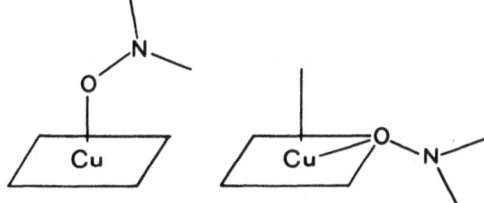

Fig. 8.3. The coordination geometries observed in copper(II)-nitroxide complexes

reported [8.5], and the two limiting structures can be schematized as shown in Fig. 8.3, i.e., with the radical occupying an axial or an equatorial coordination site. In both limits the magnetic orbital on copper(II) can be loosely described as xy, lying in the equatorial plane of the pyramid with the lobes pointing toward the ligands. When the nitroxide occupies an axial position, its π^* magnetic orbital is essentially orthogonal to the magnetic orbital on copper(II), therefore, the coupling is expected to be ferromagnetic. Indeed, this has been found to be the case for a number of complexes, in which magnetic susceptibility data showed the existence of a ferromagnetic coupling of 20–70 cm^{-1} [8.5]. On the other hand, when the nitroxide is in the equatorial plane, then there can be direct overlap with the magnetic orbital of copper(II), and consequently a large antiferromagnetic pairing of the spins. Indeed, this has been found to be the case in all the systems studied so far.

A beautiful example of a triplet spectrum has been reported [8.6] for Cu(hfac)$_2$ (TEMPOL), where TEMPOL is 4-hydroxy-2,2,6,6-tetramethyl-piperidinyl-1-oxy. This compound has been shown [8.7] to have the linear chain structure of Fig. 2.5. The EPR spectra of the solid at room temperature are shown in Fig. 8.4. They are indeed typical of a triplet with large differences between the X-band and the Q-band data, suggesting that the zero field splitting is fairly large. The Q-band spectrum shows clearly separated $\Delta M = \pm 2$ and $\Delta M = \pm 1$ transitions and at least five features, while the X-band spectrum shows only three features. This means that the zero field splitting is small compared to the microwave quantum of the Q-band and comparable to that of the X-band. Further, since at Q-band frequency the $\Delta M = \pm 1$ transitions give one feature with the simple derivative shape and three bumps, it may be assumed that the zero field splitting tensor is completely rhombic. In fact, using (7.4-12) it is easy to check that when E/D = 1/3, the two resonance fields B_{x_1} and B_{x_2} become identical. In a polycrystalline powder spectrum the corresponding feature will have a simple derivative shape.

This assignment is confirmed by single crystal spectra, which yielded $g_{xx} = 2.057(4)$, $g_{yy} = 2.013(4)$, $g_{zz} = 2.157(3)$, $D_{xx} = 0.008(1)$, $D_{yy} = -0.144(1)$, and $D_{zz} = 0.105(2)$ cm^{-1}, D = 0.171 cm^{-1}, and E/D = 0.33. The principal axes of g and D are defined in Fig. 8.5: the two sets of axes are practically parallel to each other, with z along the copper-axial oxygen bond direction, and x in between the chelate angle. The triplet EPR spectrum shows that although the structure of Cu(hfac)$_2$ TEMPOL is that of a linear chain, the magnetic properties are better approximated by a localized picture of one copper ion interacting with one

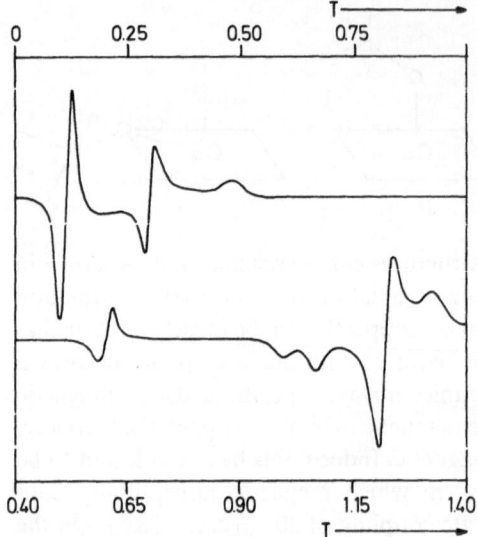

Fig. 8.4. Polycrystalline powder EPR spectra of $Cu(hfac)_2(TEMPOL)$ at room temperature. *Upper* X-band; *lower* Q-band frequency. After [8.6]

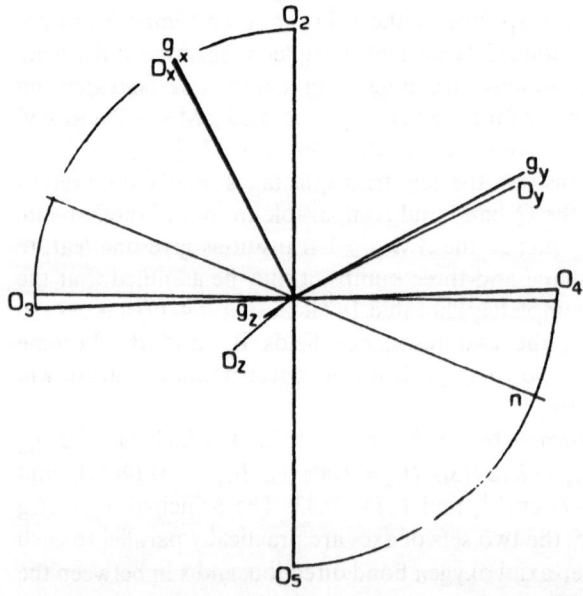

Fig. 8.5. Principal directions of the **g** and **D** tensors of $Cu(hfac)_2(TEMPOL)$ projected onto the molecular plane. After [8.6]

196

nitroxide group to which it is directly bound. Indeed, the radical interacts also with the copper atom which is bound to the OH group, but this interaction is much weaker and is not able to exchange-narrow the triplet spectra. Magnetic susceptibility measurements [8.8] suggested that the weak coupling constant is of $0.0054 \, \text{cm}^{-1}$, thus justifying the fact that a triplet spectrum is observed.

The analysis of the zero field splitting tensor is particularly problematic in this case, as was outlined in Sect. 2.2, because no simple estimation of the magnetic through-space contribution can be made. In fact, the close proximity and the delocalization of the unpaired spins makes the point-dipolar approximation in principle unsuitable here. A rough estimation of the dipolar zero field splitting can be made, however, by setting a charge in the baricenter of the NO group and using the dipolar approximation. In this way we calculate $D = 0.0955 \, \text{cm}^{-1}$, $E/D = 0$. We have found that in several dinitroxides this simple treatment gives values within 10% of the experimental values, and not too dissimilar from the values calculated with MO treatments which explicitly take into account spin delocalization. Therefore, the large observed rhombic splitting must originate from exchange contributions, since the magnetic through-space term should in any case be axial to the largest component parallel to the metal-axial oxygen direction.

Beyond copper(II) systems a rather thorough characterization has been performed also on bis-nitroxide adducts of manganese(II) hexafluoroacetylace-tonate, $Mn(hfac)_2$. Two complexes of the general formula $Mn(hfac)_2(\text{nitroxide})_2$ with the structure schematized in Fig. 8.6 were reported to show antifer-romagnetic coupling [8.9, 10] between the manganese and the nitroxides. When the nitroxide is proxyl (proxyl = 2,2,5,5,-tetramethylpyrrolidinyl-1-oxy), J was found to be $210 \, \text{cm}^{-1}$, while with TEMPO (TEMPO = 2,2,6,6,-tetra-methylpiperidinyl-1-oxy), $J = 158 \, \text{cm}^{-1}$, Manganese(II) is in the high spin d^5 ground configuration so that five magnetic orbitals must be present on the metal ion. At least one linear combination of them must have the correct symmetry for overlapping with the magnetic orbitals of the two nitroxides. A ground $S = 3/2$ state is anticipated, and this prediction is confirmed by the experiment, but the finite value of J indicates that the pairing is not complete and that manganese(II) and the radical are exchange-coupled rather than bound by a strong covalent

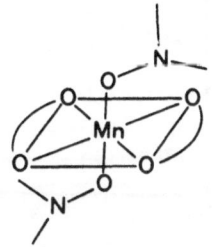

Fig. 8.6. The structure of bis-nitroxide adducts of $Mn(hfac)_2$

bond even if the Mn–O bond distances are reasonably short, 215.0(4) and 212.7(4) pm for proxyl and TEMPO, respectively. The energy separations of the excited levels are given in Fig. 8.7.

The EPR spectra of these compounds confirmed nicely the above scheme of energy levels obtained by magnetic susceptibility measurements. In Fig. 8.8 are shown the X-band spectra of Mn(hfac)$_2$(TEMPO)$_2$ at room temperature and at

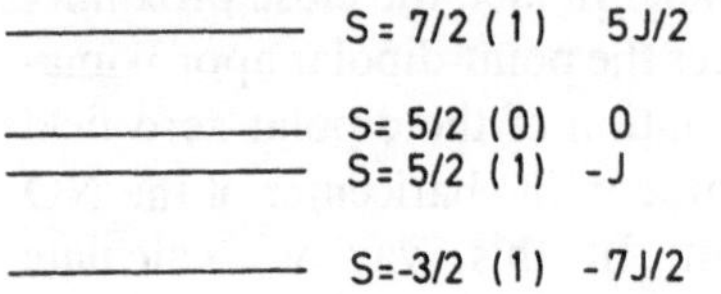

$$
\begin{array}{ll}
\text{———————} \quad S = 7/2 \; (1) & 5J/2 \\[6pt]
\text{———————} \quad S = 5/2 \; (0) & 0 \\
\text{———————} \quad S = 5/2 \; (1) & -J \\[6pt]
\text{———————} \quad S = -3/2 \; (1) & -7J/2
\end{array}
$$

Fig. 8.7. Spin energy levels for Mn(hfac)$_2$(proxyl)$_2$. After [8.8]

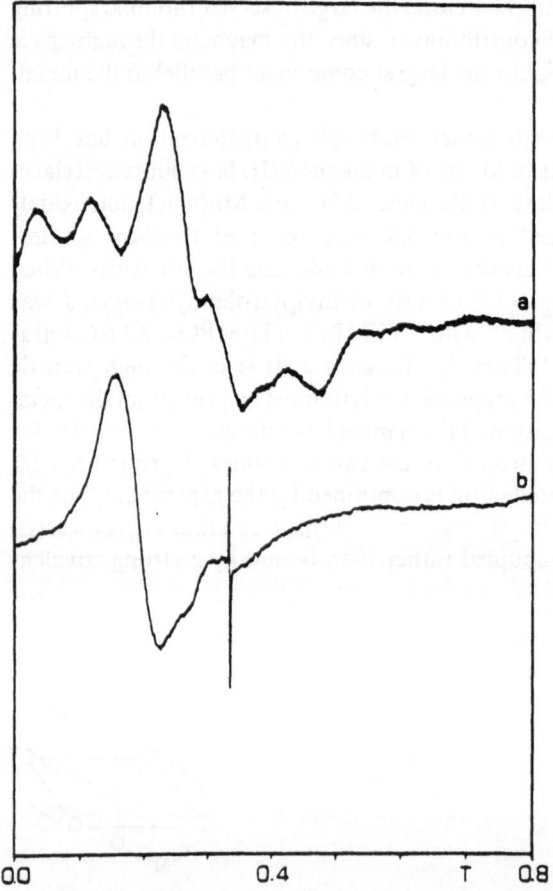

0.0   0.4   T   0.8

Fig. 8.8. Polycrystalline powder EPR spectra of Mn(hfac)$_2$(TEMPO)$_2$ at X-band frequency. *a* 300 K; *b* 4.2 K. After [8.8]

4.2 K, respectively. The low temperature spectra, confirmed also by single crystal data, have $g_{||} \approx 2$ and $g_{\perp} \approx 4$, as can be expected for a ground quartet state, with a zero field splitting which is larger than the microwave quantum. In cases like this it is extremely useful to record spectra at different frequencies, and in Fig. 8.9 we show the Q-band spectra at room temperature and at ca. 140 K. At the latter temperature the excited multiplets are essentially depopulated, so that the spectrum is again that of the ground quartet. It is apparent that they are much different in appearance from the X-band ones, showing many more transitions, in agreement with a zero field splitting which is smaller than the microwave quantum at Q-band frequency. The complete analysis of the EPR data at low and room temperature provided only an estimation of the spin hamiltonian parameters for the ground quartet and the first excited sextet, due to the broadness of the bands. For the ground quartet it was found $D = 0.63$ cm^{-1} and $E/D = 0.115$, while for the first excited sextets $D \approx 0.075$ cm^{-1}.

The experimental zero field splitting tensors were analyzed in terms of those of the individual manganese(II) ions, $\mathbf{D}_{Mn}$, and of the exchange-determined tensors, using the relations:

$$\mathbf{D}_{3/2} = 28/15\, \mathbf{D}_{Mn} - 14/30\, \mathbf{D}_{Mn-r} + 1/30\, \mathbf{D}_{r-r}; \tag{8.1}$$

$$\mathbf{D}_{5/2} = 23/35\, \mathbf{D}_{Mn} + 8/35\, \mathbf{D}_{Mn-r} - 2/35\, \mathbf{D}_{r-r}; \tag{8.2}$$

where $\mathbf{D}_{Mn-r}$ and $\mathbf{D}_{r-r}$ are the contributions determined by the manganese-radical and the radical-radical interactions, respectively. Neglecting $\mathbf{D}_{r-r}$, on the assumption that it must be much smaller than the other two, because of the

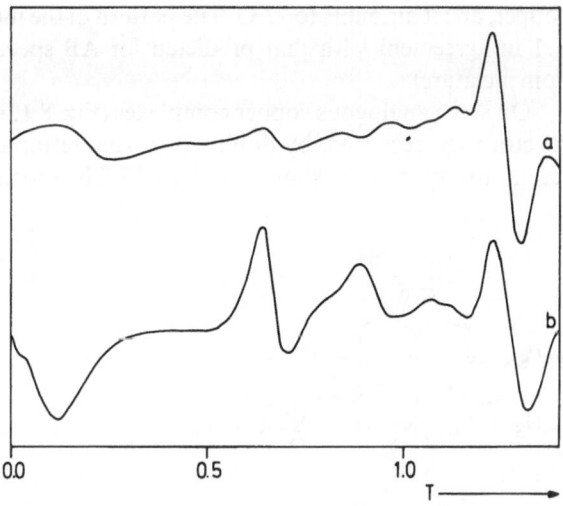

Fig. 8.9. Polycrystalline powder EPR spectra of Mn(hfac)$_2$(TEMPO)$_2$ at Q-band frequency. *a* 300 K; *b* 140 K. After [8.8]

longer radical-radical distance, the two Eqs. (8.1-2) allow us to calculate D_{Mn} and D_{Mn-r} from the experimental values observed for the quartet and the sextet levels. The values thus obtained ($D_{Mn-r} \approx -0.36$ cm^{-1} and $D_{Mn} \approx 0.25$ cm^{-1}) agree quite well with those reported for simple manganese(II) complexes with diamagnetic ligands, and with the value reported for Cu(hfac)$_2$TEMPOL.

8.3 Weak Exchange with Nitroxides

Very extensive work has been performed on transition metal complexes weakly interacting with various nitroxides [8.3]. By this we mean systems in which the distance of the metal from the NO group is too large to consider the two as directly bound, and yet they are interacting to some extent. When J is small, the separate lines of the radical and the metal ion can be observed, but the radical lines are generally broadened by the spin-spin interaction with the metal. If the latter is slowly relaxing in fluid solution, such as copper(II), oxovanadium(IV), manganese(II), gadolinium(III), etc., the spin-spin effects on the radical can be easily detected [8.10–15]. One example of such a spectrum is provided by a spin-labeled copper porphyrine, whose structure is depicted in Fig. 8.10. When R $=OC_2H_5$, the spectrum consists of four copper lines, unequally broadened by incomplete motional averaging (Fig. 8.11a). The observed splitting of the high field feature is due to the interaction with the four nitrogen atoms of the porphyrin. The spectrum of the spin-labeled derivative Fig. (8.11b, c) is much different from the sum of the spectra of the copper porphyrin and of the nitroxide, showing a doublet of triplets in the nitroxide region, and a greatly broadened copper spectrum. The splitting of the characteristic triplet of the nitroxide is clearly due to the spin-spin interaction with the unpaired electron on copper, and it amounts to 72 G. The pattern of the intensities of the two triplets is 3:1 in agreement with that predicted for AB spectra (by analogy with NMR nomenclature).

On some analogous copper complexes (Fig. 8.12), also accurate single crystal spectra were recorded, by doping into zinc tetraphenylporphin, ZnTPP. Representative spectra are shown in Fig. 8.13. The spectra were interpreted with the

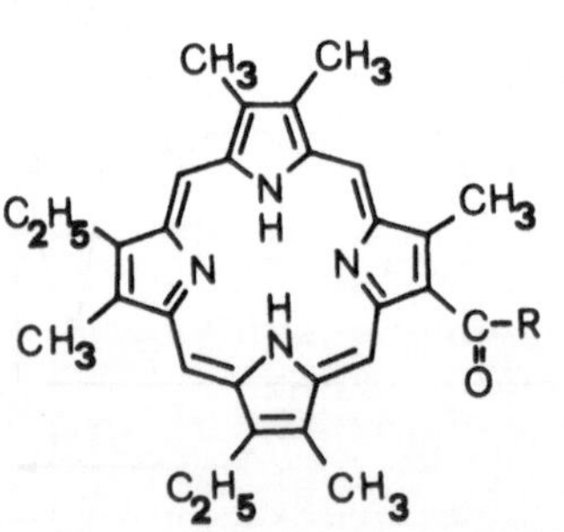

Fig. 8.10. The structure of a porphyrine ring binding to copper. R $=OC_2H_5$, tempamine

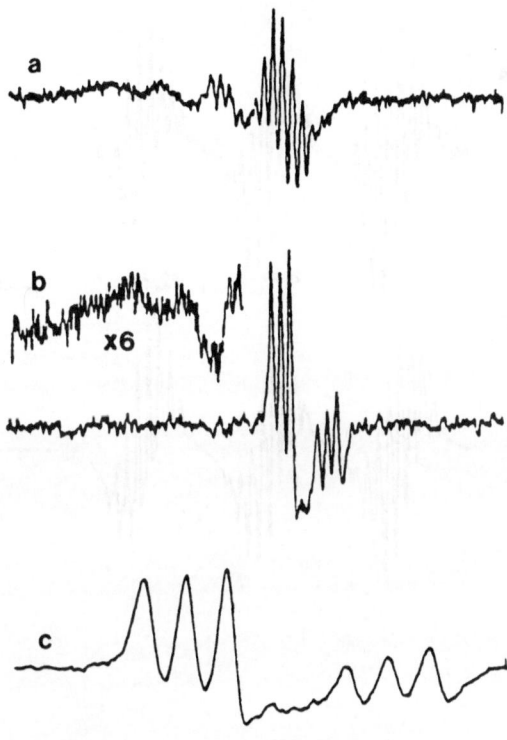

Fig. 8.11a, b. X-band EPR spectra of copper porphyrine complexes in CHCl$_3$ solution at 21°C. **a** R = OC$_2$H$_5$; **b** tempamine, scan 1000 G; **c** scan 200 G. After [8.11]

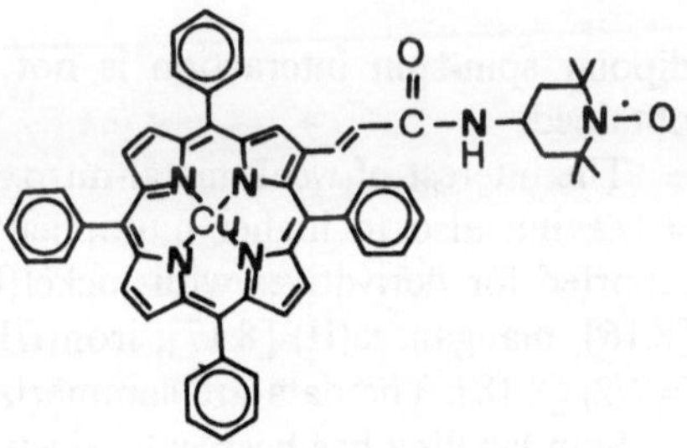

Fig. 8.12. Scheme of the structure of a spin-labeled copper porphyrine

formalism described in Sect. 3.5. They are complicated by the fact that on doping ZnTPP several different species are found. The analysis suggested that the J values in the different species range from -1 to 30×10^{-4} cm^{-1}, with metal-nitroxide distances ranging from 1300 to 1550 pm.

Spectra of frozen solutions can also be used to characterize the magnetic interaction between the metal ion and the organic radical, but in this case the

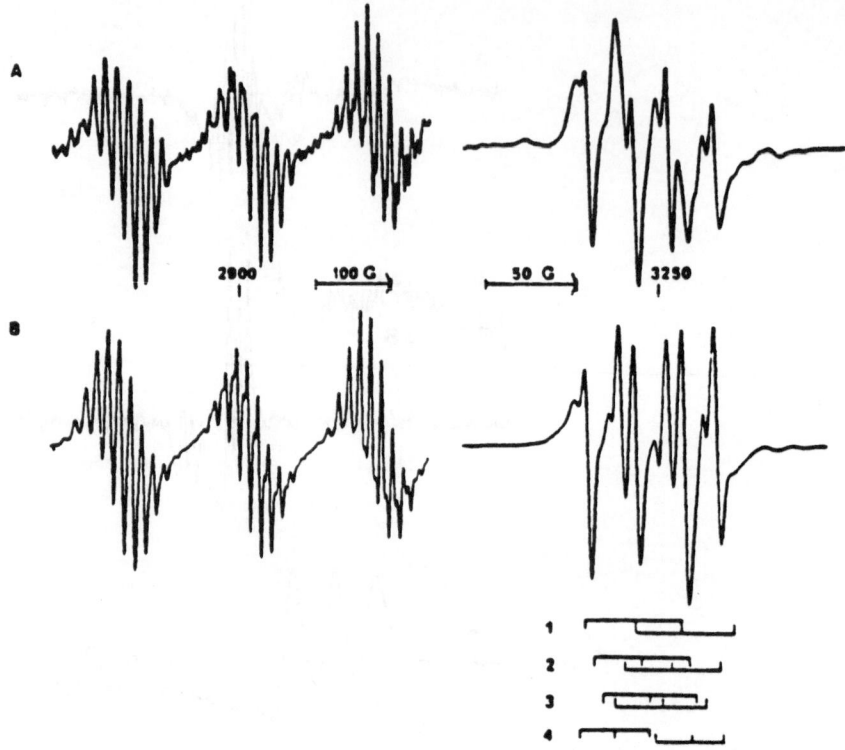

Fig. 8.13A, B. X-band single crystal spectra of the trans isomer of the copper porphyrine of Fig. 8.12 doped into ZnTPP. *Left*, copper; *right*, nitroxide regions of the spectra. **A** Experimental; **B** computer-simulated spectra. After [8.12]

dipolar spin-spin interaction is not averaged to zero and distances can be obtained.

The interest of weak metal-nitroxide interaction has not been limited to S = 1/2, but also to higher multiplicity ions. So, for instance, data have been reported for derivatives with nickel(II) (S = 1) [8.18], chromium(III) (S = 3/2) [8.16], manganese(II) [8.17], iron(III) (S = 5/2) [8.18], and gadolinium(III) (S = 7/2) [8.18]. The data are summarized also in a review article [8.3].

Spin labeling has been extensively used to investigate biological molecules, for instance, cytochrome P450. Simple model compounds have also been studied, in order to provide a firm basis for the interpretation of the spectra observed in more complicated systems [8.19]. Figure 8.14 illustrates one of these models. When two molecules of imidazole are bound, the iron(III) is in the low spin form (S = 1/2). The EPR spectra of FeTPP(Im)$_2$ and of one spin-labeled derivative are shown in Fig. 8.15. Spectrum A is that of the native porphyrin and is typical of low spin iron(III), with g = 1.52, 2.27, and 2.91. The two spin-labeled derivatives yield fairly similar spectra markedly different from those of the native species. The sharp signal at g ≈ 2 in both B and C is due to free nitroxide, which is

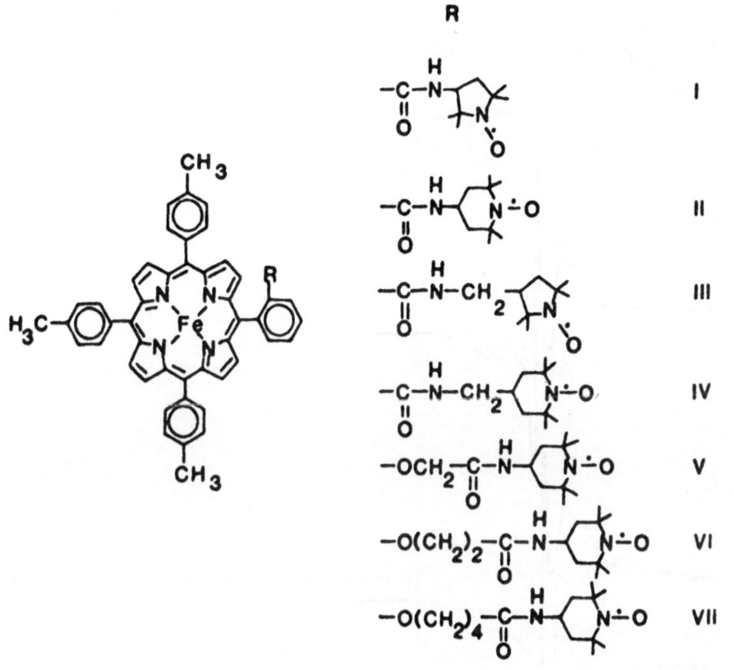

Fig. 8.14. The structure of spin-labeled iron porphyrins. After [8.19]

present as an impurity, the remaining four features being due to the iron(III) spin-labeled complex. Two signals at $g = 1.76$ and 2.14, respectively, correspond to the average between the $g = 2$ of the radical and $g = 1.52$ and 2.27 of the iron porphyrin, respectively. The other two signals at $g = 2.61$ and 2.32 correspond to the partial average of the $g = 2.29$ signal of FeTPP with $g = 2$ of the radical. This indicates that the exchange interaction is much larger than the **g** anisotropy for the former signals, but not large enough to average the latter. The spectra were simulated according to the procedure outlined in [8.13]. The best fit yielded J $= 0.28$ cm^{-1} and a distance between iron(III) and the radical of 700 pm.

8.4 Semiquinones

Orthosemiquinones are stable, negative, radical ions, which can chelate transition metal ions through two equivalent oxygen atoms (Fig. 8.16). The unpaired electron in the ligand is in a π^* molecular orbital and is spread out all over the molecule. The semiquinones can easily undergo both reduction and oxidation reactions according to the scheme shown below. Quite a few

$$\text{catecholate} \underset{+e}{\overset{-e}{\rightleftharpoons}} \text{semiquinone} \underset{+e}{\overset{-e}{\rightleftharpoons}} \text{quinone}$$

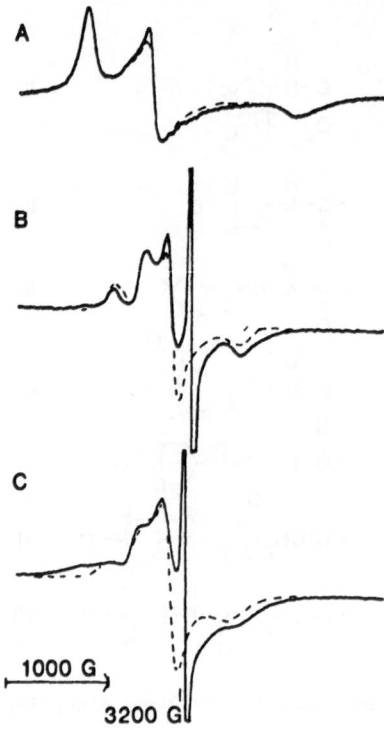

Fig. 8.15A-C. X-band spectra at $-180°C$ of (A) FeTTP(im)$_2$$^+$; (B) I-(MeIm)$_2^+$, and (C) I- (Im)$_2^+$ in toluene/chloroform glasses. I is defined in Fig. 8.14. After [8.19]

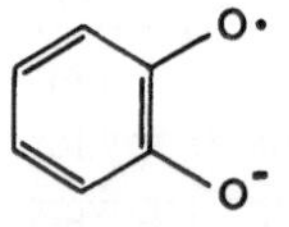

Fig. 8.16. The structure of a semiquinone

complexes have been reported with these ligands and both paramagnetic and diamagnetic metal ions [8.20–24]. In general, EPR spectra, although useful for the assignment of the formal oxidation states of both the metal and the ligand, have not been studied in detail, and often the main information obtained is that no EPR spectrum at all could be detected.

In a series of complexes of formula M(SALen)(SQ), where SALen is N,N'-ethylene bis(salicylideneiminato), SQ is an *ortho*-semiquinone and M = Fe^{III}, Mn^{III}, Co^{III}, the structure is postulated [8.25] to be as shown in Fig. 8.17. When M = iron(III) the magnetic susceptibility indicates a ground S = 2 state as a result of a strong antiferromagnetic coupling of the metal ion, S = 5/2, with the radical, but no EPR spectra could be detected, presumably due to large zero field

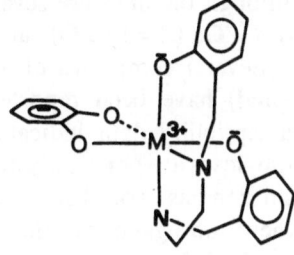

Fig. 8.17. The proposed structure of M(SALen) (SQ) complexes

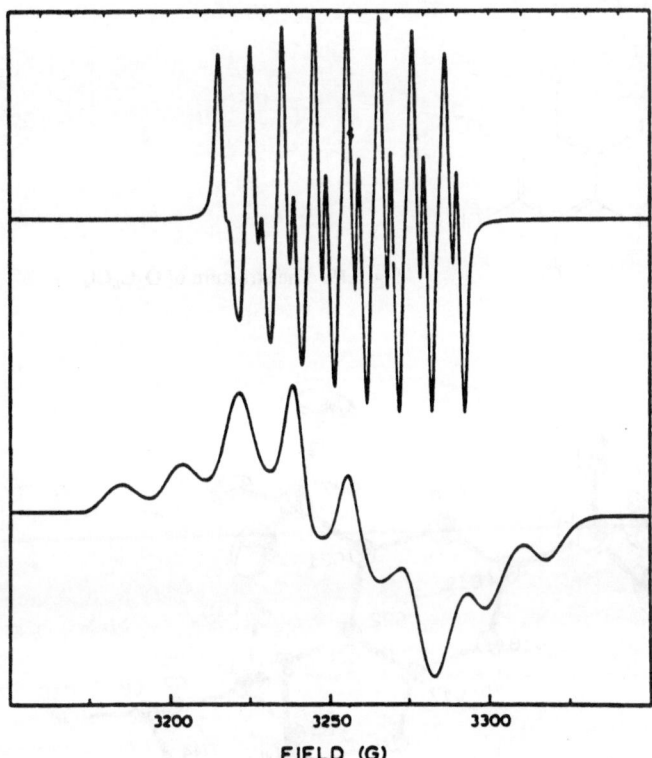

Fig. 8.18. X-band EPR spectra of Co(SALen)(3,5-DBSQ). *Upper*, room temperature; *lower*, 77 K. After [8.25]

splitting effects. When M = manganese(III), S = 2, the ground state is S = 3/2 and only a broad signal centered at $g = 2.5$ has been reported. Cobalt(III) is diamagnetic, therefore, the EPR spectrum is very simple (Fig. 8.18). The fluid solution spectrum consists of eight doublets centered at $g = 2$. The hyperfine pattern is attributed to the interaction with a ^{59}Co nucleus ($A = 10.2$ G) and a ^{1}H nucleus ($A = 3.5$ G). The spectrum remains isotropic also in glassy solution, although the lines are severely broadened, and only the hyperfine interaction with ^{59}Co ($A = 17.5$ G) can be detected.

Several complexes of general stoichiometry M(SQ)$^{n+}$ (SQ = semiquinone ligand) have been reported (M = V^{III}, Fe^{III}, Cr^{III}) [8.22–26]. The magnetic susceptibility data indicate a fairly strong coupling, so that the chromium complexes are practically diamagnetic, the vanadium has one unpaired electron, and iron has two. EPR spectra were reported only for the V(O$_2$C$_6$Cl$_4$) complex, where the ligand has the structure shown in Fig. 8.19. In fluid solution the unpaired electron appears to be located essentially on one semiquinone, a rather surprising result if one considers that two semiquinones interact strongly with the metal electrons in such a way to determine a complete pairing of them.

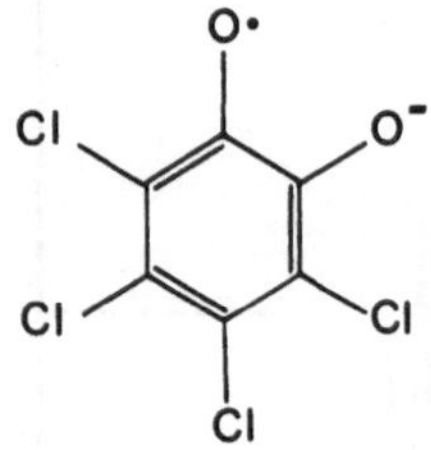

Fig. 8.19. The structure of O$_2$C$_6$Cl$_4$

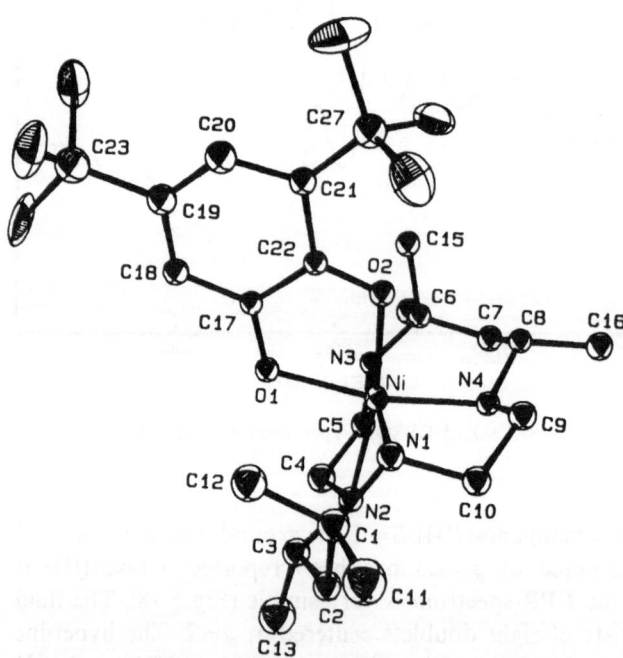

Fig. 8.20. The structure of [Ni(CTH)DTBSQ]$^+$. After [8.27]

[Ni(CTH)DTBSQ]$^+$ (CTH = dl-5,7,7,12,14,14-hexamethyl-1,4,8,11-tetra-azacyclotetradecane; DTBSQ = 3,5-di-t-butyl-semiquinone) has the structure [8.27] shown in Fig. 8.20. The nickel(II) ion is expected to have a ground $^3A_{2g}$ state, with the unpaired electrons in the $x^2 - y^2$ and z^2 orbitals. These magnetic orbitals are orthogonal to the π^* magnetic orbital of DTBSQ in this geometry and a strong ferromagnetic coupling can be anticipated. Indeed, the magnetic data show that the ground state of the complex is a quartet even at room temperature, suggesting that $J > 400$ cm^{-1}. The EPR spectra (Fig. 8.21) practically do not vary in the range 4.2–300 K. They are typical for transitions within one Kramers doublet (effective spin $S = \frac{1}{2}$), with g values: $g_1 = 5.8$, $g_2 = 2.4$, $g_3 = 1.7$. These values clearly indicate that the ground state has rhombic symmetry and that it is not a true $S = \frac{1}{2}$ state, but, in agreement with the magnetic data, it corresponds to the $M = \pm\frac{1}{2}$ levels of $S = 3/2$. In fact, the spin quartet is split in zero field and if the symmetry is axial, two Kramers doublets corresponding to $M = \pm\frac{1}{2}$ and $\pm 3/2$, respectively, are formed. If the energy separation between the two doublets is larger than the microwave quantum, then only transitions within the two doublets can be observed. The g values within the $M = \pm\frac{1}{2}$ doublet, $g^{\pm\frac{1}{2}}$, are given by: $g_{||}{}^{\pm\frac{1}{2}} = g_{||}$; $g_\perp{}^{\pm\frac{1}{2}} = 2g_\perp$, where $g_{||,\perp}$ represents the true g values for the system. Transitions within $M = \pm 3/2$ are forbidden in axial symmetry.

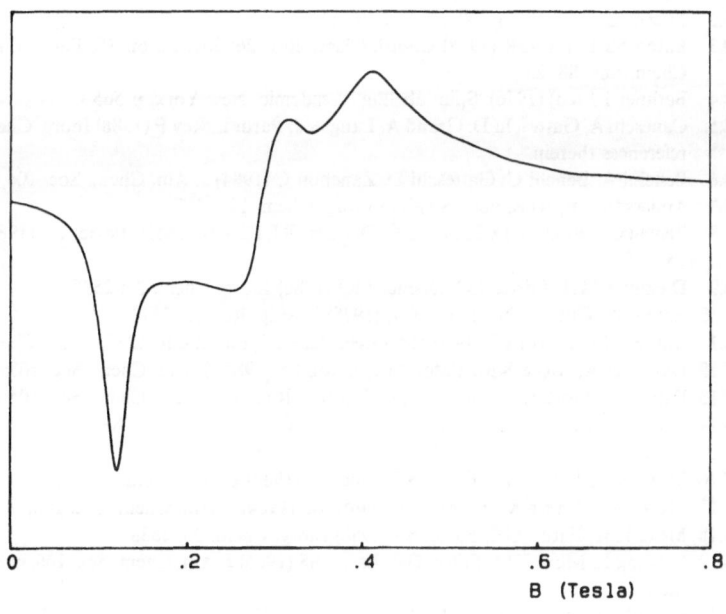

Fig. 8.21. Polycrystalline powder EPR spectrum of [Ni(CTH)DTBSQ]PF$_6$. After [8.27]

If the symmetry is lower than axial, then the g values in the two Kramers doublets are given [8.28] by:

$$g_x^{\pm\frac{1}{2}} = g_x[1+(1-3t)/\sqrt{(1+3t^2)}]; \quad g_x^{\pm 3/2} = g_x[1-(1-3t)/\sqrt{(1+3t^2)}];$$

$$g_y^{\pm\frac{1}{2}} = g_y[1+(1+3t)/\sqrt{(1+3t^2)}]; \quad g_y^{\pm 3/2} = g_y[1-(1+3t)/\sqrt{(1+3t^2)}]; \qquad (8.3)$$

$$g_z^{\pm\frac{1}{2}} = g_z[1-2/\sqrt{(1+3t^2)}]; \quad g_z^{\pm 3/2} = g_z[1+2/\sqrt{(1+3t^2)}];$$

where $t = E/D$.

If we assume that $g_x = g_y = g_\perp$ then from (8.3) and the experimental $g^{\pm\frac{1}{2}}$ values we can calculate $g_{||}$, $g_\perp$, and t. Thus, we find $g_{||} = 2.17$, $g_\perp = 2.17$, $E/D = 0.293$. In the present case the g values are the average of the values of the nickel ion and of the semiquinone. Using Eq. (3.20) and Table 3.3, assuming that g for the radical is 2.00, we finally calculate $g_{Ni} = 2.25$, in excellent agreement with the value expected for nickel(II) in octahedral symmetry.

More examples of simple model systems in which a metal ion interacts with semiquinones are given in Sect. 9.4.

References

8.1 Hyde JS, Swartz HM, Antholine WE (1979) In Berliner LJ (ed) Spin labeling II. Academic, New York, p 71.
8.2 Janzen EG (1971) Top. Stereochem. 6: 117
8.3 Eaton SS Eaton GR (1978) Coord. Chem. Rev. 26: 207, Eaton SS, Eaton GR (1988) Coord. Chem. Rev. 88: 23
8.4 Berliner LJ (ed) (1976) 'Spin labeling', Academic, New York, p 565
8.5 Caneschi A, Gatteschi D, Grand A, Laugier J, Pardi L, Rey P (1988) Inorg. Chem. 27: 1031 and references therein.
8.6 Bencini A, Benelli C, Gatteschi D, Zanchini C (1984) J. Am. Chem. Soc. 106, 5813
8.7 Anderson OP, Kuechler TS (1980) Inorg. Chem. 19: 1417
8.8 Benelli C, Gatteschi D, Zanchini C, Doedens RJ, Dickman MH, Porter LC (1986) Inorg. Chem. 25: 3453
8.9 Dickman MH, Porter LC, Doedens RJ (1986) Inorg. Chem. 25: 2595
8.10 Eaton SS, Dubois DL, Eaton GR (1978) J. Mag. Res. 32: 251
8.11 Braden GA, Trevor KT, Neri JM, Greenslade DJ, Eaton GR (1977) J. Am. Chem. Soc. 99: 4854
8.12 Damoder R, More KM, Eaton GR, Eaton SS (1983) J. Am. Chem. Soc. 105: 2147
8.13 Eaton SS, More KM, Sawant BM, Eaton GR (1983) J. Am. Chem. Soc. 105: 6560
8.14 Eaton SS, More KM, Sawant BM, Boymel PM, Eaton GR (1983) J. Magn. Res. 52: 435
8.15 Hafid S, Eaton GR, Eaton SS (1983) J. Mag. Res. 51: 470
8.16 More KM, Eaton GR, Eaton SS, Hideg K (1986) Inorg. Chem. 25: 3865
8.17 More JK, More KM, Eaton GR, Eaton SS (1984) J. Am. Chem. Soc. 106: 5395
8.18 More KM, Eaton GR, Eaton SS (1986) Inorg. Chem. 25: 2638
8.19 Fielding L, More KM, Eaton GR, Eaton SS (1986) J. Am. Chem. Soc. 108: 618 and references therein.
8.20 Kahn O, Prins R, Reedijk J, Thompson JS (1987) Inorg. Chem. 26: 3557
8.21 Lynch MW, Buchanan RM, Pierpont CG, Hendrickson DN (1981) Inorg. Chem. 20: 1038

8.22 Cass ME, Gordon NR, Pierpont CG (1986) Inorg. Chem. 25: 3962

8.23 Brown DG, Hemphill WD (1979) Inorg. Chem. 18: 2039

8.24 Buchanan RM, Fitzgerald BJ, Pierpont CG (1979) Inorg. Chem. 18: 3439

8.25 Kessel SL, Emerson RM, Debrunner PG, Hendrickson DN (1980) Inorg. Chem. 19: 1170

8.26 Buchanan RM, Kessel SL, Downs HH, Pierpont CG, Hendrickson DN (1978) J. Am. Chem. Soc. 100: 7894

8.27 Benelli C, Dei A, Gatteschi D, Pardi L (1988) Inorg. Chem. 27: 2831

8.28 Pilbrow JR (1978) J. Magn. Reson. 31: 479

9 Biological Systems

9.1 Introduction

The application of EPR to the study of magnetically coupled species in biological systems is one of the most intensively exploited fields at the moment, and almost every day new exciting examples are reported. Therefore, here more than in any other section, we will not be able to give exhaustive coverage of the area, but simply will report those examples which seem to us to be more appropriate, interesting, and worthy of attention. The whole field has already been covered by a number of review articles [9.1–4], to which the interested reader is invited to refer. Also, given the necessity of using a number of related techniques in order to obtain a good understanding of the complex matter, it may prove useful to consider other related magnetic techniques, such as nuclear magnetic resonance [9.5–7], ENDOR [9.8], magnetically perturbed Mössbauer spectroscopy [9.9–10], etc. Indeed, in the following sections we will frequently refer to the last mentioned technique, which has proved to be extremely useful for the characterization of iron proteins.

We will briefly review copper proteins, then iron proteins, and finally the photosynthetic oxygen-evolving processes, both in higher plants and bacteria.

9.2 Copper Proteins

A number of important proteins and enzymes contain copper ions at the active site [9.11–14]. The principal biological role known up to now for proteins which contain copper include oxygen transport and activation, electron transfer, iron metabolism, and superoxide dismutation. For this book the most interesting copper systems are superoxide dismutase (SOD) and the coupled dinuclear copper proteins such as hemocyanin, the oxygen-binding protein of mollusks and arthropods, and tyrosinase, an enzyme with both monooxygenase and oxidase activity.

SOD is an enzyme which is present in the erythrocytes of mammalians which catalyzes the dismutation of the toxic superoxide ions [9.15]. The enzyme from bovine erythrocytes is composed of two identical subunits, each of which contains one copper (II) and one zinc (II) ion. Frequently, it is indicated as Cu_2Zn_2SOD. The X-ray crystal structure [9.16] shows that the two metal ions are separated by about 600 pm. Copper is coordinated to four histidines, one of which is deprotonated and shared with the zinc ion. The zinc is also bonded to two additional histidines and to an aspartic acid. The copper (II) ion is in a distorted five-coordinate and the zinc in a tetrahedral environment (Fig. 9.1). The X- and Q-band EPR spectra of the native form [9.17–19], shown in Fig. 9.2, are typical of a five-coordinate copper (II) complex, intermediate between a trigonal bipyramid and a square pyramid. The zinc ion can be substituted by several other paramagnetic ions, such as copper (II) and cobalt (II). In the Cu_2Cu_2SOD and Cu_2Co_2SOD derivatives the Cu–Cu and Cu–Co pairs are antiferromagnetically coupled with $J = 52$ and 16.5 cm^{-1}, respectively [9.20–21]. The Cu_2Co_2SOD derivative is EPR silent down to liquid helium temperature, a result which is not unexpected since the coupling of two half-integer spins does not yield Kramers doublets. In fact, coupling $S_{Cu} = 1/2$ with $S_{Co} = 3/2$, two states with $S = 2$ and $S = 1$ are predicted, both largely split due to the large zero field splitting expected for the individual tetrahedral cobalt(II) ions. In fact, using Table 3.3 we expect $D_2 = 1/2\,D_{Co}$ and $D_1 = 3/5\,D_{Co}$, plus the presumably smaller term brought about by the exchange interaction. Since in distorted tetrahedral cobalt(II) complexes, D_{Co} can be as large as 10 cm^{-1} [9.22], it is easily understood that no transitions can be detected in a normal EPR experiment.

The Cu_2Cu_2SOD derivative, on the other hand, yields the EPR spectrum shown in Fig. 9.3B. Although it is typical of a triplet, with the characteristic half-field transition, it is not easily interpreted, due to the broadness of the lines, and

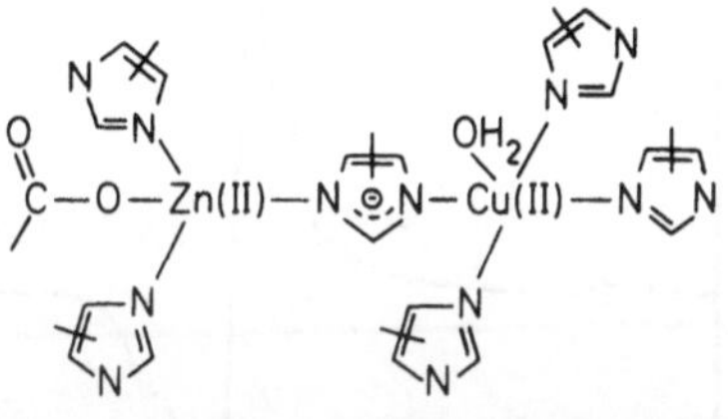

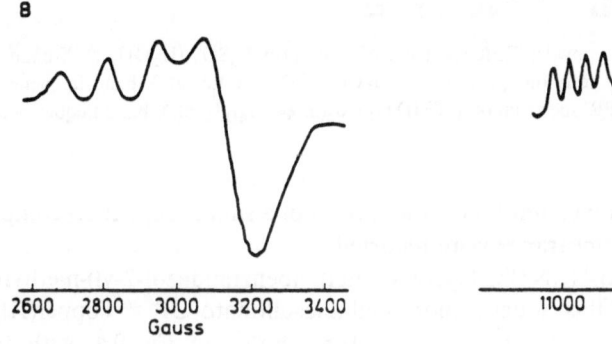

Fig. 9.1. Scheme of the active site of Cu_2Zn_2SOD

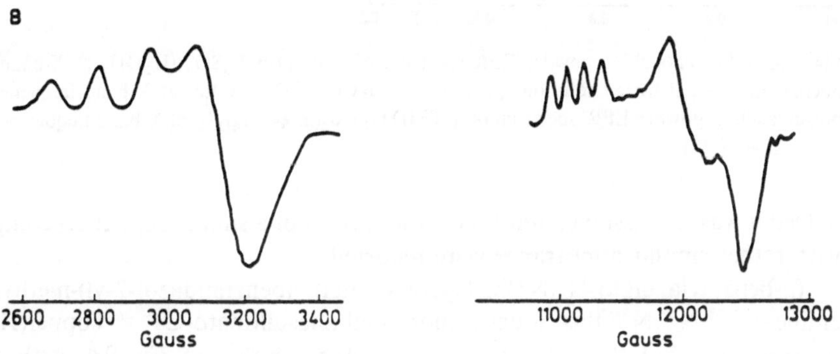

Fig. 9.2. EPR spectra of Cu_2Zn_2SOD. *Left* X-band and *right* Q-band frequency. After [9.11]

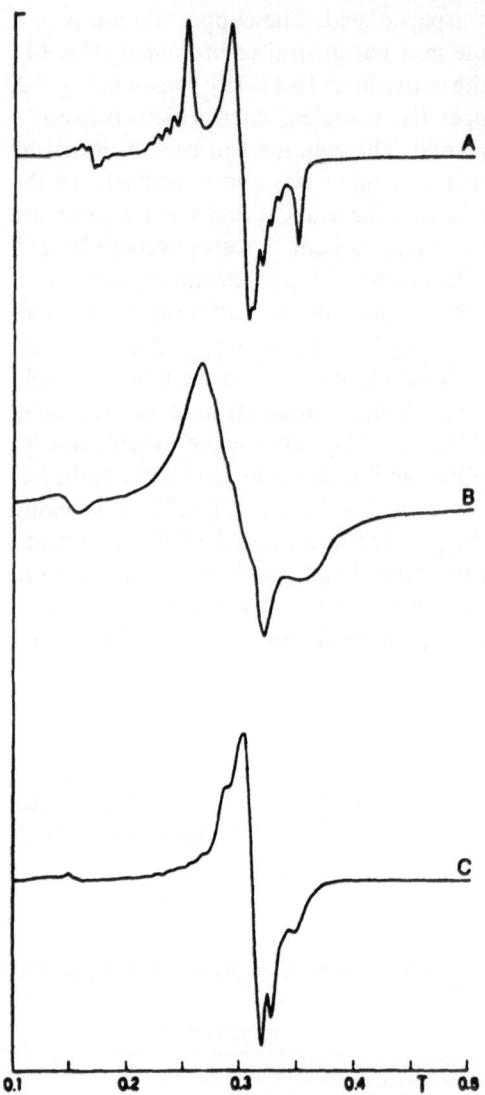

Fig. 9.3. *A* Polycrystalline powder EPR spectrum of [Cu(bpt)(CF$_3$SO$_3$)(H$_2$O)]$_2$ at X-band frequency and 7 K; *B* frozen solution spectrum of Cu$_2$Cu$_2$SOD in water at X-band frequency; *C* polycrystalline powder EPR spectrum of [(TMDT)$_2$ Cu$_2$(im)(ClO$_4$)$_2$]$_2$ at X-band frequency and 20 K. After [9.25]

indeed it was not assigned until the EPR spectra of a simple copper(II) complex with rather similar appearance were reported.

(μ-benzotriazolato-N^1,N^3)bis{ [tris(N^1-methylbenzimidazol-2-yl)-methyl)-amine-N,N^3,N$^{3'}$,N$^{3''}$]bis[aqua(trifluoromethanesulfonato-O)   copper(II)]}, [Cu(bpt) (CF$_3$SO$_3$)(H$_2$O)]$_2$ has the structure shown in Fig. 9.4, with two copper ions separated by 408.5 pm [9.24]. The coupling between the two is

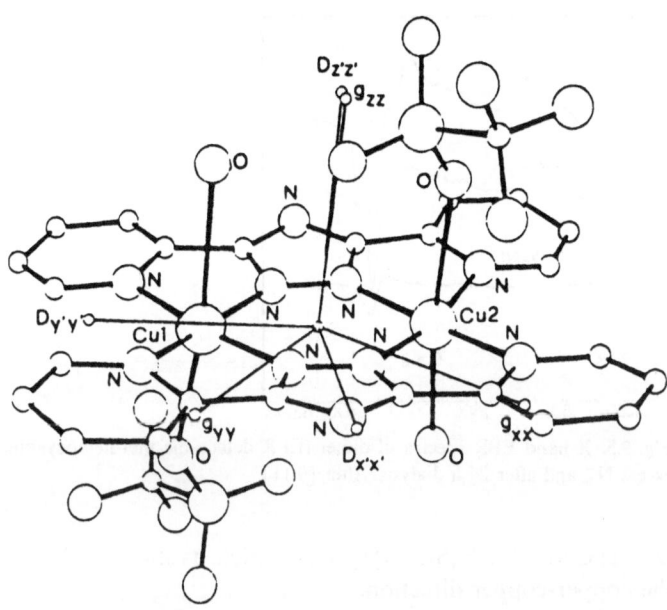

Fig. 9.4. The structure of $[Cu(bpt)(CF_3SO_3)(H_2O)]_2$ with the principal directions of the **g** and **D** tensors. After [9.25]

antiferromagnetic, with $J = 236 \text{ cm}^{-1}$. The EPR spectra [9.25] of [Cu(bpt) (CF_3SO_3) $(H_2O)]_2$ are shown in Fig. 9.3A. The features showing hyperfine splitting are assigned to the transitions parallel to z, the feature with the simple derivative shape to the two transitions parallel to x, and the remaining two features in the $g = 2$ region to the transition parallel to y. This assignment has been confirmed by single crystal spectra which yielded $g_{xx} = 2.055$ (1), $g_{yy} = 2.051$ (1), $g_{zz} = 2.232$ (1), $D_{xx} = 0.0026$ (5) cm^{-1}, $D_{yy} = 0.0338$ (3) cm^{-1}, $D_{zz} = -0.0364$ (4) cm^{-1}, with the principal axes defined in Fig. 9.4. In the two-parameter fit this corresponds to $D = -0.0546 \text{ cm}^{-1}$, $E/D = 0.29$.

The comparison of the EPR spectra of Cu_2Cu_2SOD with those of [Cu(bpt) (CF_3SO_3) $(H_2O)]_2$ suggests that the feature observed at 0.36 T, which in the literature has been used as a fingerprint of the coupled species, and another one, peaking at 0.26 T are the fine components of the transition parallel to y. According to this assignment, $g_{yy} = 2.07$ and $D_{yy} = 0.0317 \text{ cm}^{-1}$. The broad signal between 0.26 and 0.32 T must be associated with the two transitions parallel to x, which must be quasi-degenerate if $E/D \approx 1/3$. Finally, the bumps observed at fields lower than 0.26 T should correspond to the low field transitions parallel to z, the high field one being in the region obscured by mononuclear copper impurities.

The estimated zero field splitting in both [Cu(bpt) (CF_3SO_3) $(H_2O)]_2$ and Cu_2Cu_2SOD is much larger than calculated for the point-dipolar contribution, showing that even for two copper ions separated by 400–600 pm important exchange contributions can be operative. This is confirmed by the direction of

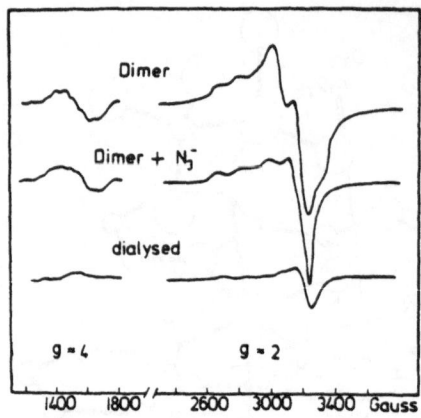

Fig. 9.5. X-band EPR spectra of dimer (EPR detectable met-hemocyanine), dimer with 100-fold excess N_3^- and after 24-h dialysis. After [9.11]

D_{zz} in $[Cu(bpt)(CF_3SO_3)(H_2O)]_2$, which, as shown in Fig. 9.4, is orthogonal to the copper-copper direction.

In hemocyanin and tyrosinase the active site contains pairs of copper ions, which are generally EPR-inactive. In fact, the deoxy form contains two copper(I) ions and is thus totally diamagnetic, but also the oxy form, which is obtained by oxidation with molecular oxygen, is diamagnetic. Spectral evidence indicates that in this case both the copper atoms are in the $+2$ oxidation state, but that a strong antiferromagnetic coupling yields a ground singlet, with no evidence for a low lying triplet state. The only EPR-active dinuclear species reported so far is the so-called dimer, or EPR-detectable met, which is obtained by NO oxidation of the deoxy form and has the EPR spectrum [9.26] shown in Fig. 9.5. It is clearly a triplet spectrum, with a well-resolved half-field transition. Temperature dependence of the signal intensity [9.27] showed that the coupling between the two copper ions must be small, $|J| < 5\ cm^{-1}$. Simulations of the spectra were performed in order to obtain structural information on the assumption that the zero field splitting is dominated by magnetic dipolar interactions. The copper–copper distance was thus estimated to be ≈ 600 pm. Addition of excess azide results in the formation of dimer N_3^-, in which the copper–copper distance was estimated to be ≈ 500 pm [9.28]. In these EPR-active, largely uncoupled derivatives, it is currently believed that the endogenous bridge responsible for the large coupling in the met form is detached. The weak EPR signal observed in met hemocyanin has been attributed to a small fraction of the sites where the endogenous bridge has a lower stability constant [9.29].

9.3 Iron Proteins

Iron proteins form a very large group, which is usually split into three classes, (1) hemoglobin and myoglobin, which are responsible for the transport and storage of oxygen; (2) iron-sulfur proteins, which are responsible for electron transport and are widely distributed in plants, animals, and microorganisms; and (3) that which contains oxo-bridged iron(III) ions, which perform a variety of biological functions, including oxygen transport, the reduction of ribo- and deoxyribonucleotides, phosphate ester hydrolysis, and iron storage. It is especially in the last two classes that proteins relevant to this book are to be found, although well-documented cases of exchange-coupled species are found also in iron porphyrins, which are structurally related to the hemoglobin class of iron proteins.

9.3.1 Iron Porphyrins and Heme Proteins

Heme is a porphinato iron complex in which the common oxidation states for the metal atom are $+2$ and $+3$, each of which can have several different spin states. Indeed, iron(II), which has a d^6 configuration, can have a ground $S = 0, 1$, and 2 state, while iron(III), which is a d^5 ion, can have $S = 1/2, 3/2$, and $5/2$. All these different spin states have been observed [9.30]. From the EPR point of view the important hemes are the iron(III) derivatives, because this ion has an odd number of electrons, and in all the possible geometries and spin states it has a ground Kramers doublet which always yields a spectrum, provided that the temperature is sufficiently low to yield a long enough relaxation time. Matters are different for the paramagnetic states of iron(II), which is an even electron ion. In this case, in fact, zero field splitting removes the spin degeneracy of the ground multiplet to such an extent that transitions between the states can often be induced only at fields much higher than those usually available.

Particularly interesting appear to be the so-called high-valent iron porphyrins [9.31, 32]. This expression is used to refer to iron porphyrin complexes more oxidized than the iron(III) oxidation state. Species of this type are, or are proposed to be, involved in various biological processes mediated by peroxidase and catalase, and by cytochrome P-450.

The peroxidases, for instance, exhibit a catalytic cycle in which the resting enzyme containing ferric heme reacts with hydrogen peroxide losing two electrons:

$$[Fe(III)P]^+ + H_2O_2 \rightarrow [FeP]^{3+} + 2H_2O,$$

Where P is just a shorthand notation for the porphinato group. The $[FeP]^{3+}$ can be formulated either as iron(V) with P^{2-}, or as iron(IV) with the P^- anion radical. In any case the $[FeP]^{3+}$ species has an odd number of electrons and should be observed in EPR. Schultz et al. [9.33] succeeded in obtaining the spectra at $T < 4 K$, and under rapid passage conditions [9.34], as shown in Fig. 9.6. The EPR data were complemented with Mössbauer spectra, which could be explained satisfactorily in terms of an iron(IV) with a ground $S = 1$ state, showing a large zero field splitting, $D = 23 cm^{-1}$. The EPR spectra were then interpreted within a model with a small anisotropic exchange, $J_{xx} = -2$, $J_{yy} = -1$, and

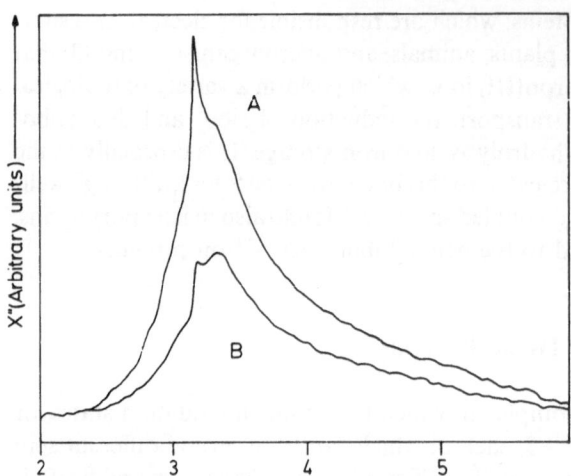

Fig 9.6. X-band EPR spectrum, of horseradish peroxidase compound I recorded under rapid passage conditions. After [9.34]

$J_{zz} = +3 \text{ cm}^{-1}$. These values in turn were explained on the basis of the dipolar coupling between the iron(IV) species and the a_{2u} radical of the porphyrin.

Recently, the oxidized species [FeCl(TPP)] (SbCl$_6$) and Fe(TPP) (ClO$_4$)$_2$ were also characterized [9.35] (TPP = tetraphenylporphinato). In this case the Fe(TPP) moiety has only two positive charges, one electron less than [FeP]$^{3+}$, and no EPR spectra could be observed. However, the Mössbauer spectra showed the presence of high spin iron(III), and the magnetic data showed that a strong antiferromagnetic coupling is operative in [FeCl(TPP)] (SbCl$_6$), $J > 500 \text{ cm}^{-1}$, while the coupling is ferromagnetic in [Fe(TPP)(ClO$_4$)$_2$, $J \approx -80 \text{ cm}^{-1}$. For the former the ground state is S = 2, and for the latter it is S = 3, both originating from the interaction of $S_{Fe} = 5/2$ and $S_r = 1/2$. It is interesting also to note that NMR spectra of the two compounds showed opposite shifts of the phenyl protons: for instance, the *ortho-* and *para* protons are shifted downfield (42 and 35 ppm, respectively) in [FeCl (TPP)] (SbCl$_6$) and upfield (-19 and -13 ppm) in Fe(TPP) (ClO$_4$)$_2$. This is a beautiful example of the alternation in sign of the hyperfine coupling constant (the EPR equivalent of the NMR isotropic shift) in the high and low total spin states obtained by coupling a spin S with another with S = $\frac{1}{2}$. Indeed, the use of Table 3.3 yields $A = +1/6 A_r$ for S = 3 and $A = -1/6 A_r$ for S = 2.

Another system which is very interesting from the EPR point of view is cytochrome-*c*-oxidase [9.36]. Although very active research is being done on this enzyme at the present time not very much is known about its structure. It is known that it is the respiratory enzyme that catalytically reduces 1 mol dioxygen to 2 mol water with the concomitant release of energy which is stored in the ADP-ATP cycle. The enzyme contains four metal centers (two irons and two

216

coppers) per functioning unit. In the oxidized (resting) form of the enzyme one iron is in a heme unit, and one copper is isolated and EPR-detectable, while at the active site there is a high spin iron(III) which is strongly coupled to a copper(II) ion to give a ground $S = 2$ state. The $S = 3$ state has been estimated to be at least $1200 \, cm^{-1}$ above the ground state. The system is EPR-silent, presumably due to large zero field splitting.

If the native enzyme is treated with cyanide, the Fe^{III}-Cu^{II} moiety remains EPR-silent, although there is no doubt that the iron(III) ion is forced into the low spin form. There are conflicting reports in the literature, but it seems now that the coupling between low spin iron(III) and copper(II) is in this case ferromagnetic, yielding a ground triplet state [9.37]. Some model compounds were also synthesized containing low spin iron(III) porphinato and a square planar copper(II) ion bridged by a cyanide ion which occupies axial positions for both the metal ions [9.38]. These species yielded EPR spectra at low temperatures with many resonances in the range 0–0.5 T. They were attributed to transitions between the levels of what was substantially described as $S' = 1$ level. The point we want to make here is that a low spin iron(III)-copper(II) pair is complicated by the fact that the iron has a quasi-degenerate orbital ground state, making the analysis of the energy levels, and of the EPR transitions, of the pair rather difficult. Indeed, if we take into account the orbital degeneracy we see that the energy levels must be expressed through a spin hamiltonian which includes the isotropic, anisotropic, and antisymmetric terms, all with very similar relative importance; therefore, discussion of ferro- or antiferromagnetic coupling may be just a semantic problem. We have already discussed this point in Sect. 2.6.

9.3.2 Iron-Sulfur Proteins

The number of proteins which contain iron atoms bound to sulfur atoms is very large, showing several different basic arrangements and many different functions [9.39, 40]. Certainly most of them are devoted to the general problem of electron transport in biology, but they are also relevant to the metabolism of H_2 and N_2. They are all named iron-sulfur proteins, and are usually classified according to the number of iron atoms which are present in the metal sites. So 1-Fe, 2-Fe, 3-Fe, and 4-Fe sites are known, whose general structures [9.40] are shown in Fig. 9.7.

Proteins containing single iron sites are of different types, but the simplest are conventional rubredoxins, which contain tetrahedrally coordinated iron(III) in the oxidized and iron(II) in the reduced form, respectively. In both oxidation states iron is in the high spin form. Also, several model compounds have been synthesized to mimic the structure and the spectral properties of the iron sites of the proteins. In view of the relevance of the single ion spin hamiltonian parameters to the interpretation of the spectra of the coupled systems, we will provide here a brief resumé of the relevant properties of both oxidized and reduced rubredoxins.

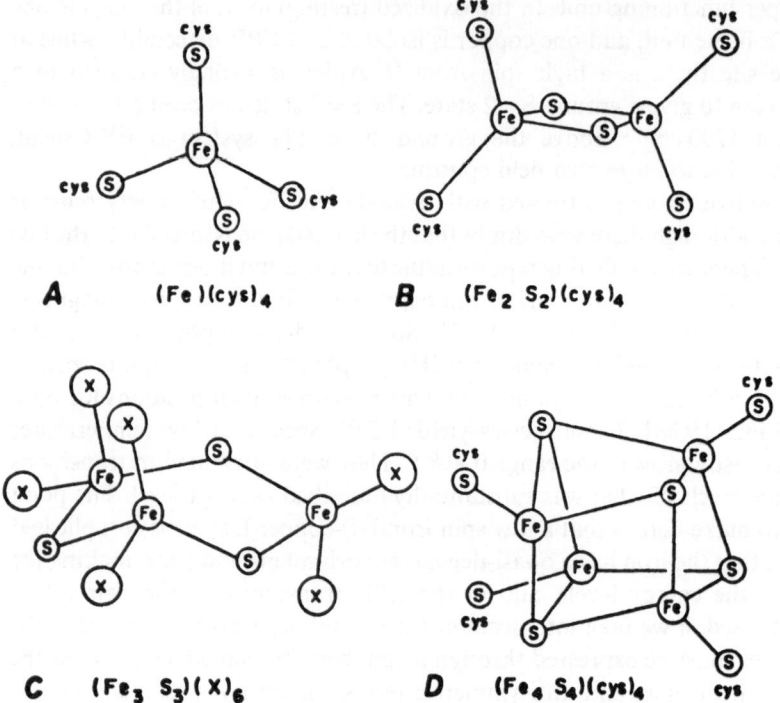

Fig. 9.7. General structures of iron-sulfur centers in proteins. After [9.40]

Rubredoxins, Rd, in their oxidized form exhibit resonances near $g = 4.3$ and $g = 9.4$ [9.41, 42]. These resonances can be explained by an $S = 5/2$ spin hamiltonian with $E/D \approx 1/3$. Mössbauer data showed [9.43] that the zero field splitting parameters in Rd from *Clostridium pasteurianum* are: $D = 1.9\ \text{cm}^{-1}$ and $E/D = 0.23$, while the hyperfine coupling constants were determined to be $A_x = -16.5\ \text{G}$, $A_y = -15.9\ \text{G}$, and $A_z = -16.9\ \text{G}$. These values are believed to be rather typical for high spin iron(III) in a tetrahedral environment of four sulfur atoms. An exception to this is the oxidized form of desulforedoxin, Dx, from *Desulfovibrio gigas*, which shows nearly axial EPR spectra [9.44] characterized by $E/D \approx 0.08$.

In the reduced forms of both Rd and Dx no EPR spectra can be detected, but the spin hamiltonian parameters could be obtained from Mössbauer data. In both cases the zero field splitting was found to be fairly large ($7.6\ \text{cm}^{-1}$ for Rd, and $-6\ \text{cm}^{-1}$ for Dx, with E/D ratios of 0.28 and 0.19, respectively) [9.45].

The 2-Fe proteins, which are generally named ferredoxins, contain in the active site dinuclear species with the structure B of Fig. 9.7. In the oxidized form the two metal atoms are high spin iron(III), while in the reduced form a trapped valence high spin iron(III)-high spin iron(II) species is present. In both the oxidized and the reduced form the metal ions are antiferromagnetically coupled to yield a ground $S=0$ and $S=1/2$ state, respectively. The former is EPR-silent, while the latter gives a characteristic spectrum, with a prominent feature at $g \approx 1.94$, which has long been used as a fingerprint of this species [9.46].

The ground state of the reduced form originates from the exchange interaction between a high spin iron(III), $S=5/2$, and a high spin iron(II), $S=2$. The coupling is of the order of 400 cm^{-1}. Typical spectra [9.47] are shown in Fig. 9.8, from which it is apparent that three g values are observed at $g_1 = 2.03$–2.04; $g_2 = 2.00$–1.90; $g_3 = 1.95$–1.80.

In the limit of large J, the g values of the ground doublet are calculated [9.46] to be given by:

$$\mathbf{g} = 7/3\ \mathbf{g}_{Fe(III)} - 4/3\ \mathbf{g}_{Fe(II)}. \tag{9.1}$$

If it is assumed that the g tensor of iron(III) is essentially isotropic and equal to 2, using (9.1) and the experimental g values it is possible to calculate the g values of the iron(II) center. Bertrand and Gayda [9.48] used Eq. (9.1) within a ligand field formalism, to show that the variations in the g values of 2-Fe ferredoxins can be justified by variations in the electronic structure of the iron(II) center. Later, the same model was applied also to a new class of 2-Fe ferredoxins, characterized by $g_{av} = 1.92$, smaller than in the other class [9.49]. Using the Angular Overlap formalism variations of the ligand field parameters have been related to the σ and π bonding characteristics of the sulfur ligands in the iron(II) environment [9.50].

The EPR spectra have been used also to determine the exchange-coupling constant, through measurements of relaxation times, as we reported in Sect. 6.3.

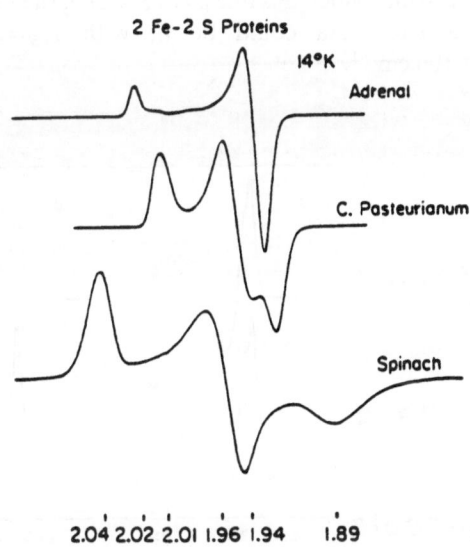

Fig. 9.8. EPR spectra of reduced iron-sulfur proteins from bovine adrenals, *Clostridium Pasteurianum*, and spinach. After [9.47]

The 3-Fe proteins were first identified in *Azotobacter vinelandi* [9.51, 52] and then were found to be present in several other ferredoxins such as in beef heart aconitase, in *Desulfovibrio gigas* ferredoxin II, FdII, hydrogenases from *D. gigas* and *D. desulfuricans, etc.* [9.39]. The EPR spectra [9.53] of three different proteins, at both X- and Q-band frequency at 14 K are shown in Fig. 9.9. At higher temperatures the spectra broaden. The EPR spectra have been described as fairly isotropic and they are centered at $g = 2.01$. This has been taken as evidence of a $S = 1/2$ ground state, and a confirmation was found in the Mössbauer studies as we reported in Sect. 4.3.1.

The pattern of energy levels was confirmed also by a study of the temperature dependence of the EPR line-width [9.54]. Indeed, it was found that the electronic relaxation time follows an exponential law:

$$T_1^{-1} \propto \exp(-\delta/kT)$$

with $\delta = 88$ cm^{-1}. This behavior agrees with an Orbach mechanism and indicates the presence of an energy level 88 cm^{-1} above the ground state. Identifying this level with the first excited doublet state, and assuming for the sake of simplicity two identical exchange constants, yielded $J_{13} = J_{23} = 40$ (6) cm^{-1}, $J_{12} = 34$ (2) cm^{-1}.

Attempts have also been made to synthesize simple molecular compounds with the same structure and properties of the proteins, but up to now only linear clusters of formula $[Fe_3S_4(SR)_4]^{3-}$ have been obtained (R = Et, Ph) [9.55]. These have been found to have a ground $S = 5/2$ state, with a fairly small zero field splitting ($\leqslant 1$ cm^{-1}) and rhombic anisotropy, as shown by the EPR spectrum, which has one $g = 4.3$. Using the same formalism as for the protein, the magnetic data could be fit with $J_{13} = J_{23} = 300$ cm^{-1} and $-100 \leqslant J_{12} \leqslant +100$ cm^{-1}.

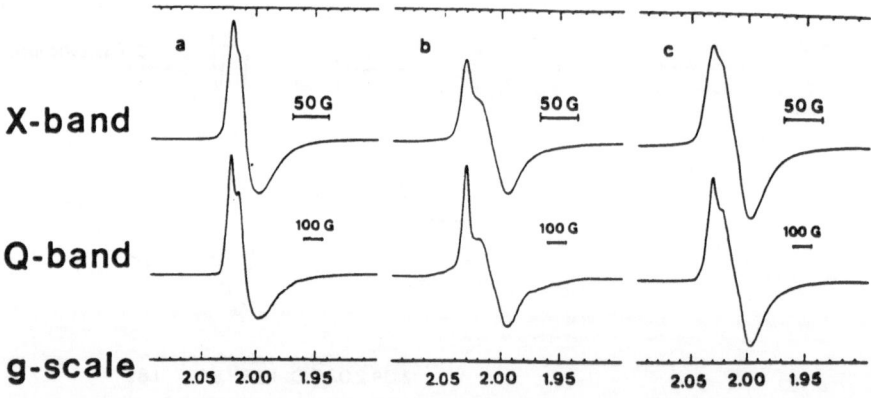

Fig. 9.9. X- and Q-band EPR spectra of 3-Fe proteins. *a A. vinelandii; b* half-reduced state of *T. commune* ferredoxin; *c* oxidized state of *T. commune* ferredoxin. After [9.53]

FdII can be isolated also in the reduced form [9.56]. The ground state has been assumed to be S = 2 on the basis of Mössbauer [9.56] and MCD [9.57] measurements. Mössbauer spectroscopy showed also that one iron has parameters typical of high spin iron(III), while the other two are equivalent, with parameters which are the average of those observed for oxidized and reduced tetrahedral sulfur sites. These data were interpreted with a model according to which the equivalent iron atoms form a completely delocalized mixed valence pair with a ground S = 9/2 state. This pair is coupled to the third high spin iron(III) ion to give a ground S = 2 state. This is largely split in zero field, with two levels separated by ≈ 0.3 cm^{-1}. This point has been confirmed by the EPR X-band spectra, which show one transition at very low field [9.58]. The ground S = 9/2 state of the delocalized pair is in agreement with a model which includes both Heisenberg exchange and a term describing valence delocalization, similar to that described in Sect. 2.7. This effect has been discussed also within the Xα-Valence Bond model by Noodleman and Baerends [9.59].

The species with four iron atoms are characterized by three different oxidation states. There are essentially two classes of proteins which have this cluster of iron atoms, namely the high potential iron protein, HiPIP, and the four-iron ferredoxins, and several model compounds have been synthesized which mimic their redox and spectral properties [9.39]. It is now generally accepted that there is a good correspondence between the states indicated below for the models and the proteins, respectively:

$$[Fe_4S_4(SR)_4]^{3-} \rightleftharpoons [Fe_4S_4(SR)_4]^{2-} \rightleftharpoons [Fe_4(SR)_4]^{-};$$

$$Fd_{red} \rightleftharpoons Fd_{ox} \rightleftharpoons (Fd_{superox});$$

$$(HiPIP_{superred}) \rightleftharpoons HiPIP_{red} \rightleftharpoons HiPIP_{ox}.$$

The states of Fd and HiPIP (in parentheses) have been observed only in vitro. For the iron-sulfur core these states correspond to:

$$[Fe_4S_4]^{+} \rightleftharpoons [Fe_4S_4]^{2+} \rightleftharpoons [Fe_4S_4]^{3+}.$$

Formally, these states correspond to $1Fe^{III} + 3Fe^{II}$, $2Fe^{III} + 2Fe^{II}$, and $3Fe^{III} + 1Fe^{II}$, respectively. The $[Fe_4S_4]^{2+}$ cluster is diamagnetic and does not show any EPR spectrum. On the other hand, both $[Fe_4S_4]^{+}$ and $[Fe_4S_4]^{3+}$ have a ground $S = \frac{1}{2}$ state and they show characteristic EPR spectra. $[Fe_4S_4]^{+}$ is characterized by $g_{av} < 2$ and the overall appearance of the spectra is similar to that of reduced 2-Fe ferredoxins, while $[Fe_4S_4]^{3+}$ is characterized by $g_{av} > 2$.

A detailed study of the EPR spectra of model compounds has been recently performed. By X-ray irradiation of the diamagnetic $(NBu)_4[Fe_4S_4(SC_6H_5)_4]$ two species were formed [9.60], which were identified with the $[Fe_4S_4]^{+}$ and $[Fe_4S_4]^{3+}$ clusters respectively. These species were found to have principal g values which correspond nicely to the values of Fd_{red} and $HiPIP_{ox}$, respectively (Table 9.1).

Table 9.1. g values of model centers compared with those of iron-sulfur proteins

		g_1	g_2	g_3	g_{av}	Ref.
A center	$[Fe_4S_4]^+$	2.089	1.969	1.877	1.978	[9.60]
Fd_{red}	$[Fe_4S_4]^+$	2.06	1.93	1.88	1.96	[9.61]
B center	$[Fe_4S_4]^{3+}$	2.108	2.006	1.987	2.034	[9.60]
HP_{ox}	$[Fe_4S_4]^{3+}$	2.12	2.04	2.04	2.07	[9.62]
		2.088	2.055	2.04	2.06	

The diamagnetism in $[Fe_4S_4]^{2+}$ can be rationalized assuming antiferromagnetic coupling between two pairs of mixed valence iron(II)-iron(III) couples [9.59]. $[Fe_4S_4]^+$ is similar to the reduced form of 2-Fe ferredoxin, because two iron(III) ions are strongly coupled to give an $S = 0$ ground state, and what is left is an iron(III)-iron(II) couple practically identical to that of the 2-Fe species.

Recently also some clusters of the type MFe_3S_4 have been isolated and studied. For instance, model compounds of formula $[MFe_3S_4(SPhCl)_3((C_3H_5)_2Cat)CN]^{3-}$, with $M = Mo$, W, ($SPhCl = p$-chlorothiophenol; Cat = catecholate) give spectra characteristic of a ground $S = 3/2$ subject to a large zero field splitting [9.63].

$CoFe_3S_4$ clusters were obtained [9.64] by incubation of *Desulfovibrio gigas* ferredoxin, which contains Fe_3S_4 clusters, with Co^{2+}. The EPR spectrum reveals the hyperfine splitting into eight components expected for ^{59}Co, and the g values, corresponding to an effective $S = \frac{1}{2}$ state, are $g_1 = 1.82$, $g_2 = 1.94$, $g_3 = 1.98$. This cluster is described as $[CoFe_3S_4]^{2+}$, isoelectronic to the $[Fe_4S_4]^+$ cluster.

9.3.3 Sulfite Reductase

The sulfite reductase isolated from *Escherichia coli* is a complex hemoflavoprotein of molecular weight 685 000 containing four sirohemes and four Fe_4S_4 clusters [9.65]. It catalyzes the six-electron reductions of SO_3^{2-} to S^{2-} and of NO_2^- to NH_3. The protein can be split and a subunit containing one siroheme and one Fe_4S_4 cluster can be isolated. This subunit is indicated as SiR. Full reduction of this unit is accomplished by two electrons. The oxidized form of the enzyme yields an EPR spectrum characteristic of a high spin iron(III) heme [9.66]. No other signal attributable to the Fe_4S_4 cluster is observed, suggesting that it is in the diamagnetic $+2$ oxidation state. Mössbauer data have confirmed that the heme is in the high spin $+3$ oxidation state, but have also shown that the iron ions of the Fe_4S_4 cluster experience a magnetic hyperfine interaction, i.e., that the spin state of the cluster is $S = 5/2$ as well [9.67, 68]. This can only occur if the two moieties, heme and 4Fe-4S, are coupled. No detailed analysis of the data has been attempted, but a model has been suggested according to which the four iron atoms of Fe_4S_4 are strongly coupled to yield the usual $S = 0$ ground state, but they are also coupled to the heme iron with smaller coupling constants.

One electron reduction of SiR causes loss of the ferriheme signals, thus showing that the iron(III) of the heme is reduced. Addition of a second electron results in the appearance of a novel EPR signal, with $g = 2.53$, 2.29, and 2.07.

9.3.4 Nitrogenase

Nitrogenase comprises two extremely air-sensitive metalloproteins (Fe protein) and one molybdenum iron protein (MoFe protein) [9.69]. The Fe protein accepts electrons from oxidative processes operating at or below -0.4 V and then specifically passes them on to the MoFe protein for use in the reduction of dinitrogen and other substrates. The MoFe protein is a tetramer of molecular weight 220 000–240 000, containing 2 mol molybdenum, approximately 30 mol iron, and approximately 30 mol acid-labile sulfur per mol peptide component.

One of the most prominent features of the MoFe protein is its EPR spectrum [9.70, 71], shown in Fig. 9.10. The spectrum is typical of an $S = 3/2$ largely split in zero field, with effective g values 4.3, 3.6, and 2. There is now evidence that this spectrum is due to a multimetal cluster which is present in the active site of the enzyme. Model compounds of formula $[\mathrm{Fe(MS_4)_2}]^{3-}$ (M = Mo, W) were found [9.72] to yield EPR spectra very similar to those of the enzyme.

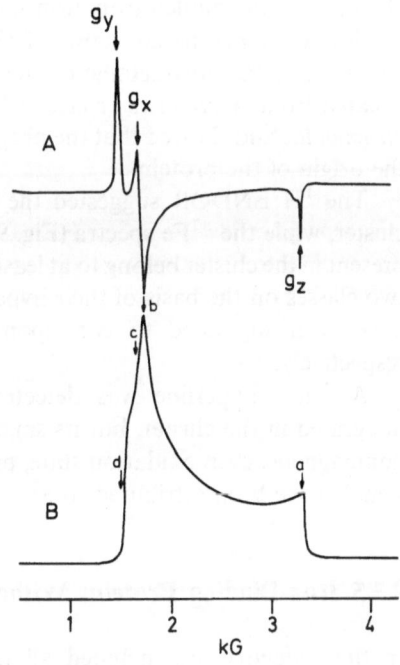

Fig. 9.10. EPR spectra of A. vinelandii MoFe protein of nitrogenase at 2 K. A absorption derivative; B rapid passage dispersion. After [9.71]

223

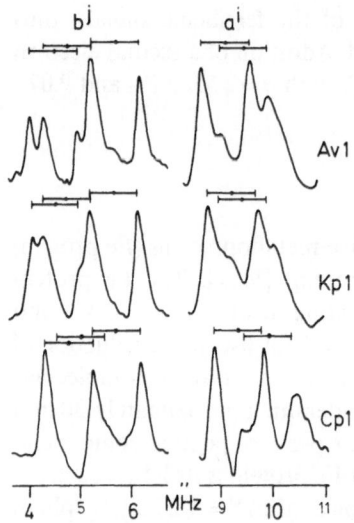

Fig. 9.11. ^{57}Fe ENDOR of MoFe protein of nitrogenase of: *Av1, Azotobacter vinelandii; Kp1, Klebsiella pneumoniae; Cp1, Clostridium pasteurianum.* After [9.71]

Mössbauer spectroscopy indicated [9.73] that the multimetal cluster includes six spin-coupled iron atoms. Crystallike ENDOR measurements [9.71] yielded the hyperfine constants of ^{1}H, ^{57}Fe, ^{95,97}Mo, and ^{33}S, and helped to characterize the polymetallic cluster. The studies were performed on proteins isolated from *Azotobacter vinelandii, Clostridium pasteurianum,* and *Klebsiella pneumoniae* and showed that the properties of the cluster are largely invariant to the origin of the protein.

The ^{1}H ENDOR suggested the presence of H_2O or OH^- bound to the cluster, while the ^{57}Fe spectra (Fig. 9.11) gave evidence that the six iron atoms present in the cluster belong to at least five different sites and can be grouped into two classes on the basis of their hyperfine coupling constants. These two types have been suggested to correspond to high spin ferric and ferrous ions, respectively.

A ^{95}Mo hyperfine was detected, showing that indeed molybdenum is integrated in the cluster, but its small value suggests that molybdenum is in a nonmagnetic, even oxidation state, probably Mo(IV). Finally, the ^{33}S ENDOR signals have been attributed to S^{2-}.

9.3.5 Iron-Binding Proteins Without Cofactors or Sulfur Clusters

In this category are included all the iron proteins which, having different functions, have in common the negative structural property that they do not have either a heme group or a sulfur cluster. They include ferritins, which are

224

widespread in plants, animals, and protista as iron storage proteins; ribonucleotide reductase, which has a central role in DNA synthesis; hemerythrin, which is the oxygen transport protein in some invertebrates; uteroferrin, and the purple acid phosphatases, which catalyze phosphoester hydrolysis. All these proteins contain μ-oxo bridged iron atoms, although additional bridging ligands may be present. For most of these enzymes EPR has provided extremely important information.

Ferritin. Ferritin consists of a multisubunit protein shell, apoferritin, surrounding a core of hydrous ferric oxide [9.74]. The core is variable in size, and may consist of as many as 4500 atoms. The average composition of the core is $(FeOOH)_8 . FeO . OPO_3H_2$. It has crystalline behavior, and its size has been estimated to be 6400 ± 2000 pm.

There are still many problems related to the structure and to the way in which the iron core is formed and depleted, and indeed ferritin is the most inorganic of the metal proteins, with a chemistry which has been related to the chemistry of rust [9.75]. Its magnetic properties are those of a superparamagnet. This means that the iron centers are strongly coupled yielding a bulk ferromagnet. However, since the domains are limited (the protein size is large but finite), the overall magnetic behavior is that of a paramagnet, in which the individual magnets are indeed the protein molecules.

Hemerythrin. Hemerythrin is an oxygen carrier protein which occurs in several phyla of marine invertebrates. It occurs as either an octamer, a trimer, or a dimer of almost identical subunits of molecular weight close to 14 000 [9.76]. The active site contains two bridged iron atoms. The iron atoms are octahedrally coordinated and they share three bridging ligands, namely an oxo group and two carboxylates belonging to aspartate and glutamate residues. The structure of the binuclear iron site in methemerythrin azide of *Themiste discretum* is shown in Fig. 9.12 [9.77] (methemerythrin is the artificially oxidized form of hemerythrin). The two iron atoms both have the oxidation number $+3$. The nonoxygenated form, deoxohemerythrin, contains two iron(II) ions, while the oxygenated form contains two iron(III) and a peroxide group. There is also a half-oxidized form of deoxyhemerythrin, semi-met-hemerythrin, which has an iron(III) and an iron(II).

Apparently, the only EPR spectra which can be readily obtained are those of the semi-met forms [9.78]. The so-called (semi-met)$_2$ is obtained by one-electron reduction of methemerythrin, while (semi-met)$_0$ is obtained by one-electron oxidation of deoxyhemerythrin. They all have principal g values which are smaller than 2, a result similar to that observed in ferredoxins. These values are readily interpreted as due to an antiferromagnetic coupling between an iron(III) ($S = 5/2$) and iron(II) ($S = 2$), yielding a ground $S = 1/2$ state.

Ribonucleotide Reductase. A similar μ-oxo-di-iron site seems to be present in the ribonucleotide reductases, enzymes which catalyze the reduction of ribonucleotides, the first step in the biochemical pathway leading to DNA synthesis

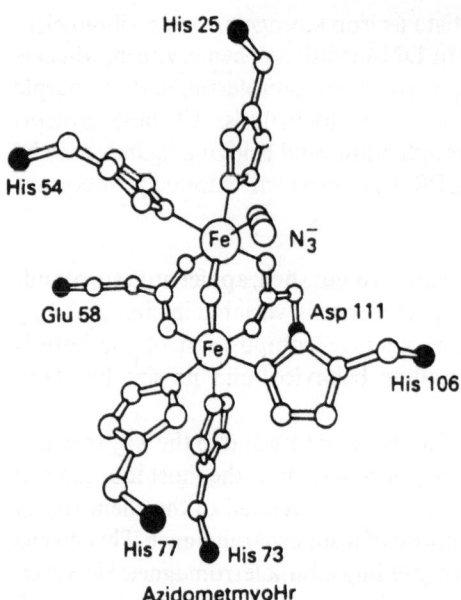

His 25

His 54

Glu 58

Fe'

N₃⁻

Asp 111

His 106

His 77

His 73

AzidometmyoHr

Fig. 9.12. Structure of the iron site of methemerythrin azide of *Themiste discretum*. After [9.75]

[9.79]. The best characterized at present is the enzyme of *Escherichia coli*, but also several studies have been performed on mammalian enzymes. The Fe-O-Fe moiety is EPR-silent, in agreement with magnetic susceptibility data, which suggest a moderate antiferromagnetic coupling between two high spin iron(III) ions. It is interesting to note that in the active site also a tyrosine radical is present, which was identified by EPR as a phenoxyl radical [9.80]. Some sort of interaction of the radical with the iron center has been evidenced by microwave saturation studies.

Uteroferrin and the Purple Acid Phosphatases. Purple acid phosphatases were first isolated from bovine spleen [9.81]. Although there are quite many phosphatases [9.82], the so-called purple acid enzymes were first isolated only after 1950. The purple acid phosphatases are characterized by their color, and by the optimum pH of function, which is below 7. They are known to contain pairs of iron atoms in the active site, and similar color and properties have been recently discovered in a protein which is present in porcine uterine flushings, which was called porcine uteroferrin [9.83]. Actually, these phosphatases can exist in two interconvertible forms: purple, which is enzymatically inactive, and pink. The former is obtained through an oxidation process, while the latter is obtained by reduction [9.82]. The EPR spectra of pink uteroferrin are characterized by a dramatic temperature dependence of the line width, and by a three-g value pattern analogous to that already described for 2-Fe sulfur proteins. In Fig. 9.13 one such spectrum is reported, showing the characteristic feature at $g \approx 1.74$.

226

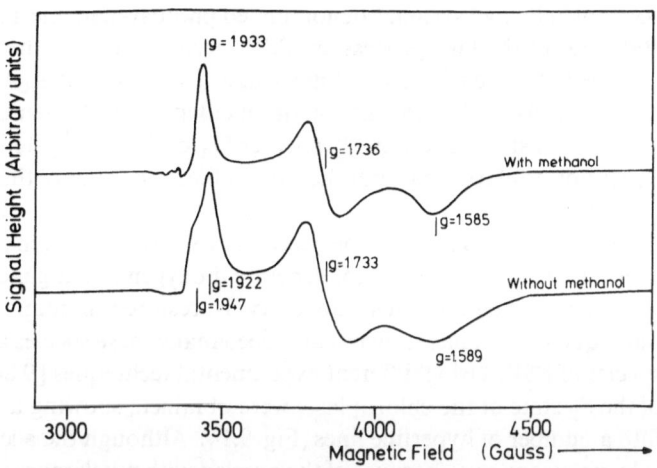

Fig. 9.13. X-band EPR spectra of pink uteroferrin at 12 K. After [9.82]

Magnetic susceptibility data indicate a ground $S = 1/2$ state for the pink form of splenic acid phosphatase, and a ground $S = 0$ for the purple form. It is thus generally accepted that in the purple form, two high spin iron(III) are antiferromagnetically coupled, while in the pink form antiparallel coupling of a high spin iron(III) and iron(II) yields the ground $S = 1/2$ state [9.82].

9.4 Exchange-Coupled Species in the Photosynthetic O_2-Evolving Complexes

We report here some of the information presently available on the photosynthetic process to show how effective EPR can be in providing hints to the nature of the active sites in biological processes, but also to show imagination must necessarily be a fundamental gift of researchers who want to study complex natural systems. In the following many conclusions will be extremely tentative and far from the elegance which can sometimes be attained in the study of simpler systems, but the importance of the processes under scrutiny will certainly induce a not too severe consideration in all the readers.

High plants can effect the photolysis of water to produce O_2 and reduced substances using sunlight. The reaction:

$$2H_2O(\text{liq}) \rightleftharpoons O_2(g) + 4H^+(aq) + 4e^- \tag{9.2}$$

evolves four electrons, but in plant photosynthesis the electrons are transferred one at a time. The overall process requires an energy transfer of 1.23 eV per electron. The process is accomplished by two light reactions in series, each of which absorbs one photon of at least 1.8 eV. The two light reactions are performed by two systems, denominated photosystem one and two, respectively, PSI and PSII. The process in PSII occurs through the generation of five intermediate oxidation states known as S_i (i = 0 to 4) states [9.84]. Both S_0 and S_1 are relatively stable in short-term incubated PSII membranes, while S_4 is relatively unstable and quickly releases O_2, cycling the system back to S_0. S_2 and S_3 are powerful oxidants, but they are stable for periods of the order of minutes at room temperature.

Manganese is known to be necessary to develop oxygen [9.85]. In fact, it has been shown that algae cannot generate dioxygen if manganese is withheld from the dietary medium; this capability is restored if manganese(II) is added. Subsequent work has shown that indeed manganese species are present in the S_2 species of PSII. Using different experimental techniques [9.86–88], EPR spectra of the S_2 state of the chloroplasts were obtained, showing a signal at $g = 2$, split into a number of hyperfine lines (Fig. 9.14). Although the accounting of the lines is by no means easy, because of the overlap with much more intense lines and the weakness of the signals in the wings of the absorption, the EPR spectra clearly indicate that the unpaired electron is interacting with more than one manganese nucleus and that the nuclei are equivalent, at least on the EPR time scale. Beyond this signal at $g \approx 2$, another one at $g = 4.1$ was also observed [9.89].

The first proposals suggested antiferromagnetically coupled mixed valence manganese dimers or tetramers [9.90, 91], but also a dimer Mn^{III}–Mn^{IV} with antiferromagnetic interaction, coupled to a third paramagnet with $S = 1$ was taken into consideration [9.88]. Up to now the most thorough attempt to justify

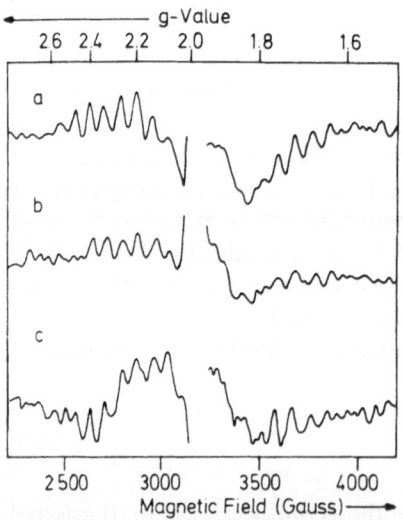

Fig. 9.14. Dependence of the hyperfine structure of the S_2 state EPR spectra of PSII membranes on temperatures of illumination: *a* 200 K illuminated, 4h dark-adapted; *b* 160 K illuminated, 6 min dark-adapted; *c* 170 K illuminated, 6 min dark-adapted. After [9.88]

228

all the known spectral features, including the temperature dependence of the signal intensities, has been made by Brudvig et al. [9.92], who have suggested a model with four manganese ions bridged by four oxygen atoms in a cubane-type structure. According to this proposal the species responsible for the EPR spectra contains three manganese(III) (S = 2) and one manganese(IV) (S = 3/2). The coupling should be strong antiferromagnetic within a manganese(III)-manganese(IV) pair; less intense antiferromagnetic between the remaining manganese(III)-manganese(III) pair, all the other interactions being ferromagnetic. Under these conditions the ground state is S = 3/2, which split in zero field, might justify the g values at 2 and 4. The suggested order of levels is shown in Fig. 9.15.

Research has been very active also on model compounds. An interesting EPR spectrum was reported [9.93] for binuclear complexes $Mn_2(ASQ)_4L_2$ (ASQ = 2-acetyl-1,4-benzo-semiquinone, L = CO) in solution (Fig. 9.16). It is apparent that the unpaired electrons interact with two equivalent ^{55}Mn nuclei. The coupling constant, $A = 45$ G, is half that expected for an isolated manganese(II) ion. The reported assignment of all the spectrum to an S = 3 state, however, is less convincing.

The mixed valence bi-μ-oxo manganese(III)-manganese(IV) dimer $(bipy)_2Mn^{III}(O)_2Mn^{IV}(bipy)_2$ is currently believed to be a reasonable model for the biological bimetallic site [9.94]. In $(bipy)_2Mn^{III}(O)_2Mn^{IV}(bipy)_2$ the two metal ions are different, with shorter bond lengths at the manganese(IV) site [9.95], as shown in Fig. 9.17. The Mn-Mn distance is 271.6 pm. The magnetic

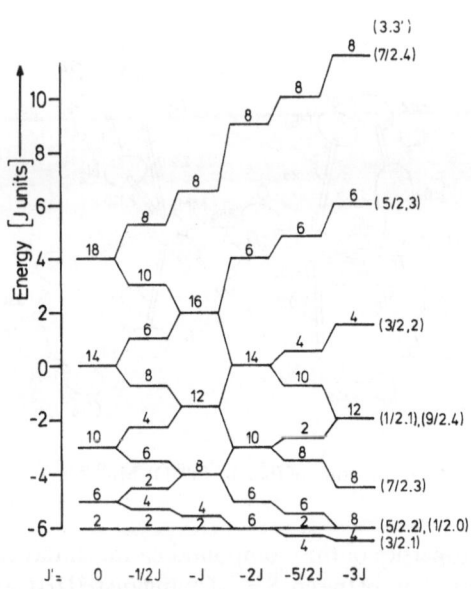

Fig. 9.15. Energy level diagrams for several exchange-coupling schemes of a 3Mn^{III}–Mn^{IV} species. After [9.92]

229

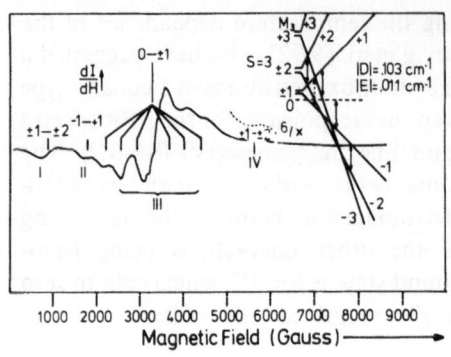

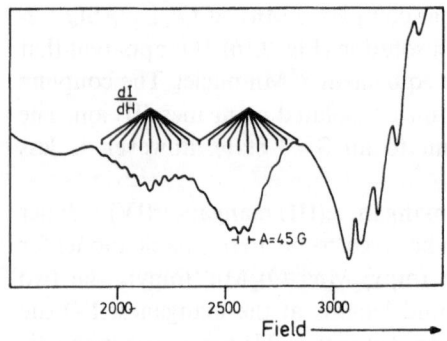

Fig. 9.16. X-band EPR spectra of $Mn_2(ASQ)_4L_2$ in dimethoxy-ethane at 11 K. After [9.93]

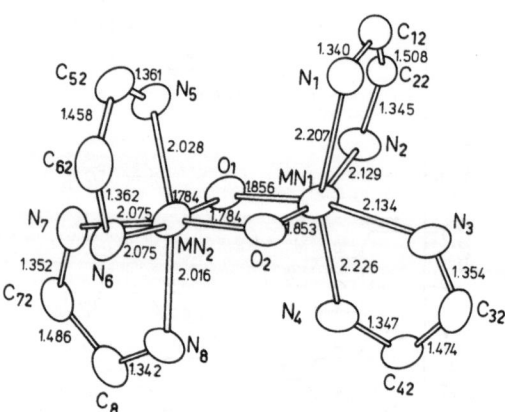

Fig. 9.17. Structure of $(bipy)_2Mn^{III}(O)_2Mn^{IV}(bipy)_2$. After [9.95]

properties of this compound can be interpreted as due to an antiferromagnetic coupling between $S = 2$ [manganese(III)] and $S = 3/2$ [manganese(IV)]. J was calculated to be $300(14)$ cm^{-1}. The frozen solution spectra of the phenanthroline derivative, which are reported to be practically identical to those of the (bipy)

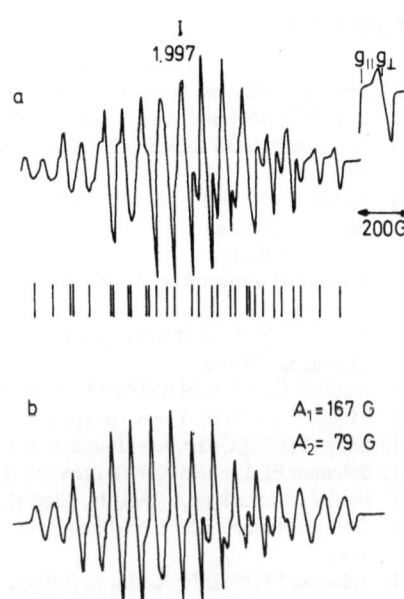

Fig. 9.18a, b. EPR spectra of (phen)$_2$Mn^{III}(O)$_2$Mn^{IV}(phen)$_2$ in CH$_3$CN at 18 K. **a** True spectrum; **b** computer simulation. After [9.94]

complex, are shown in Fig. 9.18. Due to the strong antiferromagnetic coupling, which sets the ground $S = 1/2$ state ≈ 450 cm^{-1} below the first $S = 3/2$ state, the observed spectra are only those of the doublet. The analysis of the spectra yields an average g value of 2.003, and two isotropic hyperfine coupling constants, with two unequivalent ^{55}Mn nuclei: $A_1 = 167 \pm 3$ G and $A_2 = 79 \pm 3$ G. Equations (3.20, 22, 23) indicate that the following relations must hold:

$$g = -g_{3/2} + 2g_2; \tag{9.3}$$

$$A_{Mn(IV)} = -A_{3/2}; \quad A_{Mn(III)} = 2A_2. \tag{9.4}$$

Therefore, both the g and the A values of the two individual ions must be rather close to each other. The larger hyperfine coupling in the dimer must be attributed to the interaction with manganese(III).

The mechanism of bacterial photosynthesis is completely different from that of plants. For the former the primary photochemical act involves the light-induced electron transfer from a primary electron donor, a bacteriochlorophyll dimer, to a primary electron acceptor [9.96]. The nature of the acceptor has been the matter of some controversy, however, it seems that it contains an iron-ubiquinone complex in which the metal ion and the radical are magnetically coupled. Indeed, the EPR spectra [9.97] show one signal at $g = 1.8$, which is two orders of magnitude broader than the value of the isolated ubiquinone. The g shift has been attributed to the coupling to $S = 2$ of high spin iron(II).

231

References

9.1 Smith TD, Pilbrow J (1980) Biol. Magn. Reson. 2: 85

9.2 Bray RC (1980) Biol. Magn. Reson. 2: 85

9.3 Tsvetkov YD, Dikanov SA (1987) In: Sigel H (ed) Metal ions in biological systems. Dekker 22: 207

9.4 Makinen MW, Wells GB (1987) In: Sigel H (ed) Metal ions in biological systems. Dekker 22: 189

9.5 Bertini I, Luchinat C (1986) NMR of paramagnetic species in biological systems, Benjamin Cumming, Boston

9.6 Koenig SH, Brown RD III (1983) In: Bertini I, Drago RS, Luchinat C (eds) The coordination chemistry of metalloenzymes. Reidel, Dordrecht p 19

9.7 Dwek RA (1972) 'Nuclear magnetic resonance in biochemistry: applications to enzyme systems, Clarendon, Oxford

9.8 Huttermann J, Kappl R (1987) In: Sigel H (ed) Metal ions in biological systems Dekker, 22: 1

9.9 Münck E (1979) In: The porphyrins, 4: 379

9.10 Lang G (1970) Quart. Rev. Biophys. 3, 1

9.11 Solomon EI, Penfield KW, Wilcox DE (1983) Structure and Bonding (Berlin) 53: 1

9.12 Bertini I, Scozzafava A (1981) In: (Sigel H (ed) Metal ions in biological systems. Dekker, 12: 31

9.13 Gray HB, Solomon EI (1981) In: Spiro GT (ed) Copper Proteins. Ch. 1, Wiley Interscience, New York

9.14 Solomon EI (1982) In: Karlin K, Zubieta J (eds) Copper coordination chemistry: biochemical and inorganic perspectives. Adenine, Guilderlan, NY, 1

9.15 Valentine JS, Pantoliano MW (1981) In: Spiro GT (ed) Copper proteins. Wiley Interscience, New York

9.16. Tainen JA, Getzoff ED, Been KU, Richardson JS, Richardson DC (1982) J. Mol. Biol. 18: 160

9.17 Weser U (1971) Biochim. Biophys. Acta 243: 203

9.18 Rotilio G (1972) Biochemistry 11: 2182

9.19 Liebermann RA, Sands RH, Fee JA (1982) J. Biol. Chem. 257: 336

9.20 Fee JA, Briggs RG (1975) Biochim. Biophys. Acta, 400: 439

9.21 Morgenstern-Badarau I, Cocco D, Desideri A, Rotilio G, Jordanov J, Dupre N (1986) J. Am. Chem. Soc. 108: 300

9.22 Banci L, Bencini A, Benelli C, Gatteschi D, Zanchini C (1982) Structure and Bonding (Berlin) 52: 38

9.23 Strothkamp KG, Lippard SJ (1981) Biochemistry 20: 7488

9.24 Prins R, Birker PJWML, Haasnot JG, Verschoor GC, Reedijk J (1985) Inorg. Chem. 24: 4128

9.25 Bencini A, Gatteschi D, Zanchini C, Haasnot JG, Prins R, Reedijk J (1985) Inorg. Chem. 24: 2812

9.26 Himmelwright R S, Eickman NC, Lu Bien CD, Solomon EI (1980) J. Am. Chem. Soc. 102: 5378

9.27 Schoot Uiterkamp AJM, van der Deen H, Berendsen HCJ, Boas JF (1974) Biochim. Biophys. Acta, 372: 407

9.28 Himmelwright RS, Eickman NC, Solomon EI (1979) Biochim. Biophys. Res. Commun. 86: 628

9.29 Wilcox DE, Long JR, Solomon EI (1984) J. Am. Chem. Soc. 106: 2186

9.30 Gray HB, Schugar HJ (1973) In: Eichhorn GL (ed) Inorganic biochemistry. Elsevier, Amsterdam 1: 102

9.31 Dunford HB, Stillmann JS (1976) Coord. Chem. Rev. 19: 187

9.32 Dunford HB (1982) Adv. Inorg. Biochem. 4, 41

9.33 Schulz CE, Devaney PW, Winkler H, Debrunner PG, Doan N, Chiang R, Rutter R, Hager LP (1979) FEBS Lett. 103: 102

9.34 Schulz CE, Rutter R, Sage JT, Debrunner PG, Hager LP (1984) Biochemistry 23: 4743

9.35 Gans P, Buisson G, Duee E, Marchon JC, Erler BS, Scolz WF, Reed CA (1986) J. Am. Chem. Soc. 108: 1223

9.36 Caughey WS, Wallace WJ, Volpe JA, Yoshikawa S (1976) In: Boyer PD (ed) The Enzymes 13: 299

9.37 Thomson AJ, Johnson MK, Greenwood C, Gooding PE, (1981) Biochem. J. 193: 687
9.38 Gunter MJ, Berry KJ, Murray KS (1984) J. Am. Chem. Soc. 106: 4227
9.39 Berg JM, Holm RH (1982) In: Spiro TG (ed) Iron sulfur proteins. Vol. 4; Ch. 1 Wiley, New York
9.40 Orme-Johnson WH, Orme-Johnson NR (1982) In: Spiro TG (ed) Iron Sulfur Proteins. Vol. 4, Ch. 2 Wiley, New York
9.41 Peisach J, Blumberg WE, Lode ET, Coon MJ (1971) J. Biol. Chem. 246: 5877
9.42 Rao KK, Evans MCW, Cammack R, Hall AO, Thompson CL, Jackson PJ, Johnson CE (1972) Biochem. J. 129: 1063
9.43 Debrunner PG, Münck E, Que L, Schulz CE (1977) In: Lovenberg W (ed) Iron sulfur proteins Vol. III. Academic, New York, 381
9.44 Moura I, Huynh BH, Hansinger RP, Le Gall J, Xavier AV, Münck EJ (1980) Biol. Chem. 225: 2493
9.45. Huynh BH, Kent JA (1984) Adv. Inorg. Biochem. 6: 193
9.46 Gibson JF, Hall DO, Thornley JHM, Whatley FR (1966) Proc. Natl. Acad. Sci. US 56: 987
9.47 Orme-Johnson WH, Sands RH (1972) in: Lovenberg W (ed) Iron Sulfur Proteins II, Ch. 5 Academic, New York
9.48 Bertrand P, Gayda JP (1979) Biochim. Biophys. Acta 579: 107
9.49 Bertrand P, Guigliarelli B, Gayda JP, Beardwood P, Gibson JF (1985) Biochim. Biophys. Acta 831: 261
9.50 Benelli C, Gatteschi D, Zanchini C (1987) Biochim. Biophys. Acta 1893: 365
9.51 Stout CD, Ghoshi D, Pattabhi K, Robbins AH (1980) J. Biol. Chem. 255: 1797
9.52 Emptage MH, Kent TA, Huynh BH, Rawlings J, Orme-Johnson WH, Münck E (1980) J. Biol. Chem. 255, 1793
9.53 Guigliarelli B, Gayda JP, Bertrand P, More C (1986) Biochim. Biophys. Acta 871: 149
9.54 Gayda JP, Bertrand P, Theodule FX, Moura JJG (1982) J. Chem. Phys. 77, 3387
9.55 Girerd J-J, Papaefthymiou GC, Watson AD, Gamp E, Hagen KS, Edelstein N, Frankel RB, Holm RH (1984) J. Am. Chem. Soc. 106: 5941
9.56 Huynh BH, Moura JJG, Moura I, Kent TA, Le Gall J, Xavier AV, Münck E (1980) J. Biol. Chem. 255: 3242
9.57 Thomson AJ, Robinson AE, Johnson MK, Moura JJG, Moura I, Xavier AV, Le Gall (1981) J. Biochim. Biophys. Acta 670, 93
9.58 Papaefthymiou GC, Girerd J-J, Moura I, Moura JJG, Münck E (1987) J. Am. Chem. Soc. 109: 4703
9.59 Noodleman L, Baerends EJ (1984) J. Am. Chem. Soc. 106: 1142
9.60 Gloux J, Gloux P, Lamotte B, Rins G (1985) Phys. Rev. Lett 54: 599
9.61 Stombaugh NA, Burris RH, Orme-Johnson WH (1975) J. Biol. Chem. 248: 7951
9.62 Antanaitis BC, Moss TH (1975) Biochim. Biophys. Acta 405: 262
9.63 Masharak PK, Papaefthymiou GC, Armstrong WH, Foner S, Frankel RB, Holm RH (1983) Inorg. Chem. 22: 2851
9.64 Gloux J, Gloux P, Rius G (1986) J. Am. Chem. Soc. 108: 3541
9.65 Siegel LM (1978) In: Singer JP, Ondarza RN (eds) Mechanism of oxidizing enzymes. Elsevier, North-Holland, 201
9.66 Jannick PA, Siegel LM (1982) Biochemistry 21: 3538
9.67 Christner JA, Münck E, Janick PA, Siegel LM (1983) J. Biol. Chem. 258: 11147
9.68 Christner JA, Janick PA, Siegel LM, Münck E (1983) J. Biol. Chem. 258: 11157
9.69 Nelson MJ, Lindahl PA, Orme-Johnson WH (1981) In: Eichhorn GL, Marzilli LG (eds) Advances in inorganic biochemistry. Elsevier, New York
9.70 Davis LC, Shah VK, Brill WJ, Orme-Johnson WH (1972) Biochim. Biophys. Acta 256: 512
9.71 Venters RA, Nelson MJ, McLean PA, True AE, Levy MA, Hoffman BM, Orme-Johnson WH (1986) J. Am. Chem. Soc. 108: 3487
9.72 Friesen GD, McDonald JW, Newton WE, Euler WB, Hoffman BM (1983) Inorg. Chem. 22: 2202
9.73 Huynh BH, Münck E, Orme-Johnson WH, (1979) Biochim. Biophys. Acta 576: 1929
9.74 Gabrio BW, Shoden A, Finch CA (1953) J. Biol. Chem. 204, 815

233

9.75 Lippard SJ (1986) Chem. Brit. 22, 222

9.76 Klotz IM, Keresztes-Nagy S (1961) Biochemistry 2: 445

9.77 Stenkamp RE, Sieker LC, Jensen LH (1984) J. Am. Chem. Soc. 106: 618

9.78 Wilking RG, Harrington PC (1983) Adv. Inorg. Biochem. 5: 51

9.79 Sjöberg BM, Gräslund A (1983) Adv. Inorg. Biochem. 5: 87

9.80 Ehrenberg A, Reichard P (1972) J. Biol. Chem. 247: 3485

9.81 Sundarajan TA, Sarma PS (1954) Biochem. J. 56: 125

9.82 Atanaitis BC, Aisen P (1983) Adv. Inorg. Biochem. 5: 111

9.83 Schlosnagle DC, Bazer FW, Tsibris JCM, Roberts RM (1974) J. Biol. Chem. 249: 7574

9.84 Kok B, Forbush B, McGloin M (1970) Photochem. Photobiol. 11: 457

9.85 Sauer K (1980) Acc. Chem. Res. 13: 249

9.86 Dismukes GC, Siderer Y (1981) Proc. Natl. Acad. Sci. US 78: 274

9.87 Brudvig J, Casey J, Sauer K (1983) Biochim. Biophys. Acta 723: 361

9.88 de Paula JC, Brudvig GW (1985) J. Am. Chem. Soc. 107: 2643

9.89 Casy JL, Sauer K (1984) Biochim. Biophys. Acta 767: 21

9.90 Hansson O, Andreasson LE (1982) Biochim. Biophys. Acta 679: 621

9.91 Dismukes GC, Ferris K, Watnick P (1982) Photobiochem. Photobiophys. 3: 243

9.92 de Paula JC, Beck WF, Brudvig GW (1986) J. Am. Chem. Soc. 108: 4002

9.93 Mathur P, Dismukes GC (1983) J. Am. Chem. Soc. 105: 7093

9.94 Cooper SR, Dismukes GC, Klein MP, Calvin M (1978) J. Am. Chem. Soc. 100: 7248

9.95 Plaksin PM, Stoufer RC, Mathew M, Palenick GJ (1972) J. Am. Chem. Soc. 94: 2121

9.96 Parson WW, Cogdell R (1975) J. Biochim. Biophys. Acta, 416: 105

9.97 Okamura MY, Isaacson RA, Feher, G (1975) Proc. Natl. Acad. Sci. US 72: 3491

10 Low Dimensional Materials

10.1 Linear Chain Manganese(II)

Any review of the EPR spectra of linear chain compounds must necessarily start from $[(CH_3)_4N]MnCl_3$ or, in a form more familiar in the physical literature, TMMC. Indeed, this is the first compound on which the one-dimensional properties were studied [10.1] and even after about 20 years it is still an excellent testing ground for all the refinements that new sophisticated theories are able to introduce [10.2-4]. We have already used many of the TMMC data in Chap. 6, so that we will add here only some more details.

Since manganese(II) is a ^6S ion many of its magnetic properties are particularly simple (isotropic g tensor, small single ion zero-field splitting, absence of anisotropic and antisymmetric exchange contributions). A number of manganese(II) compounds have been studied, including $CsMnCl_3.2H_2O$ (CMC) and $CsMnBr_3$. It is important to recall here that ideal one-dimensional behavior is determined by a one-dimensional structure and one-dimensional exchange pathways. In TMMC both conditions are met, while in CMC the difference between the intra- and interchain manganese-manganese contacts is much smaller, so that a less ideal behavior can be anticipated. Some relevant structural and magnetic properties of these three compounds are reported in Table 10.1. Since the ideal linear chain behavior will be attained by a given compound when the ratio of the inter- and intra-chain coupling constants J'/J tends to zero, it is apparent that TMMC is much better than the other two in this respect.

It is perhaps useful to recall here the characteristics of a well-behaved linear chain Heisenberg antiferromagnet: (1) the EPR spectrum consists of a single exchange-narrowed line with angular dependence of $|3\cos^2\Theta - 1|^{4/3}$, where Θ is the angle of the external magnetic field with the chain axis; (2) the line shape can be described by the Fourier transform of $\exp(-t^{3/2})$ at $\Theta = 0°$ and by a Lorentzian function at the magic angle $\Theta = 54.7°$; (3) the magic angle line-width is proportional to the inverse square root of the microwave frequency; (4) at $\Theta = 90°$ the line width increases with increasing frequency (inverse 10/3 effect); (5) in addition to the main line, a half-field transition can be observed. All these properties together are rarely met, and practically only TMMC behaves in this manner.

Beyond those already reported, of interest are the studies of TMMC and CMC doped with other transition metal ions. Indeed, the effect of doping must be expected to be much more intense in one-dimensional, as compared to three-

Table 10.1. Structural and magnetic properties of three manganese(II) linear chain compounds

	$[(CH_3)_4N]MnCl_3$	$CsMnBr_3$	$CsMnCl_3.2H_2O$
Structure	$P6_3/mmc$	$P6_3/mmc$	Pcca
Interchain	915.1	760.9	729∥b
distance			573∥a
Intrachain	324.5	326	473
distance			
J	9.3	13.3	4.2
T_N	0.85	8.3	4.88
J'/J	3.3×10^{-5}	1.0×10^{-3}	4.6×10^{-3}

dimensional systems, because impurities can block, or at least slow down, the effective rate of spin correlations, as was suggested in Section 6.4. Therefore, the lines must be expected to broaden and other dimensional effects such as shifts and satellite lines should be amplified by doping.

Studies of TMMC doped with both diamagnetic and paramagnetic impurities, such as copper(II), showed that the effect is essentially the same for both types of dopants, because the Mn-Cu interaction is small. In Fig. 10.1 is shown the dependence on the doping level of ΔB_{pp} of TMMC at room temperature and X-band frequency [10.14] for $\Theta = 0°$ and $90°$, respectively. It is apparent that the line width goes through a maximum at both angular settings for $x \approx 0.25$ (x is the doping level defined by the formula $[(CH_3)_4 Mn_{1-x}Cu_xCl_3)]$). This behavior is easily understood using a simple model which cuts the chain into segments of the type... Mn–Mn–Mn .., ... Cu–Mn .., ... Cu–Cu ... Within each of these segments the decay of the spin correlation function is expected to be proportional to $(Dt)^{-1/2}$, where D, the spin diffusion coefficient, is proportional to the coupling constant. For the chain an effective spin diffusion coefficient may be defined as a function of the diffusion coefficients in each different segment:

$$1/D(x) = (1-x)^2/D_{Mn} + 2x(1-x)/D_{Mn-Cu} + x^2/D_{Cu}. \qquad (10.1)$$

The linewidth under this assumption is given by:

$$\Delta B_{pp} \propto [M_2(x)D(x)^{-\frac{1}{2}}]^{2/3}, \qquad (10.2)$$

where $M_2(x)$ is the second moment, which will be a function of the composition x. Considering that the manganese spin is much larger than the copper spin, and that the second moment depends on $[S(S+1)]^2$, we see that on doping the second moment can be considered to be reduced by a factor $(1-x)$, because only the manganese segments contribute to it. As a consequence of this, on a qualitative basis, we can say that in order to have a maximum in the line width

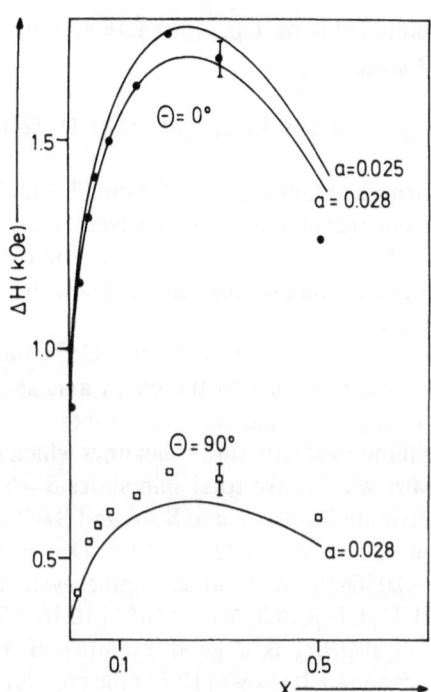

Fig. 10.1. Variation of the line width vs copper concentration in copper doped TMMC. The curves are calculated according to Eq (10.2). After [10.14]

for $x = 0.25$, $1/D(x)$ must increase on increasing x. On a quantitative basis using (10.1-2) the line width can be expressed as:

$$\Delta B_{pp}(x) = \Delta B_{pp}(0) (1-x)^{2/3} [x^2(1.5 - 2/a) - 2x(1 - 1/a) + 1]^{1/3}, \qquad (10.3)$$

where $\Delta B_{pp}(0)$ is the line width for the undoped TMMC. In order to arrive at (10.3) the spin diffusion coefficients for the manganese and copper segments have been expressed as:

$$D_{Mn} \approx 2.66 \, J_{Mn} [5/2(5/2 + 1)]^{\ddagger}; \qquad (10.4)$$

$$D_{Cu} \approx 2.66 \, J_{Cu} [1/2(1/2 + 1)]^{\ddagger}; \qquad (10.5)$$

where the exchange-coupling constant is in Kelvin. D_{Mn-Cu} is taken as $D_{Mn-Cu} = aD_{Mn}$, where a is an adjustable parameter. In (10.4) and (10.5) J_{Mn} and J_{Cu} are the coupling constants for the pure manganese and copper salts, respectively, whose experimental values are 9.3 and 63 cm^{-1}. Using (10.3 – 5) and the experimental line widths, the diffusion factor in the Mn–Cu segments has been

237

estimated to be: $D_{Mn-Cu} = 1.48$ K. If the simplest possible relation is assumed for D_{Mn-Cu}:

$$D_{Mn-Cu} = 2.66 \, J_{Mn-Cu} [5/2(5/2+1)1/2(1/2+1)]^{1/4}, \tag{10.6}$$

then we obtain $J_{Mn-Cu} = 0.6$ cm^{-1}, which is smaller than the value obtained from magnetic susceptibility and Neel temperature measurements (1.95 cm^{-1}).

This simple model, however, breaks down for measurements at the magic angle, because in this case the linewidth keeps increasing with x up to the largest available value ($x \approx 0.5$).

Doping TMMC with 20% Cd^{2+} the spectrum with the static magnetic field recorded parallel to the chain axis shows (Fig. 10.2) a central line attributed [10.15] to single ion resonances and to short chains of Mn^{2+} ions. The symmetrical structure of ten lines which is superimposed on it is attributed to ion pairs which have total spin states S = 5, 4, 3, 2, 1, and 0. The d and b lines are attributed to S = 5, c to S = 4 and S = 2, a and e to S = 3. The spectra yielded the single ion zero field splitting, D = 0.007(2) cm^{-1} and the dipolar term, D_{dip} = 0.0566(9) cm^{-1} which agree well with the values obtained by doping $[(CH_3)_4N]CdCl_3$ with Mn^{2+} [10.16, 17].

CsMnBr$_3$ is a good example of a less ideal manganese(II) linear chain compound. It shows [10.12] the angular dependence of the line width featured in Fig. 10.3. The experimental points were recorded at three different frequencies. At low frequency, 8.82 GHz, the line width has a minimum at 90°, contrary to the expected one-dimensional behavior with a minimum at the magic angle. This minimum is reestablished at higher frequencies (35.52 and 72.15 GHz), but the angular dependence can be described by the function:

$$\Delta B_{pp} = a_0 + a_1 \cos^2 \Theta + a_2 \cos^4 \Theta, \tag{10.7}$$

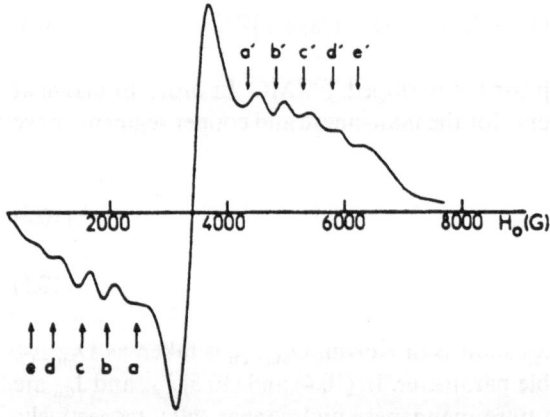

Fig. 10.2. Single crystal EPR spectrum of cadmium-doped TMMC at room temperature. The static magnetic field is parallel to the chain axis. After [10.15]

238

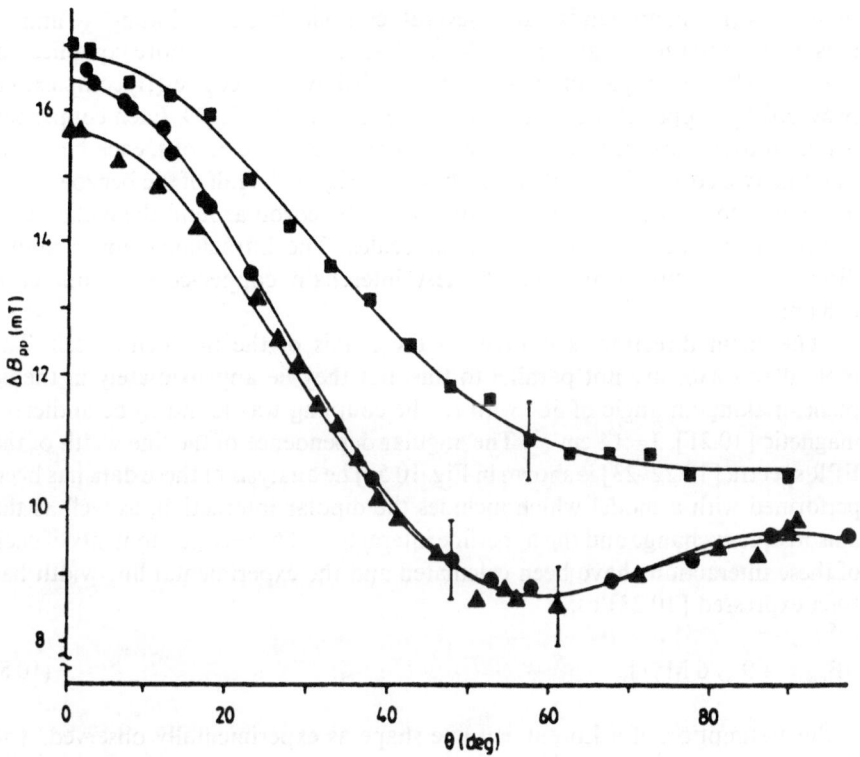

Fig. 10.3. Angular dependence of the line-width of $CsMnBr_3$ at room temperature and three different frequencies: ■ 8.82 GHZ, ○ 35.52 GHz, ▲ 72.15 GHz. After [10.12]

rather than by $|3\cos^2\Theta - 1|^{4/3}$. No satellite half-field transition could be detected, even at the lowest frequency, and the lines are much narrower in $CsMnBr_3$ than in TMMC. Therefore, from both angular dependence and line shape no support for the one-dimensional behavior of $CsMnBr_3$ is obtained. However, the frequency dependence of the line-width at the magic angle shows that it is indeed a linear chain.

Another system, which has been less well characterized, but has been reported to show the characteristic $|3\cos^2\Theta - 1|^{4/3}$ behavior is $Mn(py)_2Cl_2$ [10.18]. $[(CH_3)_3NH]MnCl_3 \cdot 2H_2O$ shows angular dependence which was interpreted as due to the sum of two contributions, one relative to the usual dipolar broadening, and the other relative to the single ion zero field splitting [10.19].

10.2 Linear Chain Copper(II)

A few systems involving copper(II) ions have been investigated. Generally, the one-dimensional behavior of these is less well established than in the manganese(II) compounds, and several complications, including **g** and **A** anisotropy, are present and make the analysis of the spectra more complicated. One of the best examples of one-dimensional behavior in copper(II) complexes is provided by copper(II) benzoate trihydrate, $Cu(benz)_2.3H_2O$. Each copper ion is bound to two oxygen atoms of two different benzoate anions, each benzoate bridging two copper ions [10.20] as shown in Fig. 10.4. Half of the benzoates are not bound to copper and the coordination polyhedron around the metal ion is completed by four bridging water molecules. The intrachain copper-copper distance is 315 pm, while the shortest interchain copper-copper contact is 698 pm.

The chain direction is parallel to the c axis of the monoclinic cell. The molecular z axes are not parallel to this, but they lie approximately in the ac plane, making an angle of 38° with c. The coupling was found to be antiferromagnetic [10.21], $J = 12$ cm^{-1}. The angular dependence of the line width of the EPR spectra [10.22–23] is shown in Fig. 10.5. The analysis of these data has been performed with a model which includes the dipolar interaction, as well as the anisotropic exchange and the hyperfine interaction. The second moments of each of these interactions have been calculated and the experimental line width has been expressed [10.23] as:

$$\Delta B_{pp} = 4/9 \sqrt{6} \, M_2^e/J, \tag{10.8}$$

in the assumption of a Lorentzian line shape as experimentally observed. The second moments were expressed as:

$$M_2^e = M_2^f + \tau M_2^{(0)}, \tag{10.9}$$

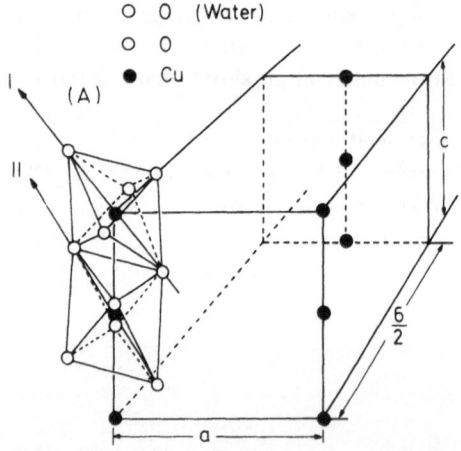

O O (Water)

O O

● Cu

Fig. 10.4. Scheme of the unit cell of $Cu(benz)_2.3H_2O$. After [10.20]

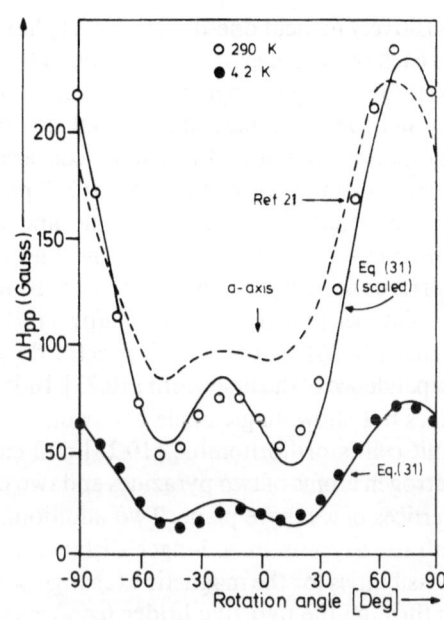

Fig. 10.5. The experimental and theoretical line width data for Cu(benz)$_2$.3H$_2$O. The rotation angle is from a in the ac plane. After [10.23]

where M_2^f and $M_2^{(0)}$ are the calculated full and secular second moments, respectively. τ is a parameter, which accounts for the increased role of secular terms in one dimension. For $\tau = 0$, Eq. (10.9) yields the usual three-dimensional behavior, while when τ is large, the angular dependence is dominated by the secular components. τ values of order of unity are consistent with Lorentzian lines, and slightly enhanced secular components.

The expressions for the various components of M_2 were given in Chap. 6. Care must be taken to consider the different principal directions of the various terms responsible for the broadening of the lines. For Cu(benz)$_2$. 3H$_2$O the angular dependence of the experimental line width was reproduced using the parameters, $D^{ex} = 0.125(15)\,cm^{-1}$ and $\tau = 1.0(2)$, where D^{ex} is the zero field splitting parameter associated with anisotropic exchange. The other parameters (A_x, A_y, A_z, D_d for the dipolar interaction, J) were fixed at the values experimentally observed.

An interesting observation to be made here is that since D^{dip} and D^{ex} are diagonal in different reference frames, the presence of anisotropic exchange limits the one-dimensional behavior. In fact, the largest deviations from Lorentzian line shapes are expected parallel to the chain direction, where the intrachain dipolar interaction reduces to the purely secular component. However, in general, D^{ex} will not have its z axis parallel to this, and nonsecular components are thus reintroduced. The only case in which the two contributions add up positively to yield one-dimensional behavior is when the two tensors are parallel to each other. An example is provided by CuSALMe, where SALMe is the Schiff base formed by methylamine and salicylaldehyde [10.24]. In this case the planar CuSALMe molecules are stacked on top of an other [10.25] so that the z molecular axes and the chain axis are practically coincident. The copper-copper distance along the chain is 333 pm, while the shortest interchain copper-copper separation is 919 pm. The line width at 76 K was found to follow a $|3\cos^2 \Theta - 1|^{4/3}$ behavior, and the line shape along the chain was found to correspond nicely to the Fourier transform of $\exp(-t^{3/2})$.

Copper pyrazine nitrate, $Cu(pyz)(NO_3)_2$, is a one-dimensional antiferromagnet [10.26], but the EPR spectra do not show the classic one-dimensional dependence of the line width [10.23]. Indeed, as shown in Fig. 10.6, the line width does not show magic angle behaviour in any of the three principal planes. The unit cell is orthorhombic [10.27] and each copper ion is coordinated by two nitrogen atoms of two pyrazines and two oxygen atoms of two nitrate ions on the vertices of a square plane. Two additional axial sites are occupied by two other nitrate oxygens at a longer distance (Fig. 10.7). In principle there are two possibilities for the magnetic exchange pathway defining the chain direction: one is through the pyrazine bridge (copper-copper separation of 671.2 pm along a), the other through nitrate ions (copper-copper separation of 514.2 pm along b). The molecular coordination plane is almost parallel to the ac plane, so that the z molecular axis is practically parallel to b. According to this structure, one can

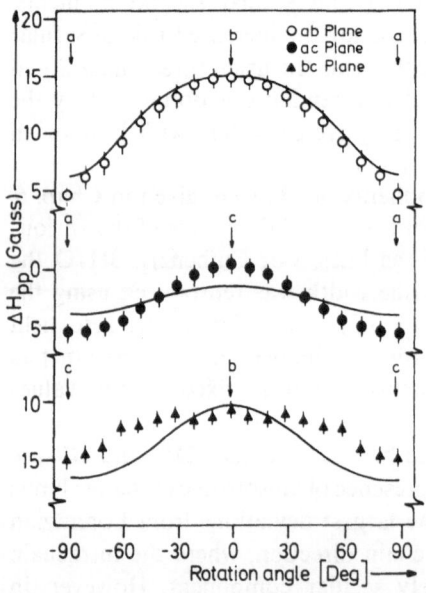

Fig. 10.6. Angular dependence of the line width of $Cu(pyz)(NO_3)_2$ at room temperature and X-band frequency. After [10.23]

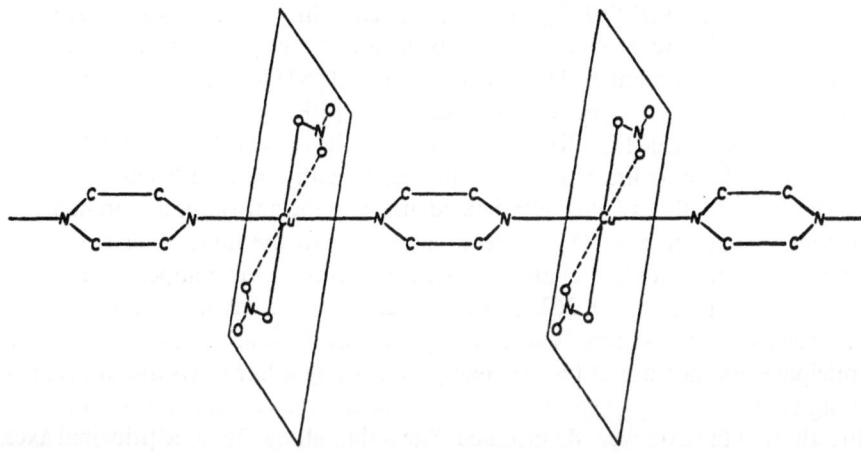

Fig. 10.7. The structure of $Cu(pyz)(NO_3)_2$

anticipate that $\mathbf{A}$, $\mathbf{D}^{dip}$, and $\mathbf{D}^{ex}$ all have the largest component parallel to b, and indeed experimentally the line width has its maximum parallel to b. The expressions for the second moment are split into two summations, one relative to the sites that are exchange-coupled to the reference spin, and the other to those which are not. For the former both dipolar and anisotropic exchange interactions must be included, while for the latter only dipolar coupling is operative. Therefore, if the strongest exchange is parallel to a, the anisotropic exchange is expected to have its largest contribution orthogonal to the chain, while if the strongest exchange is parallel to b, the largest contribution is parallel to the chain. It is possible, therefore, to differentiate the two possible exchange pathways, because inclusion of $\mathbf{D}^{ex}$ gives different values in the two cases. Indeed, for $Cu(pyz)(NO_3)_2$ the best fit to the experimental data was achieved assuming that the pyrazine bridge is most effective in transmitting the exchange interaction. Like in the $cu(benz)_2 \cdot 3H_2O$ case the experimental data were fitted with the two parameters $\mathbf{D}^{ex} = 0.128(15)$ cm^{-1} and $\tau = 0.18(8)$. Although the value of $\mathbf{D}^{ex}$ is fairly large for two copper ions separated by more than 500 pm, it is comparable to the values observed in pairs of copper ions bridged by triazolato ligands.

In $Cu(py)_2Cl_2$ (py = pyridine) the line width is dominated by the inequivalence of the magnetic chains. In fact, in the monoclinic cell there are two chains [10.23, 28], oriented in such a way that the z molecular axes form an angle of ca. 47° [10.29]. In the absence of interchain exchange two lines should be observed, but experimentally only one line is observed at both X- and Q-band frequency. Since the lines are considerably broader at Q-band, the interchain coupling constant could be estimated as $j' = 0.22$ cm^{-1}, assuming knowledge of the g anisotropy of the two chains.

Several copper halide derivatives with different cations have been reported to show one-dimensional behavior [10.30–31]. For instance in bis-piperidinium

243

tetrachlorocuprate(II) [10.32] the exchange-coupling constant was determined to be 1.1 cm^{-1}, and a $|3\cos^2\Theta - 1|^{4/3}$ behavior of ΔB_{pp} was observed at 90 K, while the complex angular dependence in $[(CH_3)_3NH]CuCl_3.2H_2O$ was interpreted within the same model described above [10.33].

$[(C_6H_{11}NH_3)CuCl_3]$, CHAC, and $[(C_6H_{11}NH_3)CuBr_3]$, CHAB, have been shown to behave as linear chain ferromagnets with $J \approx 70\text{--}100$ cm^{-1} [10.34]. The line width follows a $|\cos^2\Theta|^2$ dependence, and the line shape is Lorentzian at all the angular settings. This is presumably due to the noncollinearity of the dipolar and anisotropic exchange tensors. However, at low temperature large g shifts are observed [10.35, 36], as a consequence of the ferromagnetic nature of the intrachain interaction. Indeed, using the model outlined in Chapter 6, one anticipates an increase in the resonance field along a hard axis and a decrease along an easy axis. By experimentally measuring the resonance field shifts the directions of the axes were determined. The g shift along the three principal axes, ξ_1, ξ_2, and ξ_3 is shown in Fig. 10.8. ξ_3 corresponds to the chain axis. In CHAC the hard axes correspond to ξ_1 and ξ_3 showing that the exchange is of the Ising type, with the easy axis along ξ_2, which corresponds to the direction of the z molecular axis. For CHAB also ξ_3 is an easy axis, indicating that the anisotropy field is of the XY type.

Copper maleonitrilethiolates, $[Cu(mnt)]^{2-}$, provide interesting examples of one-dimensional $S = \frac{1}{2}$ in which EPR has been fundamental in determining the weak Heisenberg coupling. Accurate data have been reported for both the $[NMe_4]$ [10.37] and the $[N(nBu)_4]$ derivatives [10.38]. The former is monoclinic [10.39], with a copper-copper intrachain distance of 781.1 and 784.1 pm, while the latter is triclinic [10.38] with an intrachain distance of 940.3 pm. The exchange-coupling constant in $[NMe_4][Cu(mnt)]$ has been determined to be 0.112 cm^{-1} at room temperature, and to increase steadily as the temperature is lowered [10.37]. On the other hand, in $[N(nBu)_4]_2[Cu(mnt)]$ the room temperature exchange is ca. 0.04 cm^{-1} [10.38]. The smaller value can be justified on the basis of the longer metal-metal distance observed in the latter compound. However, it is interesting to note that J *decreases* with temperature for $[(N(nBu)_4]_2[Cu(mnt)]$ to a low value of 0.0107 cm^{-1} at 4.2 K. Under these conditions the exchange interaction can be smaller than the hyperfine splitting, which along z is 0.0161 cm^{-1}. Therefore, the EPR spectra at liquid helium temperature are particularly complex (Fig. 10.9). Indeed, since exchange is smaller than the hyperfine splitting, each site along the chain will be characterized by a given M_I value, and sites with different M_I must be considered as different. Therefore, the chain can be assumed to be formed of segments within which adjacent spins have the same M_I value, according to the scheme shown below:

(3/2)   (1/2   1/2   1/2   1/2) $(-3/2)$ $(-1/2$ $-1/2)$

Each segment will provide a different kind of a spectrum which can be calculated

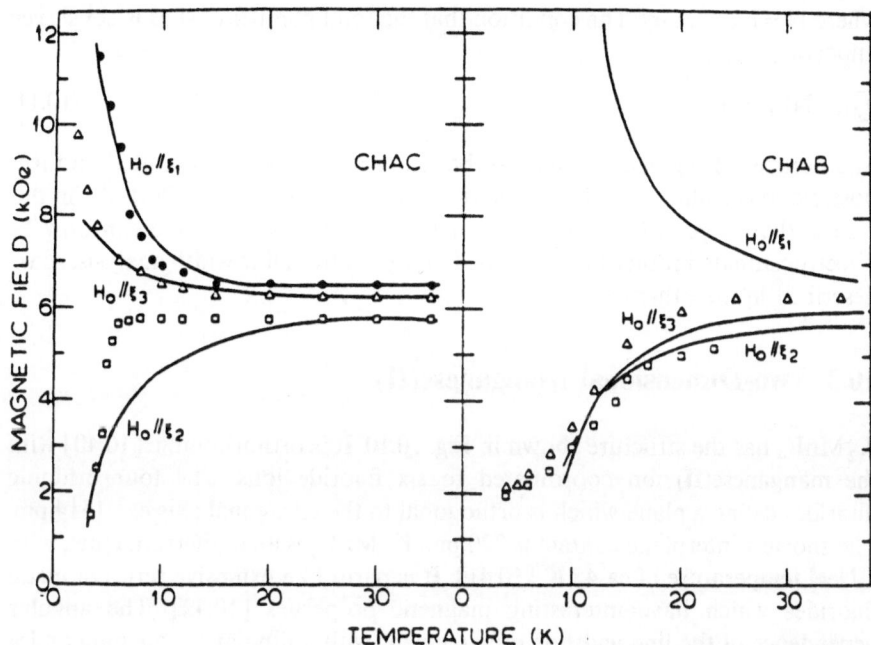

Fig. 10.8. Temperature dependence of the resonance fields for CHAC and CHAB, $\nu = 18.3$ GHz. After [10.36]

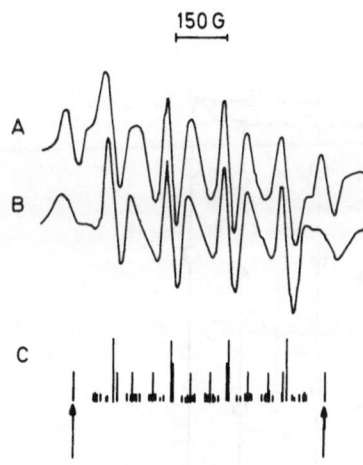

Fig. 10.9. EPR spectrum of $[N(tBu)_4]_2[Cu(mnt)_2]$ at 4.2 K with B parallel to the z molecular axis. *A* Experimental; *B* simulated; *C* stick plot. After [10.38]

assuming that the different segments are statistically determined. The probability of having a segment of length N is given by:

$$P_N = [2I/(2I+1)]^2 \, 1/(2I+1)^{N-1}, \tag{10.10}$$

245

where $N = 1, 2, \ldots \infty$. The condition that the total number of sites is conserved imposes:

$$\sum_{N=1}^{\infty} N P_N = 1. \tag{10.11}$$

The number of segments which can be relevant statistically is limited and it is possible to calculate the whole spectrum. The result is that the effect of J in this case is that of producing splittings of the spectra, and its value can now be directly estimated rather than obtained indirectly from line width analysis, as we described in the other cases.

10.3 Two-Dimensional Manganese(II)

K_2MnF_4 has the structure shown in Fig. 10.10. It is orthorhombic [10.40] with the manganese(II) ion coordinated to six fluoride ions. The four bridging fluorides define a plane which is orthogonal to the tetragonal axis $c = 1314$ pm. The shortest interplane contact is 720 pm. K_2MnF_4 is an antiferromagnet, with a Neel temperature of ca 45 K [10.41]. It is part of an extensive series of metal fluorides which have interesting magnetic properties [10.42]. The angular dependence of the line width is in agreement with a dipolar broadening and a two-dimensional behavior [10.43], obeying a law of the type:

$$\Delta B_{pp} = a + b|3\cos^2\Theta - 1|^2, \tag{10.12}$$

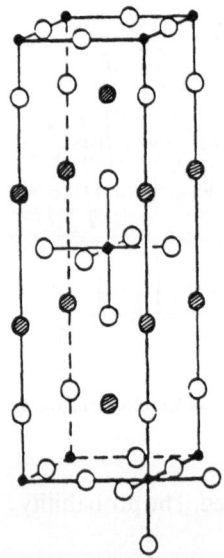

Fig. 10.10. Structure of K_2MnF_4. After [10.40]

where Θ is the angle from the perpendicular to the plane. The line shape is Lorentzian at the magic angle, while at $\Theta = 0°$ it is the Fourier transform of $\exp[-At \ln At - Bt_0]$ where A, B, and t_0 are parameters depending on the two-spin correlation function and on the correlation time. As in the one-dimensional case the line shape is intermediate between Lorentzian and Gaussian, although closer to the former. It is important to note that the actual shape strongly depends on A/B, therefore it is not possible to give a universal curve for $\Theta = 0°$.

A similar behavior has been observed also in other two-dimensional salts, such as $[CH_3NH_3]_2MnCl_4$ and $[C_2H_5NH_3]_2MnCl_4$ [10.43, 44], $[C_2H_5NH_3]_2MnBr_4$ [10.45], and $Mn(HCOO)_2 \cdot 2H_2O$ [10.46, 47].

10.4 Two-Dimensional Copper(II)

There are quite a number of two-dimensional copper(II) magnets because many copper halides, with different cations $[Cat]_2[CuX_4]$, $(X = Cl, Br)$, possess layered structures [10.48]. In general, in all the systems studied the angular dependence of the line width does not show any particular effect which can be attributed to an increased role of the secular terms, and the lines are within error lorentzians.

The most obvious reason for this is that in two-dimensional magnets it is comparatively more difficult than in one-dimensional magnets to have simplification in the expressions of, e.g. the dipolar second moment analogous to the one for which all the Θ_{ij}'s are identical $(\Theta_{ij} = \Theta)$ for one dimension. Thus, there are different angular contributions which quench the secular components.

Lacking any particular interest in this respect on the spin dynamics in these compounds, the major interest has focused on the temperature dependence of the line width. In fact, for most of these compounds, ΔB_{pp} was found to decrease linearly with temperature in the range above the transition to the ordered state. Such a behavior has been found, for instance, in $K_2CuCl_4 \cdot 2H_2O$ [10.49, 50], $Cu(HCOO)_2 \cdot 4H_2O$ [10.51], $Cu(HCOO)_2 \cdot 2(NH_2)_2CO \cdot 2H_2O$ [10.52], and in many of the layered tetrahalides [10.48].

In $Cu(HCOO)_2 \cdot 4H_2O$ each copper is coordinated to four oxygen atoms of four different formiate anions, which bridge two different copper ions [10.53]. These molecular planes define a layer parallel to the *ab* crystallographic plane of a monoclinic cell. The coordination around copper is completed by two water molecules, which are connected to other water molecules in a layer which is interposed between the copper layers. The line width decreases linearly with temperature in the range 50–300 K. This behavior has been attributed [10.50] to phonon modulation of the antisymmetric exchange, which can occur between pairs of copper ions not related by an inversion center. A quantitative analysis yielded $d/J \approx 0.005$, which is roughly 20 times smaller than one would obtain from Moryia's estimate $d \propto (\Delta g/g)J$. A similar behavior was observed in the analogous $Cu(HCOO)_2 \cdot 2(NH_2)CO \cdot 2H_2O$.

Layered $[Cat]_2[CuX_4]$ salts are the most numerous group of two-dimensional copper(II) magnets. They all have a structure formed of layers such as those depicted [10.48, 54] in Fig. 10.11. Each copper is coordinated to $4+2$ chlorides, the long bonds lying in the plane of the layer. The isotropic coupling is ferromagnetic. A detailed analysis of the EPR spectra was performed by Soos et al. [10.54] for $[Pt(NH_3)_4]CuCl_4$ and $[H_3N(CH_2)_3NH_3][CuCl_4]$. They included dipolar, anisotropic, and antisymmetric exchange interactions to analyze the line widths. Using the formulae of Sect. 6.3 they could show that dipolar, hyperfine, and **g** tensor contributions to the second moment are negligible, and included only the anisotropic and antisymmetric exchange:

$$M_2(\Theta, T) = \{M_2^A(\Theta)F_A(kT/J) + M_2^S(\Theta)F_S(kT/J)\}\chi_c/\chi(T), \qquad (10.13)$$

where A and S refer to antisymmetric and symmetric, respectively, J is the isotropic exchange constant, χ_c is the Curie susceptibility, and $\chi(T)$ is the true susceptibility at the temperature T, F_A and F_S are factors describing the static spin correlation due to the isotropic exchange, related to the two-spin correlation functions

$$F_A(kT/J) = 1 - c_1 - 2c_3 + 2c_1^2 + 2c_2^2 + 4c_3^2 \ldots ; \qquad (10.14)$$

$$F_s(kT/J) = 1 + c_1 + 4c_2\,2c_3 + 2c_1^2 + 6c_1^2 + 8c_3\,C_2 + 4c_3^2; \qquad (10.15)$$

where c_n indicates
$$c(r_{ij}, J/kT) = 3\delta_{\alpha\beta}\langle S_{i\alpha}S_{i+n\beta}\rangle/[S(S+1)]. \qquad (10.16)$$

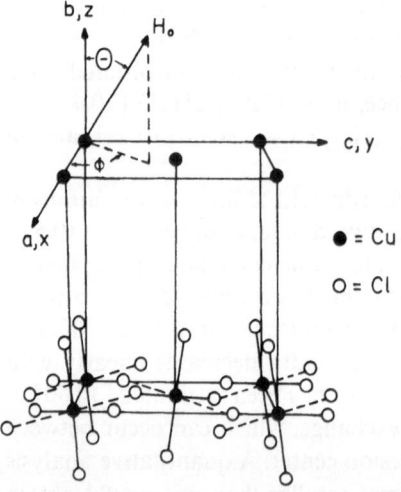

Fig. 10.11. Schematic representation of the $CuCl_4^{2-}$ layer and the polar coordinates of the magnetic field. After [10.54]

The temperature dependence of $F_I(kT/J)\chi_c/\chi(T)$, $I = A, S$ is shown in Fig. 10.12 for a square planar ferromagnet. Since for $kT > J$ the dependence is linear, the observed temperature dependence of ΔB_{pp} is justified.

The angular dependence of the anisotropic exchange part of the second moment is given by:

$$M_2^s(\Theta) = z/8(D^{ex})^2(1 + \cos^2\Theta), \tag{10.17}$$

where z is the number of nearest neighbors, and D^{ex} is the anisotropic exchange parameter. Θ is the angle from the normal to the layer. In (10.17) it was assumed that no secular enhancement is operative, in accord with experimental data.

The corresponding term for the antisymmetric exchange is given at high temperature by:

$$M_2^A(\Theta) = 1/16 zd^2(2 + \sin^2\Theta), \tag{10.18}$$

in the assumption of **d** lying in the layer, and

$$M_2^A(\Theta) = 1/16 zd^2(1 + \cos^2\Theta), \tag{10.19}$$

for **d** orthogonal to the layer.

The experimental angular dependence for $[Pt(NH_3)_4][CuCl_4]$ is shown in Fig. 10.13. It is apparent that ΔB_{pp} is smaller at $\Theta = 0°$ than at $\Theta = 90°$. Therefore, the high temperature data, where $F_A(kT/J) \approx F_S(kT/J) \approx 1$ are simply interpreted using

$$\Delta B_{pp} = 1 + c \cos^2\Theta, \tag{10.20}$$

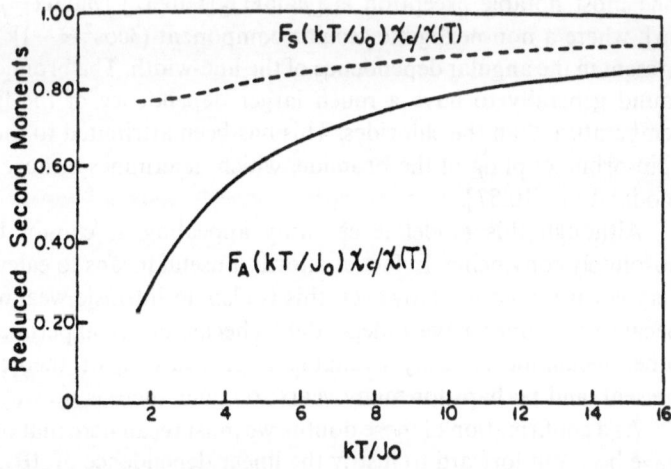

Fig. 10.12. Effect of static spin correlations in the square planar ferromagnet on the second moment of symmetric, F_S, and antisymmetric, F_A broadening perturbations. After [10.54]

249

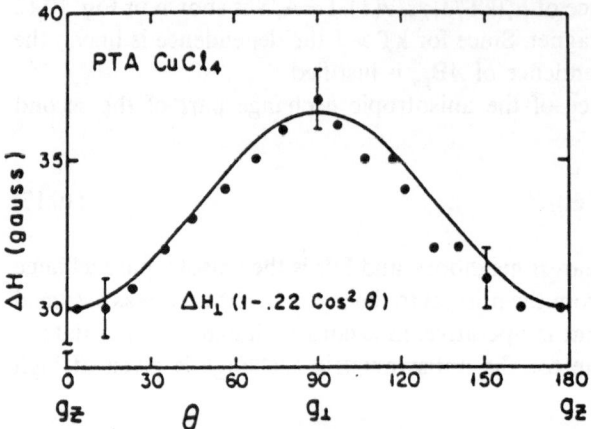

Fig. 10.13. Angular dependence of the line width of $[Pt(NH_3)_4][CuCl_4]$ at 300 K. Θ is the angle of B with the normal to the copper plane. After [10.54]

with $c > 0$ when the anisotropic exchange dominates and $c < 0$ when the antisymmetric exchange dominates. The ratio between the experimental line width at $\Theta = 0°$ and $\Theta = 90°$ yields an estimation of the d/D ratio:

$$\Delta B_\parallel / \Delta B^\perp \approx 1 + c = (4D^2 + 2d^2)/(2D^2 + 3d^2). \tag{10.21}$$

From the experimental value of $c = -0.22$ for $[Pt(NH_3)_4][CuCl_4]$ $(d/D)^2$ was estimated to be ≈ 6.

The same type of analysis has been applied to several other layered chlorocuprates. In general, the same kind of angular dependence was evidenced. The most notable exception is $[\beta alaH]_2[CuCl_4]$ ($\beta alaH = \beta$-alanine) [10.55, 56], where a non-negligible secular component $(3\cos^2 \Theta - 1)^2$ was found to be present in the angular dependence of the line-width. The bromide salts have been found generally to have a much larger dependence of the line width on the temperature than the chlorides. This has been attributed to the role of the large spin-orbit coupling of the bromide, which determines a more effective phonon modulation [10.57].

Although this model is certainly appealing, it cannot be considered as absolutely convincing. In fact, it affords a useful means to calculate d, which is a very elusive quantity. However, this is also an intrinsic weakness of the model, because it cannot have independent checks, either experimental, because no other technique is easily available to measure d, or theoretical, because at present, and perhaps for many years to come, there is no way of calculating d.

As a confirmation of these doubts we must recall here that other justifications have been put forward to justify the linear dependence of ΔB_{pp} on temperature. Among these, one which has been worked out in some detail, is the temperature dependence of the isotropic exchange. Indeed, there is evidence in the literature

250

that the interchain and the interlayer interaction may be temperature-dependent. One such example is $K_2CuCl_4.2H_2O$, which contains two magnetically non-equivalent chains, which can be characterized by two g values, g_1 and g_2. For exchange energies less than $J \approx \frac{1}{2}(g_1 - g_2)\mu_B B$, two resonances are observed. By measuring the magnetic field at which the two lines coalesce, it is possible to obtain J. For this particular compound J was found to decrease with increasing temperature in the range 77–300 K. The ratio J(300)/J(77) is $\approx 1/5$. This is not the only data of this kind, because similar results have been observed for other transition metal complexes. On this basis it was concluded that also phonon modulation of the isotropic exchange can be responsible for the temperature dependence of the line width. The effect of phonon modulation was estimated through a simplified model, which assumes an exponential dependence of J on the distance between the paramagnetic centers. The model has been applied with some success to some layered copper(II) compounds [10.58].

References

10.1 Dietz, RE, Merritt FR, Dingle R, Hone D, Silbernagel, BG, Richards PM (1971) Phys. Rev. Lett. 26: 1186
10.2 Lagendijk A (1978) Phys. Rev. B 18: 1322
10.3 Benner H, Broedehl M, Seitz W, Wiese J (1983) J. Phys. C 16: 6011
10.4 Siegel E, Mosebach H, Pauli N, Lagendijk A (1982) Phys. Lett. 90A: 309
10.5 Morosin B, Graeber EJ Acta Cryst. 23: 766
10.6 Dupas C, Renard JP (1976) Solid state commun. 20: 581
10.7 Goodyear J, Kennedy D (1972) Acta Cryst. B28: 1640
10.8 Jensen SJ, Andersen P, Rasmussen SE (1962) Acta Chem. Scand. 16: 1890
10.9 de Jonge WJM, Swüste CHW, Kopinga K, Takeda K (1975) Phys. Rev. B 12: 5858
10.10 Eibschütz M, Sherwood RC, Hsu FSL, Cox DE 1972 AIP Conf. Proc. N10 P1, Graham CD, Rhyme JR, Rhyme IJ, eds. p 684
10.11 de Jonge WJM, Hijmans JPAM, Boersma F, Schouten JC, Kopinga K (1978) Phys. Rev. B17: 2922
10.12 Kalt H, Siegel E, Pauli N, Mosebach H, Wiese J, Edgar A (1983) J. Phys. C 16: 6427
10.13 Allroth E (1980) J Phys. C13: 383.
10.14 Clement S, Dupas C, Renard JP, Cheikh-Rouhou A (1983) J. Physique 43: 767
10.15 Clement S, Renard JP, Ablart G (1984) J. Magn. Reson. 60: 46
10.16 McPherson GL, Henling LM, Koch RE, Quarls HF. (1977) Phys. Rev. B16: 1983
10.17 Tazuke Y (1977) J. Phys. Soc. Jpn.42: 1617
10.18 Richards PM, Quinn RK, Morosin B (1973) J. Chem. Phys. 59: 4474
10.19 Iio K, Isobe M, Nagata K (1975) J. Phys. Soc. Jpn. 38: 1212
10.20 Koizumi H, Osaki K, Watanabe T (1963) J. Phys. Soc. Jpn. 18: 117
10.21 Date M, Yamazaki H, Motokawa M, Tazawa S (1974) Progr. Theor. Phys. Suppl. No 405, 194
10.22 Okuda K, Hata H, Date M (1972) J. Phys. Soc. Jpn. 33: 1574
10.23 McGregor KT, Soos ZG (1976) J. Chem. Phys. 64: 2506
10.24 Bartkowski RR, Morosin B (1972) Phys. Rev. B̂, 4209
10.25 Lingafelter EC, Simmons GL, Morosin B, Scheringer S, Freinburg C (1961) Acta Cryst. 14: 1222
10.26 Losee DB, Richardson HW, Hatfield WE, (1973) J. Chem. Phys. 59: 3600
10.27 Santoro A, Mighell AD, Reimann CW (1970) Acta Cryst. 26: 979

10.28 Hughes RC, Morosin B, Richards PM (1975) Phys. Rev. B11: 1795

10.29 Morosin B (1975) Acta Cryst. 31: 362

10.30 Karra JS, Kemmerer G E (1979) Phys. Lett 73A: 63

10.31 Ikeke M, Date M (1971) J. Phys. Soc. Jpn 30: 93

10.32 Drumheller JE, Seifert PL, Zaspel CE, Snively LO (1980) J. Chem. Phys. 73: 5830

10.33 Ritter MB, Drumheller JE, Kite TM, Snively LO (1983) Phys. Rev. B28: 4949

10.34 Groenedijk HA, Blöte HWJ, van Duyneveldt AJ, Gaura RM, Landee CP, Willett RD (1981)
       Physica 106B: 47

10.35 Poertadji S, Ablart G, Pescia J, Clement S, Cheikh-Rouhou AJ (1983) J. Physique Lett 44:
       L561

10.36 Hoogerbeets R, van Duyneveldt AJ (1983) Physica 121B: 233

10.37 Kuppusamy P, Manoharan PT (1985) Inorg. Chem. 24: 3053

10.38 Plumlee KW, Hoffmann BM, Ibers JA (1975) J. Chem. Phys. 63: 1926

10.39 Mahadevan C, Seshasayee M (1984) J. Crystallogr. Spectrosc. Res. 14: 213

10.40 Richards PM In. Local Properties at Phase Transitions, Editrice Compositori, Bologna, 1975
       p 539

10.41 Breed DJ (1967) Physica 37, 35

10.42 De Jongh LJ, Block R (1975) Physica 79: 568

10.43 Richards PM, Salamon MB (1974) Phys. Rev. B9, 32

10.44 Boesch HP, Schmocker U, Waldner F, Emerson K, Drumheller JE (1971) Phys. Lett. A36: 461

10.45 Riedel EF, Willett RD (1975) Solid State Commun. 16: 413

10.46 Lynch M, Kokoszka G, Szydlik P (1979) J. Phys. Chem. Solids 40: 79

10.47 Shia L, Kokoszka G (1974) J. Chem. Phys. 60: 1101

10.48 Willett RD In. Magneto Structural Correlations in Exchange Coupled Systems, Willett RD,
       Gatteschi D, Kahn O. (eds) Dordrecht, 1985 p 269

10.49 Zaspel CE, Drumheller JE (1977) Phys. Rev. B16: 1771

10.50 Okuda T, Date M (1970) J. Phys. Soc. Jpn. 28: 308

10.51 Castner TG, Jr, Seehra MS (1971) Phys Rev B4: 38

10.52 Shimizu M, Ajiro Y (1980) J. Phys. Soc. Jpn. 48: 414

10.53 Kiriyama R, Ibamoto H, Matsuo K (1954) Acta Cryst 7: 482

10.54 Soos ZG, McGregor KT, Cheung TTP, Silverstein AJ (1977) Phys. Rev. B16: 3036

10.55 Willett RD, Jardine FH, Rouse I, Wong RJ, Landee CP, Numata M (1981) Phys. Rev. B 24:
       5372

10.56 Kite TM, Drumheller JE, Emerson K (1982) J. Magn. Res. 48: 20

10.57 Willett RD, Extine M (1981) Phys. Letters 81A: 536

10.58 Kite TM, Drumheller JE (1983) J. Magn. Reson. 54: 253

11 Excitons

11.1 Introduction

In the previous chapters we have taken into consideration magnetic interactions in insulators. The EPR spectra of conductors are much different, often corresponding to Pauli paramagnetism, and they cannot be interpreted within the scheme of weak magnetic interactions. Between insulators and conductors there are, of course, semiconductors. They often show EPR spectra which are not too dissimilar from those of insulators, although their detailed interpretation shows the role of electron delocalization. In particular, many organic molecular semiconductors at sufficiently low temperatures show triplet spectra, which in a first approximation can be considered as arising from the exchange interaction between neighboring centers [11.1–3]. These interactions, however, are not always localized between adjacent centers and their accurate description requires the use of a number of concepts on the band theory of solids. In the language of the band theory of solids these phenomena take the collective name of *excitons*, and they are not restricted to semiconductors but can be found also in insulators [11.4–5]. In the next section we will briefly introduce the concept of exciton using elementary band theory concepts, and in the following part of the chapter we will work out some examples where the close similarity of the spectra of some excitonic systems to the spectra of extended lattices reported in Chaps. 6 and 10 will be apparent.

11.2 Excitons

Among the elementary excitations in semiconducting solids, the one which involves the transition of an electron from the filled valence band to the empty conduction band requires the creation of an electron-hole pair. The electron is in the conduction band and the hole is left in the valence band. When the electron-hole interaction is taken into account, a number of energy levels in the energy gap between the valence and the conduction band are generated. These levels are called *exciton states*, and the interacting electron-hole pair is called an *exciton*.

In a number of cases the electron and the hole are separated by several lattice constants and their interaction is conveniently represented as a purely electrostatic interaction between particles with opposite charges. These types of excitons are called Wannier-Mott excitons and their motion is well represented as the motion of a quasi-particle of effective mass μ^* moving in a homogeneous dielectric [11.6–18].

In the case where the electron and the hole are localized on the same or on an adjacent lattice position we have a strong electron-hole interaction and excitons of this type are called Frenkel excitons. The most important property that Frenkel and Wannier excitons have in common is the possibility of successive transfer of energy from one lattice site to another, as a consequence of the translational symmetry of the wave function. We can say that the excitons move through the solid.

Excitons are a collective excited state of a solid and, in principle, one can expect an excitonic system when in the solid molecules are present with available excited states which can also interact with each other and transmit the excitation through the entire crystal. Common examples of excitonic systems are given by organic crystals and the excitons in these systems take the name of molecular excitons [11.4]. Examples of molecular organic crystals are the crystal of benzene or naphtalene or other aromatic systems. These are insulating solids in which the organic molecules weakly interact with each other through van der Waals forces. If one can excite in a molecule an electron from the ground into an excited state, this excitation is not localized on one particular site of the crystal, but due to the interaction between molecules in the crystal, it is delocalized over the entire crystal. In other words, if N is the number of molecules forming the crystals with one molecule in the unit cell and we have the possibility of forming only one excited state per molecule, we have N excited crystal states. These states form a band of states and the width of this exciton band depends on the strength of the intermolecular interactions. When more than one molecule is present in the unit cell, it has been shown that more than one band corresponds to any excited molecular state. In the particular case of two independent molecules per unit cell, the exciton bands are split into two bands separated in energy. This energy separation is known as Davidov splitting [11.8]. In benzene crystals the Davidov splitting of the singlet excitons was found to be $\approx 40 \, \text{cm}^{-1}$.

Up to this point we have never explicitly considered the spin of the excitons. In molecular crystals one can have both singlet and triplet excited states. For example in benzene or other aromatic molecules, transitions between the ground singlet state and excited triplet states are allowed by spin-orbit coupling and, therefore, triplet states can be populated via photoexcitation. Since direct transitions toward the ground singlet states are spin-forbidden, these triplet states can have a lifetime longer than that of the singlet-singlet excitations. The decay of the molecule from one triplet excited state into the singlet ground state can occur radiatively by emission of radiation. This phenomenon is known as phosphorescence. The lifetime of phosphorescent organic crystals can range from 10 to 10^{-3} s. EPR spectra from the triplet states of photoexcited organic systems have been measured since 1959 and are still being extensively studied [11.9–12].

Other systems in which triplet excitons have been investigated through EPR spectroscopy are the segregated stack charge transfer (CT), organic solids and the organic free radical systems. Organic CT solids which present exciton states are generally formed by π molecular organic acceptors like tetracyanoquinodimethane (TCNQ), chloranil, tetracyanobenzene (TCNB), and donors like tetrathiafulvalene (TTF) or N,N,N',N'-tetramethyl-p-phenylenediamine (TMPD). The acceptors (A) and donors (D) in the crystals, which give rise to triplet exciton spectra, are disposed into alternating (dimerized, trimerized, tetramerized) segregated stacks containing all A's or D's. In these systems triplet exciton spectra have been observed both from free excitons [11.1] and from excitons trapped in impurity centers (*exciton traps*) [11.13].

It is apparent from the above introduction that a detailed description of the exciton nature and dynamics is beyond the scope of the present book. In the following we will limit ourselves to a qualitative description of the effect of the exciton motion on the EPR spectra of triplet excitons in those systems where the triplets can be viewed as generated by exchange interaction between neighboring centers.

11.3 EPR of Triplet Excitons in Linear Organic Radical Systems

A number of solids containing linear chains of planar organic radicals stacked face to face have been synthesized from the beginning of 1960 up to now. These compounds range from insulators to magnetic semiconductors and spin correlation in these systems has been extensively studied. Some review articles dealing with the structure and properties of linear organic radical solids can be found in Refs. [11.1, 11.14–17].

One of the most thoroughly studied systems is that formed by the TCNQ molecules, whose structure is illustrated in Fig. 11.1. The π molecular electronic system can easily accept one electron to form the stable radical anion TCNQ⁻. These anions tend to stack one over an other to form segregated stacks of planar molecules. The negative charge is stabilized by either simple M^+ cations ($M = Na$, Cs, Rb, NH_4) or more complex molecular systems like cationic complexes.

The TCNQ-containing compounds show a large variety of structural motifs and exhibit a number of different magnetic and electrical properties. They range from metallike substances with temperature independent paramagnetism to essentially diamagnetic materials of low electrical conductivity. In a number of these compounds a paramagnetism attributable to thermally accessible triplet

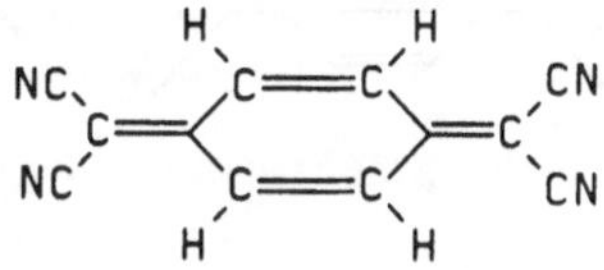

Fig. 11.1. Structure of tetracyanoquinodimethane, TCNQ

states was observed and EPR spectra characteristic of mobile S=1 spin excitations have been measured [11.18–26]. As a consequence of the large variability in the physical properties, these compounds have been studied by a number of investigators and several theoretical models have been suggested to account for their properties. Among these models we may cite the singlet-triplet model [11.1, 23], the collective electronic state theory [11.22, 27, 28], the linear Heisenberg antiferromagnet [11.29], the Hubbard model [11.30], and the band structure model [11.31]. It is apparent that we cannot discuss the details of the above models here or make a full comparison among the different theories, but we will limit ourselves to a description of the EPR spectra observed in these compounds in order to show the nature of the information which can be derived from the electronic structure using EPR spectroscopy. The connection between the measured quantities and the electronic nature of the materials will, of course, depend on the particular model used.

In the following we will focus our attention mainly on salts of the $TCNQ^-$ anion, which show most of the EPR characteristics of triplet excitons. In the crystal structures of $TCNQ^-$ salts two main stacking motives (Fig. 11.2) have been observed, namely the ring-to-ring (R–R) and the ring-external bond (R–B) stacking. More complex moieties have also been found, e.g., stacks of $(TCNQ)_3^{2-}$ radicals, etc. In these complex salts the stacking units contain both charged and neutral TCNQ molecules. In Fig. 11.3 we schematically show the linear arrangement of planar organic molecules. In Fig. 11.3a the molecules stack face to face in a regular arrangement and each molecule equally interacts with two adjacent molecules. This arrangement is generally less stable than that shown in Fig. 11.3b, which can be described as a stacking of dimeric units in which each molecule interacts more strongly with one of its nearest neighbors. A typical example of regular stacking is given by the room temperature structure [11.32] of $Rb[TCNQ]_2$, while at temperatures lower than 213 K a dimerized structure is preferred [11.33].

The TCNQ salts, and generally organic molecular crystals, can be described as π molecular crystals with tightly bound electrons and small π electron intersite differential overlap. Not all of the TCNQ salts show triplet exciton spectra: excitonic states generally occur when the crystal lattice geometry favors a singlet pairing of electrons on neighboring molecules. In this situation an excited triplet state is available on each pair of molecules as a consequence of the exchange

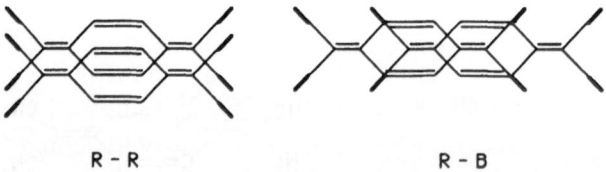

R – R   R – B

Fig. 11.2. Two typical arrangements of the TCNQ molecules found in the crystal structure of $TCNQ^-$ salts

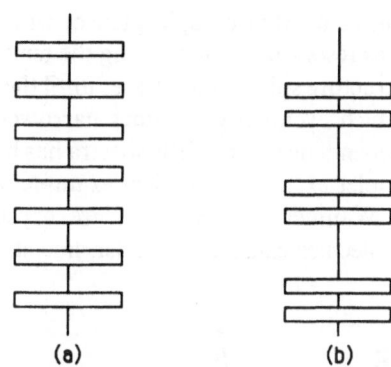

Fig. 11.3a, b. Schematic view of two possible arrangements of planar organic molecules to form a linear chain. **a** Regular arrangement; **b** dimerized stacking

interaction. The following general features are common to most compounds as well as to CT and molecular crystals: (1) the triplet exciton band-widths are large compared to nuclear hyperfine interactions and the EPR spectra will not show any nuclear hyperfine structure; (2) the Davidov-type splitting of the triplet exciton bands is large compared to the fine structure splitting of the $S = 1$ spin state; (3) there is only a weak dependence of the fine structure splitting on the **k** vector of the exciton band and the observed splitting is a thermal average over the possible **k** states.

One of the fundamental points in choosing the theoretical model to account for the magnetic properties of TCNQ salts is the amount of energy required to form the excited charge transfer state $TCNQ^{2-}$. This state is formally obtained through the reaction

$$TCNQ^- + TCNQ^- \rightarrow TCNQ^0 + TCNQ^{2-}, \tag{11.1}$$

in which one electron is removed from one $TCNQ^-$ anion radical onto the nearest neighbor ion.

In the case of zero or a small energy difference a band model or a Hubbard hamiltonian should be appropriate, while in the opposite case a Heisenberg linear chain model could be applied. Kommandeur et al. concluded that in the alkali salts of $TCNQ^-$ this energy should be nearly zero and developed a band model to account for the observed magnetism [11.31–34]. EPR spectra were generally interpreted within the framework of an approximate Hubbard model [11.30–35]. Soos et al. applied the linear Heisenberg theory quite successfully to interpret both the EPR spectra and magnetic properties of a number of $TCNQ^-$ salts [11.29, 36].

As long as the EPR spectra are concerned, at sufficiently low temperature, they show a fine splitting and can be interpreted using the $S = 1$ spin hamiltonian

$$H_{S=1} = \mu_B \mathbf{B} \cdot \mathbf{g} \cdot \mathbf{S} + \mathbf{S} \cdot \mathbf{D} \cdot \mathbf{S}, \tag{11.2}$$

with a nearly isotropic **g** tensor with $g \approx 2.00$. In any case no hyperfine structure was resolved. On increasing the temperature the EPR lines broaden and the fine structure splitting decreases until the spectrum collapses into a single broad line, which eventually becomes narrower at higher temperature. This temperature dependence of the EPR spectra has been ascribed to the presence of Frenkel-type triplet excitons. A typical example of the spectrum is shown in Fig. 11.4.

Wannier spin excitons are expected to be delocalized onto nonadjacent molecules causing a smaller, fine structure splitting which cannot be observed.

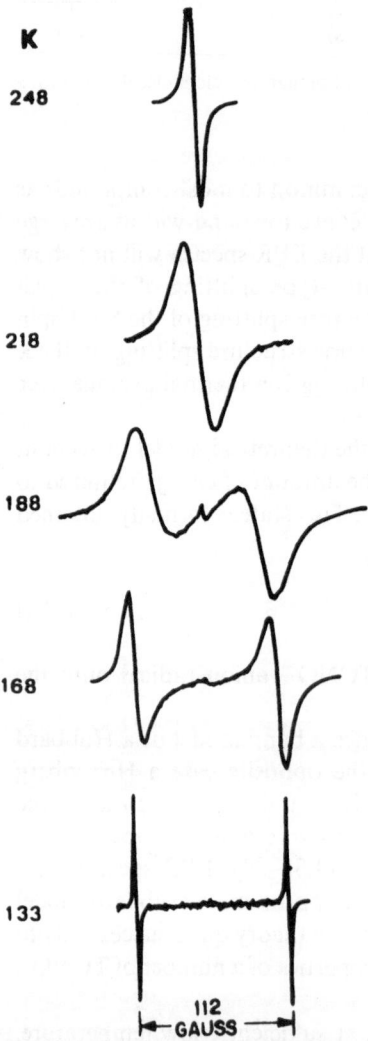

Fig. 11.4. Temperature dependence of the triplet exciton EPR spectra in a single crystal of $(Ph_3AsCH_3)(TCNQ)_2$. The orientation of the crystal in the magnetic field is fixed, but the spectrometer gain is varied. After [11.24]

258

A strong $g \approx 2$ line is often present at any temperature, the nature of this signal will be briefly discussed at the end of the chapter.

A property of the Frenkel spin excitons is that they are self-trapped, i.e., the triplet excitation is accompanied by a local lattice distortion. The observation of triplet resonances in Wannier-type triplet excitons as well as in certain charge transfer organic solids, often requires the presence of a trapping center which could stop the excitation in a particular point of the crystal.

In Fig. 11.5 we show an energy level diagram computed by Hibma et al. using an approximate Hubbard hamiltonian for a chain of interacting $TCNQ^-$ dimers for $k = 0$ [11.35]. Using a Hubbard hamiltonian the energies of the states are parameterized by three parameters: t_1, t_2, and U. The t_i's are nearest neighbor transfer integrals and U is the effective repulsive energy between two electrons on the same molecule to form a $TCNQ^{2-}$ ion. An analytical expression for the eigenvalues of the Hubbard hamiltonian was found for $t_1 \gg t_2$, i.e., for a chain of interacting dimers. In Fig. 11.5 E_T is the energy separation between the ground singlet state and the exciton states, E_g is the energy gap for electrical conduction, I_T is the binding energy of the triplet excitons, and I_S is the energy difference between a free electron-hole pair state and the first excited singlet exciton state. Analytical expressions for the above quantities as a function of the Hubbard hamiltonian parameters can be found in Ref. [11.35].

The intensity I of the EPR spectra depends on E_T through the equation

$$I \cdot T = a \cdot \exp(-E_T/kT), \tag{11.3}$$

where a is a proportionality constant and k is the Boltzmann constant. The excitation energies E_T measured following the temperature dependence of the

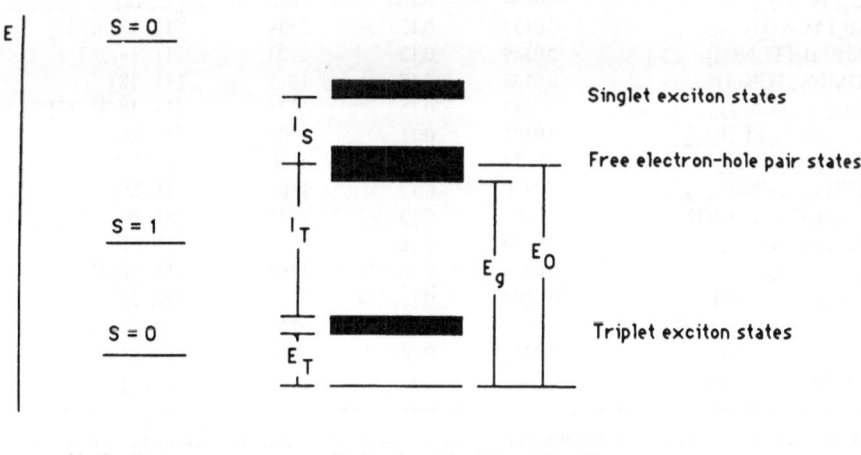

Fig. 11.5. Energy level diagram for an isolated couple of 1/2 spins (*left*) and for a chain of weakly interacting dimers (*right*) as computed in [11.35]

EPR spectra with (11.3) for a number of TCNQ salts are reported in Table 11.1 together with the measured zero field splitting parameters D and E/D of the triplet spin hamiltonian. In Table 11.1 the data for $[Ni(mnt)_2][TMPD]_2$ (TMPD = tetramethylphenylendiaminium) are also included. Here, the excitonic electrons are delocalized onto the chain of $TMPD^+$ cation radicals. D is of the same order of magnitude for all of the complexes with the same stoichiometry and decreases on passing from $[TCNQ]_2^-$ to $[TCNQ]_4^-$, while the anisotropy of the D tensor, E/D, remains almost constant throughout the entire series. The direction of maximum zero field splitting was always found to be parallel, within experimental error, to the line connecting the centers of the molecular planes.

The zero field splitting tensor in comparison to the experimental one has been generally estimated using a generalized point-dipolar approximation. In this framework the tensor takes the form

$$D_{ij} = 1/2\, g^2\, \mu_B^2\, \Sigma \rho_\alpha \rho_\beta (r_{\alpha\beta}^2 \delta_{ij} - 3i_{\alpha\beta} j_{\alpha\beta}) r_{\alpha\beta}^{-5}, \tag{11.4}$$

where i, j = x, y, z and ρ_α and ρ_β are the spin densities for atom α on molecule 1 and for atom β on molecule 2. In (11.4) $r_{\alpha\beta}$ is the distance between atoms α and β. The use of ρ_α and ρ_β is the generalization of the point dipolar model. A number of calculations have been performed on the $TCNQ^-$ anion to compute the spin densities on the nuclear centers. In Table 11.2 we present some of these results and report the D and E values for $Rb_2[TCNQ]_2$ computed by (11.4).

Table 11.1. Activation energy E_T, and zero field splitting parameters for selected triplet excitonic systems

Compound[a]	$|D|$ (cm^{-1})	E/D	E_T (cm^{-1})	Ref.
$Li_2[TCNQ]_2$	0.0154	0.15	1855	[11. 18]
$Rb_2[TCNQ]_2$	0.0133	0.12	2339	[11. 18, 31, 35]
$(MPM)_2[TCNQ]_2$	0.0149	0.12	2903	[11. 19, 20]
$(DMB)_2[TCNQ]_2$	0.0133	0.17	1613	[11. 18]
$(TMB)_2[TCNQ]_2$	0.0144	0.15	1371	[11. 18, 35, 37]
$(DMCHA)_2[TCNQ]_2$	0.0141	0.11	2500	[11. 18]
$(NBP)_2[TCNQ]_2$	0.0124	0.16	1130	[11. 21]
$(NBP)_2[TCNQF_4]_2$	0.0120	0.17	1209	[11. 22]
$[Ni(mnt)_2][TMPD]_2$	0.020	0.13	1936	[11. 38]
$Rb_2[TCNQ]_3$	0.0094	0.16	——	[11. 18]
$Cs_2[TCNQ]_3$	0.0094	0.16	1290	[11. 18, 23]
$(MPM)_2[TCNQ]_3$	0.0094	0.16	2500	[11. 20]
$(Ph_3PCH_3)_2[TCNQ]_4$	0.0063	0.16	524	[11. 20, 24, 39]
$(Ph_3AsCH_3)_2[TCNQ]_4$	0.0063	0.16	524	[11.24]
$(Et_3NH)_2[TCNQ]_4$	0.0041	0.12	290	[11.25]

[a]Abbreviations: *MPM* = morpholinium; *DMB* = 1,3-dimethylbenzimidazolinium; *TMB* = 1,3,5-trimethylbenzimidazolinium; *DMCHA* = diethylmethylcyclohexyl ammonium; *NBP* = N-butylphenazinium; mnt = maleonitriledithiolato; *TMPD* = N,N,N',N'-tetramethyl-p-phenylenediamine; *TCNQF₄* = 2,3,5,6-tetrafluoro tetracyanoquinodimethane; *Ph₃PCH₃* = triphenylmethylphosphonium; *Ph₃AsCH₃* = trimethylphenylarsonium; *Et₃NH* = triethylammonium.

Table 11.2. Computed spin densities ρ_i for an isolated TCNQ—ion radical and fine structure parameters calculated for $Rb_2[TCNQ]_2^a$

$$\text{>C}\overset{-1}{\underset{}{\text{C}}}\langle\,\,^{2}\!-\!3^{/}_{\,\,4}\!-\!N$$

ρ_1	ρ_2	ρ_3	ρ_4	ρ_N	$D(cm^{-1})$	$E(cm^{-1})$	Ref.
0.045	0.056	0.225	−0.003	0.067	0.0175	0.0018	[11.40]
0.057	0.067	0.219	0.009	0.040	0.0198	0.0022	[11.41]
0.059	0.109	0.191	0.002	0.039	0.0202	0.0023	[11.42]
0.061	0.132	0.180	0.001	0.031	0.0217	0.0025	[11.43]

[a] The experimental values are $D = 0.0133$ cm^{-1}, $E = 0.0016$ cm^{-1} (see Table 11.1).

$Rb_2[TCNQ]_2$ is a 1:1 salt which presents two stable crystal phases. In one of these phases the TCNQ$^-$ ions are arranged in a regular stack of crystallographically equivalent molecules and no fine structure splitting was detected in the EPR spectra. In the other phase the stack can be described as formed by a stacking of $[TCNQ]_2^{2-}$ dimers and the EPR spectra are typical triplet exciton spectra. The shortest and the longest interplanar distances are 315 and 348 pm, respectively [11.33].

It is apparent from Table 11.2 that the computed zero field splitting parameters merely reproduce the order of magnitude of the measured zero field splitting tensor. This can be due to either the inadequacy of the dipolar model to account for the zero field splitting or to an inaccurate computation of the spin densities. Since the computed zero field splitting parameters are always larger than the experimental ones, another possibility for explaining the discrepancy between computed and observed parameters is that the zero field splitting is determined by interactions between more distant centers [11.18, 35]. In the hypothesis that the strongest interaction is between far-neighbor centers, Silverstein and Soos [11.44] computed $D = 0.0126$ cm^{-1} and $E = 0.0015$ cm^{-1} for $Rb_2[TCNQ]_2$ using the charge densities of reference [11.40] in Table 11.2. In fact, distant interactions have been observed in TCNQ salts as weak, fine structure lines flanking the more intense ones. These lines have been attributed to excitons larger than the normal ones, i.e., delocalized on more distant molecules [11.18, 35]. In Fig. 11.6 the variation of the zero field splitting parameters as a function of the interplanar distance R is shown [11.18].

Several mechanisms contribute to the line width and to the temperature dependence of the line widths of the triplet exciton spectra [11.28]. When two (or more) crystallographically independent chains are present in the crystal they become magnetically equivalent along appropriate crystallographic directions and it is experimentally found that the line width of the signals observed when the static magnetic field is along these directions are much sharper than in a general orientation where two independent molecules are seen. This sudden change in the line-width is due to interchain jumping of the excitons between the two crystallographically nonequivalent chains. It is well known that in the slow

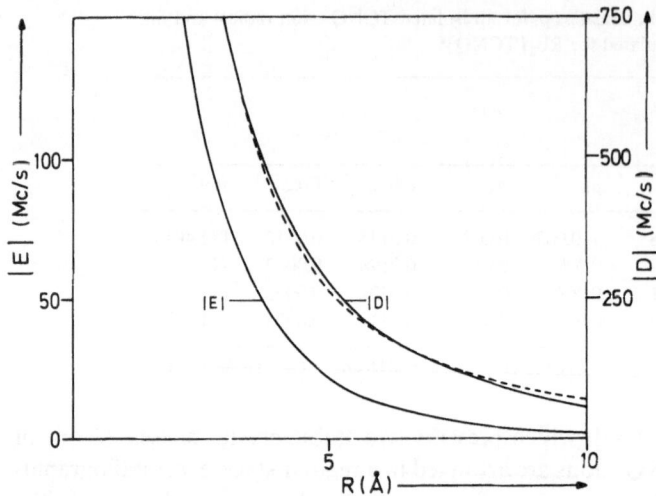

Fig. 11.6. Calculated zero field splitting parameters for a $[TCNQ]_2^{2-}$ dimer as a function of the interplanar distance R. The *dotted curve* is computed using the equation $D = 6660/R^2$. After [11.18]

jumping limit, when two separate resonances are seen, the line width of each resonance is given by

$$\Gamma = \Gamma_{A,B} + v_t \, (v_t \ll v_A - v_B),$$ (11.5)

where Γ_A and Γ_B are the line widths of chains A and B, respectively, without hopping, v_t is the hopping frequency, and v_A and v_B are the resonance frequencies of A and B. The separation between the A and B lines is slightly smaller than $v_A - v_B$. In the fast jumping limit, where only a single signal is observed, the line width is given by

$$\Gamma = 1/2(\Gamma_A + \Gamma_B) + (v_A - v_B)^2/8v_t \, (v_t \gg v_A - v_B).$$ (11.6)

Equations (11.5–6) show that when $v_t \ll v_A - v_B$, the lines are broadened by v_t, the hopping frequency, vice versa when the lines coincide, $v_A = v_B$, the width is equal to $1/2(\Gamma_A + \Gamma_B)$, i.e. the width without hopping. Single crystal measurements thus allow one to measure v_t at different temperatures. These measurements [11.28] for $Rb_2[TCNQ]_2$ and $(TMB)_2[TCNQ]_2$ showed that v_t versus T has the form

$$v_t = v_0 + \exp(-E_L/kT),$$ (11.7)

where $v_0 = 1.1 \times 10^2$ MHz, $E_L = 645$ cm^{-1} for $Rb_2[TCNQ]_2$, and $v_0 = 2.0 \times 10^3$ MHz, $E_L = 1210$ cm^{-1} for $(TMB)_2[TCNQ]_2$. E_L in Eq. (11.7) can be seen as an activation energy for the hopping of the exciton between translationally nonequivalent chains. This energy should be nearly equal to the height of the barrier of the trap depth associated with the local lattice distortions due to the exciton formation.

Hibma and Kommandeur [11.28] have shown that the line width anisotropy associated with v_t can originate extra peaks in the polycrystalline powder spectra.

Low temperature line widths contain information about the motion of the excitons along the chain. This motion depends on the relative values of E_L, the self-trapping energy, and t_i, the transfer integrals. When t_i is much larger than E_L, the exciton motion will be wavelike with no activation energy and the lattice distortion will be distributed over a large number of centers. When E_L is larger than t_i, the exciton will be self-trapped and a diffusional motion with an activation energy results. The main consequences of the intrachain motion is the averaging out of any hyperfine interaction even at low temperature. Soos [11.45] derived the following equation for the contribution of the hyperfine interaction to the line width

$$v_{av} = v_{hf}^2 / \sqrt{(2v_j \, v_r)}, \tag{11.8}$$

where v_{hf} is the width of the hyperfine structure for a localized exciton, v_j is the hopping frequency along the chain, and v_r is the characteristic frequency for three-dimensional randomization processes. For a purely one-dimensional motion (11.8) would be

$$v_{av} \approx (v_{hf}^4 / v_j)^{1/3}. \tag{11.9}$$

Typical values of v_{av} are of the order of a few MHz. It can also happen that at low temperature the hyperfine splitting is only partially averaged out giving rise to an anisotropic, inhomogeneously broadened (Gaussian) line shape. A typical example is $(TMB)_2 [TCNQ]_2$.

The broadening and eventual collapsing of the fine structure lines with increasing temperature has been attributed mainly to interactions between excitons. These exchange interactions change the spin state of the interacting excitons thus producing an averaging of the dipolar interactions. The dipolar splitting is therefore dependent on the collision frequency: as the collision frequency increases, the splitting becomes smaller and the lines broaden until eventually they merge into a single line. Expressing the zero field splitting and line widths in MHz we have the following equations

$$\Delta v - \Delta v_0 = 2/(3\sqrt{3})v; \quad d_0^2 - d^2 = 8v^2; \tag{11.10}$$

$$\Delta v - \Delta v_0 = 1/\sqrt{3} \, d_0^2/v; \qquad d_0^2 - d^2 = 0; \tag{11.11}$$

where Δv and Δv_0 are line widths in the presence and in the absence of the interaction, respectively, and d and d_0 are the line separations. Equation (11.10)

Table 11.3. Activation energies and preexponential factors for the line broadening processes due to exciton-exciton interactions

Compound	E_0(MHz)	$E(cm^{-1})$	Ref.
$(Ph_3AsMe)_2[TCNQ]_2$	1.4×10^4	1290	[11.46]
$(Ph_3PMe)_2[TCNQ]_2$	5×10^5	1530	[11.47]
$Cs_2[TCNQ]_3$	4×10^4	1774	[11.47]
$(MPM)_2[TCNQ]_2$	1×10^4	4275	[11.47]
$Rb_2[TCNQ]_2$	2×10^4	2903	[11.28]
$K_2[TCNQ]_2$	3×10^5	3065	[11.28]
$(TMB)_2[TCNQ]_2$	8×10^3	3226	[11.28]

holds in the slow exchange limit $v \ll d_0$ and Eq. (11.11) holds in the fast exchange limit $v \gg d_0$. The interaction frequencies measured in this way show a temperature dependence of the form (11.7). The preexponential and exponential factors measured in a number of cases are reported in Table 11.3.

It should be noted that the values measured from Δv or d values are generally different. Chesnut and Meinholtz suggested a procedure to overcome this anomaly [11.46].

In many examples of EPR spectra of TCNQ compounds one or two strong central lines at $g \approx 2.00$ are present. The nature of this band has been widely investigated in the system $(MPM)_2 [TCNQ_2]$ [11.20, 26, 46, 48]. Here, the band has a larger and a narrower component. The presence of this band is independent of the synthetic method and the temperature dependence of the intensity has an activated form like Eq. (11.7). The narrower line is Lorentzian with $H_{pp} = 0.5 - 0.8$ gauss and isotropic. The broad line has an anisotropic line width with H_{pp} in the range 3–10 gauss and has a Gaussian lineshape. A number of explanations have been suggested to account for the presence of these bands which include the formation of molecular defects or the presence of free electron-hole pairs in the conduction band. In the MPM salt ENDOR spectra have shown that the broad line is due to isolated $TCNQ^-$ ions exactly in the same position as at the end or the beginning of a TCNQ chain [11.49].

References

11.1   Nordio PL, Soos ZG, McConnell HM Ann. Rev. Phys. Chem. (1966) 17: 237
11.2   Soos ZG Ann. Rev. Phys. Chem. (1974) 25: 121
11.3   Soos ZG, Klein DJ (1975) In: Foster R (ed) Molecular associations, Academic, New York, vol 1
11.4   Davidov AS 1971 'Theory of molecular excitons', Plenum, New York
11.5   Knox RS, 1963 'Excitons', Academic, New York
11.6   Wannier GH (1937) Phys. Rev. 52: 191
11.7   Mott NF (1938) Trans. Faraday Soc. 34: 500
11.8   Davidov AS (1948) J. Expl. Theoret. Phys. (U.S.S.R.) 18: 210
11.9   Hutchinson CA, Magnum BW (1958) J. Chem. Phys. 29: 952
11.10 van der Waals JH, de Groot MS (1959) Mol. Phys. 2: 333

11.11 Sternlicht H, McConnell HM (1961) J. Chem. Phys. 35: 1793

11.12 Pasimeni L, Corvaja C, Agostini G, Giacometti G (1985) Chem. Physics 73: 537

11.13 Pasimeni L, Guella G, Corvaja C (1982) J. Magn. Reson. 47: 240

11.14 Endres H (1983) In: Miller JS (ed) 'Extended linear chain compounds' Plenum, New York, vol 3, p 263

11.15 Hoffman BM, Martinsen J, Pace LJ, Ibers JA (1983) In: Miller JS (ed) 'Extended linear chain compounds' Plenum, New York, vol 3, p 459

11.16 Zeller HR (1973) Adv. Solid State Phys. 13: 31

11.17 Garito AF, Heeger A J (1974) Acc. Chem. Res. 7: 232

11.18 Hibma T, Dupuis P, Kommandeur J (1972) Chem. Phys. Letters 15: 17

11.19 Marechal MA, McConnell HM (1965) J. Chem. Phys. 43: 497

11.20 Bailey JC, Chestnut DB (1969) J. Chem. Phys. 51: 5118

11.21 Harms RH, Keller HJ, Nöthe D, Wehe D, Heimer NE, Metzger RM, Gundel D, Sixl H (1982) Mol. Cryst. Liq. Cryst. 85: 249

11.22 Metzger RM, Heimer NE, Gundel D, Sixl H, Harms RH, Keller HJ, Nöthe D, Wehe D (1982) J. Chem. Phys. 77: 6203

11.23 Chestunt DB, Arthur P (1962) J. Chem. Phys. 36: 2969

11.24 Chestnut DB, Phillips WD (1961) J. Chem. Phys. 35: 1002

11.25 Flandrois S, Amiell J, Carmona F, Delhaes P (1975) Solid State Commun. 17: 287

11.26 Hoffmann SK, Corvan PJ, Singh P, Sethulekshmi CN, Metzger RM, Hatfield WE (1983) J. Am. Chem. Soc., 105: 4608

11.27 Winkler RG, Reineker P (1987) Mol. Phys. 60: 1283

11.28 Hibma T, Kommandeur J (1975) Phys. Rev. B 12: 2608

11.29 Soos ZG (1965) J. Chem. Phys. 43: 1121

11.30 Soos ZG, Klein DJ (1971) J. Chem. Phys. 55: 3284

11.31 Vegter JG, Kommandeur J (1975) J. Mol. Cryst. Liq. Cryst., 30: 11

11.32 Shirotani I, Kobayashi H (1973) Bull. Chem. Soc. Jpn. 46: 2595

11.33 Hoekstra A, Spoelder T, Vos A (1972) Acta Cryst. B28: 14

11.34 Vegter JG, Kommandeur J, Fedders PA (1973) Phys. Rev. B 7: 2929

11.35 Hibma T, Sawatzky GA, Kommandeur J (1977) Phys. Rev. B 15: 3959

11.36 Soos ZG, Hughes RCJ. (1967) Chem. Phys. 46: 253

11.37 Dupuis P, Flandrois S, Neel J (1969) C.R. Acad Sci. (Paris) C269: 1091

11.38 Hove MJ, Hoffman BM, Ibers JA (1972) J. Chem. Phys. 56: 3490

11.39 Thomas DD, Merkl AW, Hildebrandt AF, McConnell HM (1964) J. Chem. Phys. 40: 2588

11.40 Lowitz DA (1967) J. Chem. Phys. 46: 4698

11.41 Bieber A, André JJ, (1974) Chem. Phys. 5: 166

11.42 Jonkman HT, van der Velde GA, Nieuwpoort WC (1974) Chem. Phys. Letters 25: 62

11.43 Johansen H (1975) Int. J. Quantum Chem., 9: 459

11.44 Silverstein AJ, Soos ZG (1976) Chem. Phys. Letters, 39: 525

11.45 Soos ZG (1965) J. Chem. Phys. 44: 1729

11.46 Chestnut DB, Meinholtz DC (1984) J. Chem. Phys. 80: 3540

11.47 Jones MT, Chestnut DB (1963) J. Chem. Phys., 38: 1311

11.48 Kepler RG (1963) J. Chem. Phys. 39: 3528

11.49 Maniero AL, Priolisi O, Corvaja CJ. Chem. Soc. Faraday Trans. (in press)

Appendix A. Second Quantization

The formalism of second quantization, or the occupation number representation for fermions, allows one to write multi electron wave functions and operators more simply than using the Slater antisymmetrized product of one-electron functions [2.1]. The main advantage of this method compared to that of Slater determinants is that it leads to a compact notation, while the latter is rather cumbersome even for small numbers of electrons. However, there is really no result which can be proved within the second quantization formalism which cannot be obtained using the Slater formalism.

In the one-electron approximation a quantum state of a system is defined by an antisymmetrized product of the occupied one-electron spin orbitals. A typical form of an N electron system is:

$$\psi = A | \Phi_{k1}(r_1) \Phi_{k2}(r_2) \ldots \Phi_{kN}(r_N) \rangle, \tag{A.1}$$

where A is an antisymmetrizer, r_i represents space and spin variables of electron i, and k_i is a set of quantum numbers defining the state of the electron i.

When we have chosen a complete orthonormal set of one-electron spin orbitals we can associate to each state (A.1) a vector formed by the occupation number of each spin orbital of the complete set. This number, due to the antisymmetry requirement of the wave function, can be either 0 or 1. Equation (A.1) gives the occupation number representation of the system.

Let us consider a three-electron state as an example. In second quantization it is represented by:

$$\psi_{klm} = |0_1, 0_2 \ldots 1_k \ldots 1_l \ldots 1_m \ldots 0_n \rangle, \tag{A.2}$$

which indicates that one electron is present in Φ_k, one in Φ_l, and one in Φ_m, while all the others are vacant. ψ_{klm} is equivalent to the Slater determinant:

$$\psi_{klm} = 1/\sqrt{3!} \begin{vmatrix} \Phi_k(1) & \Phi_l(1) & \Phi_m(1) \\ \Phi_k(2) & \Phi_l(2) & \Phi_m(2) \\ \Phi_k(3) & \Phi_l(3) & \Phi_m(3) \end{vmatrix}. \tag{A.3}$$

It should be noted that while in the Slater representation ψ_{klm} is given by a 3×3 determinant, (A.2) is a vector in an infinite dimensional space.

A general state vector can be written as:

$$|n\rangle = |n_1, n_2 \ldots n_k \ldots \rangle \tag{A.4}$$

with $n_k = 0, 1$. The state vector for which $n_k = 0$, for all k, is a vacuum state. The state (A.4) is orthonormalized by the conditions:

$$\langle n'_1, n'_2, \ldots n'_k, \ldots | n_1, n_2, \ldots n_k \ldots \rangle = \delta_{n1'n1} \delta_{n2'n2} \cdots \delta_{nk'nk} \cdots \qquad \text{(A.5)}$$

and it forms a vector of a Hilbert space.

All the possible state vectors can be constructed starting from the vacuum state using creation operators, a_k^*, i.e., an operator which creates an electron in the state Φ_k. In the Slater formalism a creation operator transforms an $N \times N$ Slater determinant into an $(N+1) \times (N+1)$ one. Of course, this $(N+1)(N+1)$ determinant will be nonzero if Φ_k is different from any other state already present, otherwise the determinant will have two equal rows, i.e., a_k^* creates a particle in the k-th orbital provided it is initially vacant

$$a_k^* | n_1, n_2 \ldots 0_k \ldots n_N \rangle = | n_1, n_2 \ldots 1_k \ldots n_N \rangle. \qquad \text{(A.6)}$$

If n_k in the initial state is equal to 1, the result of the creation operator is zero:

$$a_k^* | n_1, n_2 \ldots 1_k \ldots n_N \rangle = 0. \qquad \text{(A.7)}$$

Conversely, an annihilation operator a_k removes an electron in the k-th orbital provided that the orbital in question initially contained an electron:

$$a_k | n_1, n_2 \ldots 1_k \ldots n_N \rangle = | n_1, n_2 \ldots 0_k \ldots n_N \rangle. \qquad \text{(A.8)}$$

If it operates on a vacant orbital, the result is zero:

$$a_k | n_1, n_2 \ldots 0_k \ldots n_N \rangle = 0. \qquad \text{(A.9)}$$

In order to pass from the second quantization to the Slater formalism it is important to arrange the one-electron orbitals in a given order and then to employ the same order in the Slater determinant.

The creation and annihilation operators have several interesting properties:

1. $a_k^* a_k^* = 0.$ \qquad (A.10)

This is obvious if initially $n_k = 1$. On the other hand, the first operator creates an electron in k, if it is initially vacant, and the second will find $n_k = 1$. Analogously one has:

2. $a_k a_k = 0.$ \qquad (A.11)

3. $a_k^* a_j^* + a_j^* a_k^* = 0.$ \qquad (A.12)

In fact, if n_k and n_j are zero in $|n\rangle$, $a_j^* a_k^* |n\rangle = |n_1, n_2 \ldots 1_k \ldots 1_j \ldots \rangle =$

$-|n_1, n_2 \ldots 1_j \ldots 1_k \ldots\rangle = -a_k^* a_j^* |n\rangle$. Condition 3 indicates that $a_k^* a_j^*$ anti-commute. This can be symbolically written as:

$$\{a_k^*, a_j^*\} = 0. \tag{A.13}$$

4. $a_k a_j + a_j a_k = 0;$ (A.14)

$$\{a_k, a_j\} = 0. \tag{A.15}$$

5. $\{a_k^* a_j\} = \delta_{kj}.$ (A.16)

It also follows from above:

6. $a_k^* a_k = n_k.$ (A.17)

In fact, let $|n_j\rangle = |n_1, n_2 \ldots 1_j \ldots 0_k \ldots\rangle$, then $a_k^* a_j |n_j\rangle$ $= |n_1, n_2 \ldots 0_j \ldots 1_k \ldots\rangle = -|n_1, n_2 \ldots 1_k \ldots 0_j \ldots\rangle = -a_j a_k^* |n_j\rangle$. For $k = j$ $a_j^* a_j |n_j\rangle = |n_j\rangle$ and $a_j a_j^* |n_j\rangle = 0$, so that $(a_j^* a_j + a_j a_j^*)|n_j\rangle = |n_j\rangle$, or equivalently $\{a_k^*, a_k\} = 1$. The eigenvalue of $a_k^* a_k$ is the occupation number of the k-th orbital. $a_k^* a_k$ defines a number operator for the k-th orbital:

$$N_k = a_k^* a_k. \tag{A.18}$$

One- and two-electron operators can be expanded in terms of the creation and annihilation operators. Let us start from one-electron operators, which we indicate as $\Sigma_i f_i$. The matrix element of this operator between two states $|a\rangle$ and $|b\rangle$ is zero, if the two vectors differ in more than one component orbital. If they differ in just one orbital, e.g., k_1 in $|a\rangle$ and k_2 in $|b\rangle$, then:

$$\langle a|\Sigma_i f_i|b\rangle \equiv \langle \Phi_{k1}|f|\Phi_{k2}\rangle. \tag{A.19}$$

In the right term we have omitted the subscript indicating the electron, because the matrix element is taken between one-electron states.

If the states are identical, then

$$\langle a|\Sigma_i f_i|a\rangle \equiv \langle \Phi_1|f|\Phi_1\rangle + \langle \Phi_2|f|\Phi_2\rangle + \ldots + \langle \Phi_N|f|\Phi_N\rangle. \tag{A.20}$$

If we define an operator:

$$h = \Sigma_{k1,k2}\langle \Phi_{k1}|f|\Phi_{k2}\rangle a_{k1}^* a_{k2}, \tag{A.21}$$

then it is easy to show that this operator leads to the same matrix elements between the $|a\rangle$ and $|b\rangle$ states as the $\Sigma_i f_i$ operator.

268

The procedure can be extended to two-electron operators, yielding for a general operator $\Sigma_{i \neq j} g_{ij}$ to the identity:

$$\Sigma_{i \neq j} g_{ij} \equiv \Sigma_{k1,k2,k3,k4} \langle \Phi_{k1} \Phi_{k2} | g | \Phi_{k3} \Phi_{k4} \rangle a_{k1}^* a_{k2}^* a_{k4} a_{k3}. \tag{A.22}$$

Among one-electron operators an interesting special case is that in which each $f_i = 1$, so that the total operator, N, represents the total number of electrons and is called the *number operator*:

$$N = \Sigma_{k1} \langle k | 1 \rangle a_k^* a_1 = \Sigma_{k1} a_k^* a_1 \delta_{k1} = \Sigma_k a_k^* a_k = \Sigma_k N_k. \tag{A.23}$$

For our purposes it is important to establish the relations between second quantized operators and angular momentum operators. Let us start from an intuitively clear example and then obtain general formulae. Let us consider $a_m^* a_{m'}$, and suppose that m and m' are two orbitals on the same center, both with spin up. Let us further assume that both orbitals belong to the same configuration, for instance of d orbitals, and are characterized by two different values of the magnetic quantum number, m and m', respectively. The operator removes an electron from m' and creates another one in m, keeping the spin up. The effect of the operator has been that of changing the overall $M = \Sigma_i m_i$ value by $m' - m$. This is the same effect of angular momentum operators, therefore, it can be hoped to relate $a_k^* a_1$ to angular momentum operators. This indeed has been done. For orbitals belonging to the same 1 manifold, the following relations must hold [A.2]:

$$a_{m\sigma}^* a_{m'\sigma'} = \sum_{n=0}^{2l} \sum_{\omega=0}^{1} A_{n\omega} \langle lm | O_{m-m'}{}^n | lm' \rangle . \tag{A.24}$$
$$\langle \tfrac{1}{2}\sigma | O_{\sigma-\sigma'}{}^{\omega} | \tfrac{1}{2}\sigma' \rangle \Sigma_i O_{m-m'}{}^n (l_i) O_{\sigma-\sigma'}{}^{\omega}(S_i)$$

where

$$A_{n\omega} = (2n+1)(2\omega+1)[\langle 1 \| O^n \| 1 \rangle \langle \tfrac{1}{2} \| O^{\omega} \| \tfrac{1}{2} \rangle]^{-2} \tag{A.25}$$

and

$$\langle j \| O^n \| j \rangle = 2^{-n} \sqrt{[2j+n+1)!/(2j-n)!]}. \tag{A.26}$$

O^n and O^{ω} are angular momentum operator equivalents [A.3], i is the number of electrons. A list of operator equivalents in common use is given in Table A.1.

Equation (A.24) looks formidable, but its use is actually easier than it might be supposed. If, for instance, we consider states with no orbital angular momentum localized on one center A, only the spin variables will be important in (A.24). We find

$$a_{0\sigma}^* a_{0\sigma'} = [\tfrac{1}{2}]^{\sigma\sigma'}, \tag{A.27}$$

Table A.1. Relations between some common operator equivalents and angular momentum operators

n	$m-m'$	$O^n_{m-m'}$
0	0	1
1	± 1	$\pm 1/\sqrt{2}\, j_\pm$
1	0	j_z
2	± 2	$\sqrt{6}/4\, j_\pm^2$
2	± 1	$\pm\sqrt{6}/4\,(j_z j_\pm + j_\pm + j_\pm j_z)$
2	0	$\frac{1}{2}[3j_z^2 - j(j+1)]$
4	± 4	$\sqrt{70}/16\, j_\pm^4$
4	± 3	$-\sqrt{35}/8\,[\,j_z j_\pm^3 + j_\pm^3 j_z\,]$
4	± 2	$\sqrt{10}/6[\{7j_z^2 - j(j+1) - 5\}j_\pm^2$ $+ j_\pm^2 \{7j_z^2 - j(j+1) - 5\}]$
4	± 1	$\pm\sqrt{5}/8\,[\{7j_z^3 - 3j(j+1)j_z - j_z\}j_\pm +$ $+ j_\pm \{7j_z^3 - 3j(j+1)j_z - j_z\}]$
4	0	$1/8[35j_z^4 - \{30j(j+1)-25\}j_z^2 - 6j$ $(j+1) + 3j^2(j+1)^2]$

Table A.2. Values of $[\tfrac{1}{2}]^{\sigma\sigma'}$

σ	σ	$[\tfrac{1}{2}]^{\sigma\sigma'}$
+	+	$(1 + 2S_{Az})/2$
−	−	$(1 - 2S_{Az})/2$
+	−	S_{A+}
−	+	S_{A-}

where $[\tfrac{1}{2}]^{\sigma\sigma'}$ is a shorthand notation for $\sum\limits_{\omega=0}^{1} \langle \tfrac{1}{2}\sigma | O_{\sigma-\sigma'}{}^{\omega} | \tfrac{1}{2}\sigma' \rangle$. The actual values are given in Table A.2. The expressions for a different center B are identical. An orbitally nondegenerate state corresponds to a configuration of half-filled orbitals, therefore, we find that $\alpha = \alpha'$ and $\beta = \beta'$. Using Table A.2 and Eqs. (2.49–52) we find that the following relations must hold:

$$a^*_{\alpha+} a_{\alpha-} a^*_{\beta-} a_{\beta+} = S_{\alpha+} S_{\beta-};$$

$$a^*_{\alpha-} a_{\alpha+} a^*_{\beta+} a_{\beta-} = S_{\alpha-} S_{\beta+};$$

$$a^*_{\alpha+} a_{\alpha+} a^*_{\beta+} a_{\beta+} = 1/4[1 + 2S_{\alpha z} + 2S_{\beta z} + 4S_{\alpha z}S_{\beta z}];$$

$$a^*_{\alpha-} a_{\alpha-} a^*_{\beta-} a_{\beta-} = 1/4[1 - 2S_{\alpha z} - 2S_{\beta z} + 4S_{\alpha z}S_{\beta z}]. \tag{A.28}$$

By summing over all the spin states we obtain Eq. (2.53).

Table A.3. Values of $[1]^{mm'}$

m	m'	$[1]^{mm'}$
0	0	$1/6 - 1/6(3l_z^2 - 2)$
1	1	$1/6 + 1/4 l_z + 1/12(3 l_z^2 - 2)$
-1	-1	$1/6 - 1/4 l_z + 1/12(3l_z^2 - 2)$
± 1	∓ 1	$1/4 l_\pm^2$
0	± 1	$\sqrt{2}/8(1_\mp \pm 1_z 1_\mp \pm 1_\mp 1_z)$
± 1	0	$\sqrt{2}/8(1_\pm \pm 1_z 1_\pm \pm 1_\pm 1_z)$

The second case which is worth considering is that of a system which has one orbitally degenerate and one orbitally non-degenerate state. With the former we may associate an angular momentum $1 = 0$, while with the latter, for instance, $1 = 1$, as shown in Sect. 2.6. Then using (A.24–26):

$$a_{m\sigma}^* a_{m'\sigma'} = [\langle \tfrac{1}{2}\sigma | O_{\sigma-\sigma'}{}^0 | \tfrac{1}{2}\sigma' \rangle O_{\sigma-\sigma'}{}^0(S) + 4\langle \tfrac{1}{2}\sigma | O_{\sigma-\sigma'}{}^1 | \tfrac{1}{2}\sigma' \rangle O_{\sigma-\sigma'}{}^1(S)]$$
$$\Sigma_n A_{n0} \langle 1m | O_{m-m'}{}^n | 1m' \rangle O_{m-m'}{}^n(L) \tag{A.29}$$

for the degenerate center, and
$$a_\sigma^* a_{\sigma'} = [\tfrac{1}{2}]^{\sigma\sigma'} \tag{A.30}$$

for the non-degenerate one. (A.30) can be written as:
$$a_{m\sigma}^* a_{m'\sigma'} = 2[\tfrac{1}{2}]^{\sigma\sigma'}[1]^{mm'}, \tag{A.31}$$

$[1]^{mm'}$ is given in Table A.3 We may perform the summations on the spin variables first and we find:

$$\Sigma_{\sigma\sigma'} a_{m\sigma}^* a_{m'\sigma'} a_\sigma^* a_\sigma = 2[1]^{mm'}(\tfrac{1}{2} + 2S_A \cdot S_B); \tag{A.32}$$

$$\Sigma_\sigma a_{m\sigma}^* a_{m'\sigma} = 2[1]^{mm'}; \tag{A.33}$$

$$\Sigma_\sigma a_\sigma^* a_\sigma = 1, \tag{A.34}$$

Therefore (2.53) can be rewritten as:
$$H_{eff} = \Sigma_{mm'}[4/U A_{om}A_{om'}^* - 2B_{om'mo}][1]^{mm'}(2S_A \cdot S_B). \tag{A.35}$$

By writing out explicitly $[1]^{mm'}$, and passing from the imaginary basis $|1m\rangle$ to the real one, we finally obtain the expression (2.57).

References

A.1 Weissbluth M (1978) Atoms and molecules, Academic, New York
A.2 Elliott RJ, Thorpe MF (1968) J. Appl. Phys. 39: 202
A.3 Buckmaster HA (1962) Can. J. Physics 40: 1670

Appendix B. Properties of Angular Momentum Operators and Elements of Irreducible Tensor Algebra

B.1 Properties of Angular Momentum Operators

The angular momentum operator of a quantum system, $\mathbf{J}$, is defined by three cartesian components J_x, J_y, and J_z having the following commutation relationships (units: $\hbar = 1$):

$$[J_x, J_y] = iJ_z; \ [J_y, J_z] = iJ_x; \ [J_z, J_x] = iJ_y \tag{B.1}$$

or, in terms of the operators $J_\pm = J_x \pm iJ_y$

$$[J_z, J_\pm] = \pm J_\pm; \ [J_+, J_-] = 2J_z. \tag{B.2}$$

Basis vectors can be represented by the quantum numbers, J and M, plus any other set of quantum numbers, α, which must be added to J and M to form a complete set. They have the following properties:

$$\mathbf{J}^2 |\alpha JM\rangle = J(J+1)|\alpha JM\rangle; \tag{B.3}$$

$$J_z |\alpha JM\rangle = M|\alpha JM\rangle; \tag{B.4}$$

$$J_\pm |\alpha JM\rangle = [J(J+1) - M(M \pm 1)]^{1/2} |\alpha JM \pm 1\rangle; \tag{B.5}$$

$$\langle \alpha JM | \alpha'J'M' \rangle = \delta_{\alpha\alpha'} \delta_{JJ'} \delta_{MM'}. \tag{B.6}$$

Equation (B.5) has been written following the Condon and Shortley phase convention. The $|\alpha JM\rangle$ kets form a $(2J+1)$-dimensional vector space. J in (B.3–6) can take any integer or half-integer positive value and $-J \leq M \leq J$.

B.2 Addition of Two Angular Momenta. Clebsch-Gordon Coefficients and "3j" Symbols

B.2.1 Definitions

Let j_A and j_B be angular momenta of quantum systems A and B, respectively. The total angular momentum of quantum systems A and B together is

$$\mathbf{J} = \mathbf{j}_A + \mathbf{j}_B. \tag{B.7}$$

The tensor product of the $(2j_A + 1)$ kets of A and $(2j_B + 1)$ kets of B gives a space of $(2j_A + 1)(2j_B + 1)$ kets which are simultaneous eigenvectors of j_A^2, j_B^2, j_{Az}, and j_{Bz}:

$$|j_A j_B m_A m_B\rangle = |j_A m_A\rangle |j_B m_B\rangle. \tag{B.8}$$

The basis vectors of (B.7) can be obtained from (B.8) by a unitary transformation as simultaneous eigenvectors of j_A^2, j_B^2, J^2, J_z, which are all commuting operators, as

$$|j_A j_B JM\rangle = \sum_{m_A m_B} |j_A j_B m_A m_B\rangle \langle j_A j_B m_A m_B | JM\rangle. \tag{B.9}$$

The coefficients of the unitary transformation $\langle j_A j_B m_A m_B | JM\rangle$ are called *Clebsch-Gordon coefficients*.

B.2.1.1 Equivalent Definitions

1. $\langle j_A m_A j_B m_B | j_A j_B JM\rangle$ In: Condon EU, Shortley GH (1935) The theory of atomic spectra, Cambridge University Press, New York
2. $C^{j_A j_B J}_{m_A m_B M}$ In: Biedenharn LC (1952) Tables of the Racah coefficients, Oak Ridge National Lab., Physics Division, ORNL-1098
3. $C(j_A j_B J; m_A m_B)$ where it has been considered that $M = m_A + m_B$; In: Rose ME Elementary theory of angular momentum, (1957) Wiley, New York
4. $S^{j_A j_B}_{Jm_A m_B}$ In: Wigner EP (1958) Group theory and its application to the quantum mechanics of atomic spectra, Academic, New York

B.2.1.2 Phase Convention and Properties

A commonly used convention to fix the relative phase of kets (B.9) is
1. $\langle j_A j_B j_A (J - j_A) | JJ\rangle$ is real and positive.
2. $|j_A j_B JM\rangle$ follows Eq. (B.5).
   The Clebsch-Gordon coefficients have the following main properties:

1. They are real and form an orthogonal matrix;
2. The coefficients are different from zero, if $m_A + m_B = M$ and $|j_A - j_B| \leq J \leq j_A + j_B$;
3. $\langle j_A j_B m_A m_B | JM\rangle = (-1)^{j_A + j_B - J} \langle j_B j_A m_B m_A | JM\rangle$.

B.2.1.3 The "3j" Symbols

Wigner defined the following quantities which are proportional to the Clebsch-

Gordon coefficients in order to better exploit the symmetry properties:

$$\begin{pmatrix} j_A & j_B & J \\ m_A & m_B & -M \end{pmatrix} = (-1)^{j_A - j_B + M}(2J+1)^{-\frac{1}{2}} \langle j_A j_B m_A m_B | JM \rangle. \tag{B.10}$$

The 3j symbols have the following properties:
1. Even permutations of the columns leave the numerical value unchanged:

$$\begin{pmatrix} j_A & j_B & J \\ m_A & m_B & M \end{pmatrix} = \begin{pmatrix} j_B & J & j_A \\ m_B & M & m_A \end{pmatrix} = \begin{pmatrix} J & j_A & j_B \\ M & m_A & m_B \end{pmatrix}.$$

2. Odd permutations of the columns multiply the value by $(-1)^{j_A + j_B + J}$:

$$\begin{pmatrix} j_A & j_B & J \\ m_A & m_B & M \end{pmatrix} = (-1)^{j_A + J_B + J} \begin{pmatrix} 3j_B & j_A & J \\ m_B & m_A & M \end{pmatrix}.$$

3. When the signs of m_A, m_B, and M are simultaneously changed, the value is multiplied by $(-1)^{j_A + J_B + J}$:

$$\begin{pmatrix} j_A & j_B & J \\ m_A & m_B & M \end{pmatrix} = (-1)^{j_A + J_B + J} \begin{pmatrix} j_B & j_A & J \\ m_B & m_A & M \end{pmatrix}.$$

4. Regge symmetry: a 3j symbol can be transformed in the square matrix:

$$\begin{pmatrix} j_A & j_B & J \\ m_A & m_B & M \end{pmatrix} \rightarrow \begin{pmatrix} j_B + J - j_A & j_A + J - j_B & j_A + j_B - J \\ j_A - m_A & j_B - m_B & J - M \\ j_A + m_A & j_B + m_B & J + M \end{pmatrix}.$$

If this matrix is reflected through its diagonals or the rows or columns are cyclically permuted, the 3j symbol which is associated with the resultant matrix has the same value as the original one. These rules add 72 new symmetries to the symmetries in 1, 2, and 3.

B.2.2 Methods of Calculations

A general formula for the calculation of the 3j symbols has been developed by Racah:

$$\begin{pmatrix} a & b & d \\ \alpha & \beta & \delta \end{pmatrix} = (-1)^{a-b-\delta}\, \Gamma(abd)^{1/2} \times$$

$$[(a+\alpha)!(a-\alpha)!(b+\beta)!(b-\beta)!(d+\delta)!(d-\delta)!]^{1/2} \times \tag{B.11}$$

$$\Sigma_t(-1)^t[t!(d-b+t+\alpha)!(d-\alpha+t-\beta)!(a+b-d-t)!(a-t-\alpha)!(b-t-\beta)!]^{-1}$$

with $\alpha + \beta + \delta = 0$ and $|a-b| \le d \le (a+b)$ and
$\Gamma(abd) = (a+b-d)!(b+d-a)!(d+a-b)!/(a+b+d+1)!$

In (B.11) Σ_t extends over all integral values of t for which the factorials are ≥ 0. The number of terms in the summation is $\mu+1$, where μ is the smallest of the nine numbers:

$$a\pm\alpha; \, b\pm\beta; \, d\pm\delta; \, a+b-d; \, b+d-a; \, d+a-b.$$

B.2.2.1 Special Formulae

1. $\begin{pmatrix} j & j & 1 \\ m & -m & 0 \end{pmatrix} = (-1)^{j-m} \, m \, [(2j+1)(j+1)j]^{-1/2}.$

2. $\begin{pmatrix} j_1 & j_2 & j_3 \\ 0 & 0 & 0 \end{pmatrix} = (-1)^{J/2} \, [(j_1+j_2-j_3)!(j_1+j_3-j_2)!(j_2+j_3-j_1)!/$
$$(j_1+j_2+j_3+1)!]^{1/2} \times$$
$$(J/2)!/[(J/2-j_1)!(J/2-j_2)!(J/2-j_3)!]$$

   where $J = j_1 + j_2 + j_3$.

3. $\begin{pmatrix} j_1 & j_2 & 0 \\ m_1 & -m_2 & 0 \end{pmatrix} = (-1)^{j_1-m_1}(2j_1+1)^{-1/2}\delta_{j_1j_2}\,\delta_{m_1m_2}.$

B.3 The "6j" Symbols and the Racah W Coefficients

B.3.1 Definitions

The 6j symbols appear in the case of the coupling of three angular momenta $j_1, j_2,$ and j_3 as the coefficients of the unitary transformation which connects two different coupling schemes. From the direct product of the vector spaces of the three angular momenta j_1, j_2, and j_3 we get a $(2j_1+1)(2j_2+1)(2j_3+1)$-dimensional space spanned by the kets $|j_1j_2j_3m_1m_2m_3\rangle \equiv |j_1m_1\rangle|j_2m_2\rangle|j_3m_3\rangle$. Defining a total angular momentum of the system as $\mathbf{J}=\mathbf{j_1}+\mathbf{j_2}+\mathbf{j_3}$, the subspace of angular momentum (JM) can be defined according to:

$$\min|j_1\pm j_2\pm j_3|\leq J\leq(j_1+j_2+j_3); \; -J\leq M\leq J.$$

In general, different systems of basis vectors can be defined according to the coupling scheme of the angular momenta:

1. Coupling scheme $\mathbf{j_1}+\mathbf{j_2}= \mathbf{j_{12}}; \mathbf{j_{12}}+\mathbf{j_3}=\mathbf{J}$: vectors $|j_1j_2j_{12}j_3JM\rangle$

$$|j_1j_2j_{12}j_3JM\rangle= \sum_{m_1m_2m_3m_{12}} |j_1j_2j_3m_1m_2m_3\rangle\langle j_1j_2m_1m_2|j_{12}m_{12}\rangle \times$$
$$\times \langle j_{12}j_3m_{12}m_3|JM\rangle \quad \text{(B.12)}$$

2. Coupling scheme $j_2 + j_3 = j_{23}$; $j_1 + j_{23} = J$: vectors $|j_1 j_2 j_3 j_{23} JM\rangle$

$$|j_1 j_2 j_3 j_{23} JM\rangle = \sum_{m_1 m_2 m_3 m_{23}} |j_1 j_2 j_3 m_1 m_2 m_3\rangle \langle j_2 j_3 m_2 m_3 | j_{23} m_{23}\rangle \times$$
$$\times \langle j_1 j_{23} m_1 m_{23} | JM\rangle \quad \text{(B.13)}$$

Since the kets in (B.12) and (B.13) span the same vector space as those in (B.8) and (B.9), they are related by a unitary transformation

$$|j_1 j_2 j_{12} j_3 JM\rangle = \sum_{j_{23}} |j_1 j_2 j_3 j_{23} JM\rangle \langle j_1 j_2 j_3 j_{23} JM | j_1 j_2 j_{12} j_3 j_{23} JM\rangle \quad \text{(B.14)}$$

The coefficients in (B.14) can be arranged in a 6j symbol using the following definition:

$$\begin{Bmatrix} j_1 & j_2 & j_{12} \\ j_3 & J & j_{13} \end{Bmatrix} = (-1)^{j_1 + j_2 + j_3 + J} [(2j_{12} + 1)(2j_{13} + 1)]^{-1/2} \times$$
$$\times \langle j_1 j_2 j_{12} j_3 JM | j_1 j_2 j_3 j_{23} JM\rangle \quad \text{(B.15)}$$

Symbols differing from (B.15) by a phase factor have been defined by Racah and are still used. They are defined by:

$$\begin{Bmatrix} j_1 & j_2 & j_3 \\ j_4 & j_5 & j_6 \end{Bmatrix} = (-1)^{j_1 + j_2 + j_3 + j_4 + j_5} W(j_1 j_2 j_5 j_4; j_3 j_6). \quad \text{(B.16)}$$

B.3.2 Properties

1. The 6j symbols are real.
2. A 6j symbol is left unchanged by interchange of any two rows or by switching the upper and lower members of any two rows.
3. A 6j symbol is nonzero if the elements of each of the triads
   (j_1, j_2, j_3), (j_1, j_5, j_6), (j_4, j_2, j_6), (j_4, j_5, j_3). $\quad$ (B.17)
a) Satisfy the triangular condition $(j_i, j_k, j_l) = |j_i - j_k| \le j_l \le j_i + j_k$;
b) Have an integral sum.
4. The 6j symbols are related to the 3j symbols associated with the triad (B.17) by the equation:

$$\begin{Bmatrix} j_1 & j_2 & j_3 \\ j_4 & j_5 & 6 \end{Bmatrix} = \sum_{\text{all } M, \, m_1 m_2} (-1)^{j_4 + j_5 + j_6 + M_1 + M_2 + M_3} (2j_3 + 1) \times$$

$$\times \begin{pmatrix} j_1 & j_2 & j_3 \\ m_1 & m_2 & m_3 \end{pmatrix} \begin{pmatrix} j_1 & j_5 & j_6 \\ m_1 & M_5 & -M_6 \end{pmatrix} \begin{pmatrix} j_4 & j_2 & j_6 \\ -M_4 & m_2 & m_6 \end{pmatrix}$$

$$\times \begin{pmatrix} j_4 & j_5 & j_3 \\ m_4 & -M_5 & m_3 \end{pmatrix}. \quad \text{(B.18)}$$

5. Orthogonality relation:

$$\sum_{j_3}(2j_3+1)\begin{Bmatrix} j_1 & j_2 & j_3 \\ j_4 & j_5 & j_6 \end{Bmatrix}\begin{Bmatrix} j_1 & j_2 & j_3 \\ j_4 & j_5 & j_6' \end{Bmatrix}=\delta_{j_6 j_6'}(2j_6+1)^{-1}.$$

6. The Regge symmetry:

$$\begin{Bmatrix} j_1 & j_2 & j_3 \\ j_4 & j_5 & j_6 \end{Bmatrix}=\begin{Bmatrix} (j_1+j_2+j_4-j_5)/2 & (j_1+j_2+j_5-j_4)/2 & j_3 \\ (j_1+j_4+j_5-j_2)/2 & (j_4+j_2+j_5-j_1)/2 & j_6 \end{Bmatrix}.$$

B.3.3 Methods of Calculation

A general formula for the calculation of the 6j symbols has been reported by Racah:

$$\begin{Bmatrix} j_1 & j_2 & j_3 \\ j_4 & j_5 & j_6 \end{Bmatrix}=[\Gamma(j_1j_2j_3)\Gamma(j_1j_5j_6)\Gamma(j_4j_2j_6)(j_4j_5j_6)]^{1/2} \times$$
$$\times \Sigma_t(-1)^t(t+1)!/[(t-j_1-j_2-j_3)!(t-j_1-j_5-j_6)! \times$$
$$\times (t-j_4-j_2-j_6)!(t-j_4-j_5-j_3)!(j_1+j_2+j_4+j_5-t)! \times$$
$$\times (j_2+j_3+j_5+j_6-t)!(j_3+j_1+j_6+j_4-t)! \qquad (B.19)$$

where the sum over t and the Γ symbols are the same as in (B.11). The number of terms in the summation is $\mu+1$, where μ is the smallest of the 12 numbers:

$$j_1+j_2+j_3; \ j_1+j_5-j_6; \ j_4+j_2-j_6; \ j_4+j_5-j_3; \ j_2+j_3-j_1; \ j_5+j_6-j_1;$$
$$j_2+j_6-j_4; \ j_5+j_3-j_4; \ j_3+j_1-j_2; \ j_6+j_1-j_5; \ j_6+j_4-j_2; \ j_3+j_4-j_5.$$

B.3.3.1 Special Formulae

1. One of the j is zero:

$$\begin{Bmatrix} j_1 & j_2 & 0 \\ j_4 & j_5 & j_6 \end{Bmatrix}=(-1)^{j_1+j_4+j_6}[(2j_1+1)(2j_4+1)]^{-1/2}\delta_{j_1j_2}\delta_{j_4j_5}$$

2. The smallest j is equal to 1/2:

$$\begin{Bmatrix} j_1 & j_2 & j_3 \\ 1/2 & j_3-1/2 & j_2+1/2 \end{Bmatrix}=(-1)^{j_1+j_2+j_3}[(j_1-j_2+j_3)(j_1+j_2-j_3+1)/$$
$$[(2j_2+1)(2j_2+2)2j_3(2j_3+1)]\}^{1/2}$$

$$\begin{Bmatrix} j_1 & j_2 & j_3 \\ 1/2 & j_3-1/2 & j_2-1/2 \end{Bmatrix}=(-1)^{j_1+j_2+j_3}\{(j_1+j_2+j_3+1)(j_2+j_3-j_1)/$$
$$[2j_2(2j_2+1)2j_3(2j_3+1)]\}^{1/2}$$

277

3. The smallest j is equal to 1:

$$\begin{Bmatrix} j_1 & j_2 & j_3 \\ 1 & j_3-1 & j_2-1 \end{Bmatrix} = (-1)^{j_1+j_2+j_3}$$

$$\{(j_1+j_2+j_3)(j_1+j_2+j_3+1)(j_2+j_3-j_1-1)$$
$$(j_2+j_3-j_1)/$$
$$[(2j_2-1)2j_2(2j_2+1)(2j_3-1)2j_3(2j_3+1)]\}^{1/2}$$

$$\begin{Bmatrix} j_1 & j_2 & j_3 \\ 1 & j_3-1 & j_2 \end{Bmatrix} = (-1)^{j_1+j_2+j_3}$$

$$\{2(j_1+j_2+j_3+1)(j_2+j_3-j_1)(j_1-j_2+j_3)(j_1+j_2-j_3+1)/$$
$$[2j_2(2j_2+1)(2j_2+2)(2j_3-1)2j_3(2j_3+1)]\}^{1/2}$$

$$\begin{Bmatrix} j_1 & j_2 & j_3 \\ 1 & j_3-1 & j_2+1 \end{Bmatrix} = (-1)^{j_1+j_2+j_3}$$

$$\{(j_1-j_2+j_3-1)(j_1-j_2+j_3)(j_1+j_2-j_3+1)$$
$$(j_1+j_2-j_3+2)/$$

$$[(2j_2+1)(2j_2+2)(2j_2+3)(2j_3-1)2j_3(2j_3+1)]\}^{1/2}$$

$$\begin{Bmatrix} j_1 & j_2 & j_3 \\ 1 & j_3 & j_2 \end{Bmatrix} = (-1)^{j_1+j_2+j_3}$$

$$\{2[j_1(j_1+1)-j_2(j_2+1)-j_3(j_3+1)]/$$
$$[2j_2(2j_2+1)(2j_2+2)2j_3(2j_3+1)(2j_3+2)]\}^{1/2}$$

B.4 The "9j" Symbols

B.4.1 Definition

The 9j symbols can be defined for the coupling of four angular momenta like the 6j symbols have been defined for the coupling of three angular momenta.

The total angular momentum of a 4j quantum system is $J = \Sigma_i\, j_i$. The subspace of angular momentum (JM) can be defined in the $\Pi_i\,(2j_i+1)$-dimensional space spanned by the vectors:

$$|j_1j_2j_3j_4m_1m_2m_3m_4\rangle = \Pi_i|j_im_i\rangle,$$

according to different coupling schemes. Let us consider for example the two coupling schemes

1. $j_1=j_2=j_{12}$; $j_3+j_4=j_{34}$; $j_{12}+j_{34}=J$: vectors $|j_1j_2j_{12}j_3j_4j_{34}JM\rangle$.
2. $j_1+j_3=j_{13}$; $j_2+j_4=j_{24}$; $j_{13}+j_{24}=J$: vectors $|j_1j_3j_{13}j_2j_4j_{24}JM\rangle$.

The coefficient of the unitary transformation which connects the basis vectors of 1 and 2 define the following 9j symbols

$$\langle j_1 j_2 j_{12} j_3 j_4 j_{34} JM | j_1 j_3 j_{13} j_2 j_4 j_{24} JM \rangle =$$
$$= [(2j_{12}+1)(2j_{34}+1)(2j_{13}+1)(2j_{24}+1)]^{1/2} \begin{pmatrix} j_1 & j_2 & j_{12} \\ j_3 & j_4 & j_{34} \\ j_{13} & j_{24} & J \end{pmatrix}. \qquad (B.20)$$

B.4.2 Properties

1. The 9j symbols are real.
2. A 9j symbol is invariant in the even permutation of rows and columns and in a reflection through one of the diagonals.
3. A 9j symbol is multiplied by $(-1)^r$, where $r = \Sigma_{all\ j} j_i$ in the odd permutation of rows and columns.
4. A 9j symbol can be expressed in terms of 3j symbols according to:

$$\begin{pmatrix} j_1 & j_2 & j_3 \\ j_4 & j_5 & j_6 \\ j_7 & j_8 & j_9 \end{pmatrix} = \Sigma_{all\ m} \begin{pmatrix} j_1 & j_2 & j_3 \\ m_1 & m_2 & m_3 \end{pmatrix} \begin{pmatrix} j_4 & j_5 & j_6 \\ m_4 & m_5 & m_6 \end{pmatrix} \begin{pmatrix} j_7 & j_8 & j_9 \\ m_7 & m_8 & m_9 \end{pmatrix} \times$$
$$\times \begin{pmatrix} j_1 & j_4 & j_7 \\ m_1 & m_4 & m_7 \end{pmatrix} \begin{pmatrix} j_2 & j_5 & j_8 \\ m_2 & m_5 & m_8 \end{pmatrix} \begin{pmatrix} j_3 & j_6 & j_9 \\ m_3 & m_6 & m_9 \end{pmatrix} \qquad (B.21)$$

5. Orthogonality relation:

$$\sum_{j_{13} j_{24}} (2j_{13}+1)(2j_{24}+1) \begin{pmatrix} j_1 & j_2 & j_{12} \\ j_3 & j_4 & j_{34} \\ j_{13} & j_{24} & J \end{pmatrix} \begin{pmatrix} j_1 & j_2 & j'_{12} \\ j_3 & j_4 & j'_{34} \\ j_{13} & j_{24} & J \end{pmatrix} =$$
$$= \delta_{j_{12} j'_{12}} \delta_{j_{34} j'_{34}} [(2j_{12}+1)(2j_{34}+1)]^{-1}$$

B.4.3 Special Formulae

1. One of the j is zero: the 9j symbol collapses to a 6j symbol:

$$\begin{pmatrix} j_1 & j_2 & j_3 \\ j_4 & j_5 & j_6 \\ j_7 & j_8 & 0 \end{pmatrix} = (-1)^{j_2+j_4+j_3+j_7} \delta_{j_3 j_6} \delta_{j_7 j_8} [(2j_3+1)(2j_7+1)]^{-1/2} \times$$
$$\times \begin{Bmatrix} j_1 & j_2 & j_3 \\ j_5 & j_4 & j_7 \end{Bmatrix}$$

B.5 Irreducible Tensor Operators

B.5.1 Definitions

A tensor operator can be defined as a set of operators that transform linearly one into another under rotation of the coordinate system. A set of $(2k + 1)$ operators which transform linearly one into the other is called a *kth-rank* tensor operator or a *tensor operator of order k*. A *scalar* operator, O, is a zero-rank tensor operator, a *vector* operator, O, is a first-rank tensor operator having cartesian components O_x, O_y, and O_z. A tensor operator is generally indicated by the symbol $T_{kq}(O)$, where k is the rank, q is in the range $\pm k$, and O indicates the type of quantum mechanical operator used. Unless it is explicitly necessary we will omit in the following the O symbol, meaning any tensor operator of rank k.

A tensor operator $T_{kq}(q = -k, \ldots, k)$ is called *irreducible tensor* operator if its $(2k + 1)$ components transform under a rotation R of the coordinate system according to:

$$RT_{kq}R^{-1} = \sum_{-k}^{k} {}_{q'} T_{kq'}R^k_{qq'},\tag{B.22}$$

where R^k is the Wigner rotation matrix and spans the k-th irreducible representation of the real orthogonal rotation group SO(3). Through (B.22) the components of an irreducible tensor operator can be expressed as a linear combination of the more common cartesian components. Some examples are shown in Table 3.1.

B.5.1.1 Compound Irreducible Tensor Operators

The tensor product of two irreducible tensor operators of rank k_1 and k_2, T_{k_1} and T_{k_2}, can be defined as the set of $(2k_1 + 1)(2k_2 + 1)$ operators:

$$T_{k_1} \otimes T_{k_2} = \{T_{k_1 q_1} T_{k_2 q_2}\}.\tag{B.23}$$

The set of operators (B.23) is in general reducible and we can construct from it an irreducible tensor operator of oder k called the *tensor product of* T_{k_1} and T_{k_2} of order k according to:

$$\{T_{k_1} \otimes T_{k_2}\}_{kq} = \sum_{q_1 q_2} \langle k_1 k_2 q_1 q_2 | kq \rangle T_{k_1 q_1} T_{k_2 q_2}.\tag{B.24}$$

When not explicitly stated, T_{k_1} and T_{k_2} are commuting tensor operators such as operators acting on two vector spaces 1 and 2 separately, like S_A and S_B in Chap. 3, or operators acting separately on spin and space coordinates, like the spin-orbit coupling operator of one particle l·s.

According to (B.10) we can rewrite (B.24) as

$$\{T_{k_1} \otimes T_{k_2}\}_{kq} = \sum_{q_1 q_2} (2k+1)^{1/2} (-1)^{k_2 - k_1 - q} \begin{pmatrix} k_1 & k_2 & k \\ q_1 & q_2 & -q \end{pmatrix} \times$$
$$\times T_{k_1 q_1} T_{k_2 q_2} \tag{B.25}$$

where one necessarily has $|k_1 - k_2| \le k \le (k_1 + k_2)$.

It is an obvious extension of the above definition to construct tensor operators as products of more than two irreducible tensor operators and the relative formulae will not be explicitly given here.

B.5.1.2 Tensor Product of Two First-rank Irreducible Tensor Operators

From the tensor product of two irreducible tensor operators with $k_1 = k_2 = 1$ we get three sets of irreducible tensor operators corresponding to $k = 0, 1$, and 2. For $k = 0$ from (B.25) one has

$$\{T_1 \otimes T_1\}_{00} = (3)^{-1/2} [T_{11} T_{1-1} - T_{10} T_{10} + T_{1-1} T_{11}]. \tag{B.26}$$

The usual scalar product of two vector operators $(\mathbf{T} \cdot \mathbf{T}) = T_x T_x + T_y T_y + T_z T_z$ is related to (B.26) by

$$(\mathbf{T} \cdot \mathbf{T}) = -(3)^{1/2} \{T_1 \otimes T_1\}_{00}, \tag{B.27}$$

or more generally

$$(\mathbf{T_k} \cdot \mathbf{T_k}) = (-1)^k (2k+1)^{1/2} \{T_k \otimes T_k\}_{00}. \tag{B.28}$$

For $k = 1$ one has

$$\{T_1 \otimes T_1\}_{11} = (2)^{-1/2} [T_{11} T_{10}] - (2)^{-1/2} [T_{10} T_{11}];$$
$$\{T_1 \otimes T_1\}_{10} = (2)^{-1/2} [T_{11} T_{1-1}] - (2)^{-1/2} [T_{1-1} T_{11}]; \tag{B.29}$$
$$\{T_1 \otimes T_1\}_{1-1} = (2)^{-1/2} [T_{10} T_{1-1}] - (2)^{-1/2} [T_{1-1} T_{10}].$$

For $k = 2$ one has

$$\{T_1 \otimes T_1\}_{22} = [T_{11} T_{11}];$$
$$\{T_1 \otimes T_1\}_{21} = (2)^{-1/2} [T_{11} T_{10}] + (2)^{-1/2} [T_{10} T_{11}];$$
$$\{T_1 \otimes T_1\}_{20} = (2/3)^{1/2} [T_{10} T_{10}] + (6)^{-1/2} [T_{11} T_{1-1} + T_{1-1} T_{11}]; \tag{B.30}$$
$$\{T_1 \otimes T_1\}_{2-1} = (2)^{-1/2} [T_{1-1} T_{10}] + (2)^{-1/2} [T_{10} T_{1-1}];$$
$$\{T_1 \otimes T_1\}_{2-2} = [T_{1-1} T_{1-1}].$$

It is interesting to note that (B.30) applied to two tensor operators acting on the same set of coordinates, such as two spin operators, gives identically zero, as a consequence of the commutation relations. This can be easily verified by the reader.

B.5.2 Properties of Irreducible Tensor Operators

1. Commutation relations with the total angular momentum operator $\mathbf{J}$ (Racah's commutation relations):

$$[J_\pm, T_{kq}] = T_{kq\pm1}[(k\mp q)(k\pm q+1)]^{1/2}$$
$$[J_z, T_{kq}] = qT_{kq}$$

2. Adjoint tensor operators.
   The adjoint or Hermitean conjugate, $\bar{T}_{kq}$, of a tensor operator T_{kq} is defined by:

$$\bar{T}_{kq} = (-1)^q T_{k-q}^+ \tag{B.31}$$

where $+$ denotes the Hermitean conjugate. Using (B.31) the spherical harmonics Y_{lm} form an Hermitean tensor operator. Definitions equivalent to (B.31) are:

$$\bar{T}_{kq} = (-1)^{k\pm q} T_{k-q}^+. \tag{B.32}$$

With respect to (B.32) the spherical harmonics of odd order are anti-Hermitean.

3. Wigner-Eckart theorem:

$$\langle \tau JM|T_{kq}|\tau'J'M'\rangle = (-1)^{J-M}\langle \tau J\|T_k\|\tau'J'\rangle \begin{pmatrix} J & k & J' \\ -M & q & M' \end{pmatrix}, \tag{B.33}$$

where $\langle \tau J\|T_k\|\tau'J'\rangle$ is called *reduced matrix element*. τ indicates any set of quantum numbers required to completely specify the quantum system. The following form of the theorem is sometimes used:

$$\langle \tau JM|T_{kq}|\tau'J'M'\rangle = (-1)^{k-J'+M}(2J+1)^{1/2}\langle \tau J\|T_k\|\tau'J'\rangle \times$$
$$\times \begin{pmatrix} J & k & J' \\ -M & q & M' \end{pmatrix} \tag{B.34}$$

The conjugate of a reduced matrix element is (k integral)

$$\langle \tau J\|T_k\|\tau'J'\rangle^* = (-1)^{J'-J}\langle \tau'J'\|T_k^+\|\tau J\rangle.$$

Equation (B.33) or (B.34) can be used to compute the matrix element of any tensor operator provided the reduced matrix elements are known.

B.5.3 Reduced Matrix Elements of Special Operators

1. Identity operator:

$$\langle \tau J \| 1 \| \tau' J' \rangle = \delta_{\tau\tau'} \delta_{JJ'} [(2J+1)]^{1/2},$$

2. Total angular momentum:

$$\langle \tau J \| J \| \tau' J' \rangle = \delta_{\tau\tau'} \delta_{JJ'} [J(J+1)(2J+1)]^{1/2}.$$

3. Tensor product of two irreducible tensor operators.
   Let T_{k_1} be an irreducible tensor operator acting on the vector space $|\tau_1 j_1 m_1\rangle$ and T_{k_2} an irreducible tensor operator acting on $|\tau_2 j_2 m_2\rangle$, then

$$\langle \tau_1 j_1 \tau_2 j_2 J \rangle \| \{ T_{k_1} \otimes T_{k_2} \}_k \| \tau_1' j_1' \tau_2' j_2' J' = \langle \tau_1 j_1 \| T_{k_1} \| \tau_1' j_1' \rangle$$

$$\langle \tau_2 j_2 \| T_{k_2} \| \tau_2' j_2' \rangle [(2J+1)(2J'+1)(2k+1)]^{1/2} \begin{Bmatrix} j_1 & j_1' & k_1 \\ j_2 & j_2' & k_2 \\ J & J' & k \end{Bmatrix} \tag{B.35}$$

There are some particular cases in which the 9j symbol in (B.35) reduces to a 6j symbol:

$$T_{k_2} = 1 \quad \text{and} \quad k = k_1,$$
$$\langle \tau_1 j_1 \tau_2 J \| T_{k_1} \| \tau_1' j_1' \tau_2' j_2' J' \rangle = \delta_{\tau_2 \tau_2'} \delta_{j_2 j_2'}$$

$$\langle \tau_1 j_1 \| T_{k_1} \| \tau_1' j_1' \rangle \times$$
$$\times (-1)^{j_1 + j_2 + J' + k_1} [(2J+1)(2J'+1)]^{1/2} \begin{Bmatrix} j_1 & J & j_2 \\ J' & j_1' & k_1 \end{Bmatrix} \tag{B.36}$$

b) $T_{k_1} = 1$ and $k = k_2$,
$$\langle \tau_1 j_1 \tau_2 j_2 J \| T_{k_2} \| \tau_1' j_1' \tau_2' j_1' J' \rangle = \delta_{\tau_1 \tau_1'} \delta_{j_1 j_1'}$$
$$\langle \tau_2 j_2 \| T_{k_2} \| \tau_2' j_2' \rangle$$
$$\times (-1)^{j_1 + j_2' + J + k_2} [(2J+1)(2J'+1)]^{1/2} \begin{Bmatrix} J & k_2 & J' \\ j_2' & j_1 & j_2 \end{Bmatrix} \tag{B.37}$$

c) Scalar product ($k = 0$ and $k_1 = k_2$).
$$\langle \tau_1 j_1 \tau_2 j_2 JM | (T_{k_1} \cdot T_{k_2}) | \tau_1' j_1' \tau_2' j_2' J'M' \rangle = \delta_{JJ'} \delta_{MM'}$$
$$\langle \tau_1 j_1 \| T_{k_1} \| \tau_1' J_2 \rangle \langle \tau_2 j_2 \| T_{k_2} \| \tau_2' j_2' j_1' \rangle (-1)^{J + j_2 + j_1} \begin{Bmatrix} j_1 & k & j_1' \\ j_2' & J & j_2 \end{Bmatrix} \tag{B.38}$$

4. A particular case often met in practice arises when the states are classified by only one total angular momentum, J, and one magnetic quantum number, M. In this situation both T_{k_1} and T_{k_2} act on the same vectors and the following equation holds:

$$\langle J\|\{T_{k_1}\otimes T_{k_2}\}_k\|J'\rangle = (-1)^{k+J+J''}[2k+1]^{1/2}\Sigma_{J''} \begin{Bmatrix} k_1 & k_2 & k \\ J' & J & J'' \end{Bmatrix} \times$$

$$\times \langle J\|T_{k_1}\|J''\rangle\langle J''\|T_{k_2}\|J'\rangle \tag{B.39}$$

5. Particular expressions.

a) $T_{k_1} = T_1(j_1)$

$$\langle j_1 j_2 J + 1\|T_1(j_1)\|j_1 j_2 J\rangle = 1/2\{[(j_1+j_2+1)^2 - (J+1)^2][(J+1)^2 + -(j_1-j_2)^2]/(J+1)\}^{1/2}$$

$$\langle j_1 j_2 J\|T_1(j_1)\|j_1 j_2 J\rangle = 1/2[(j_1(j_1+1)+J(J+1)-j_2(j_2+1)] \times \times (2J+1)^{1/2}[J(J+1)^{-1/2}$$

$$\langle j_1 j_2 J - 1\|T_1(j_1)\|j_1 j_2 J\rangle = -1/2\{[(j_1+j_2+1)^2 - J^2][J^2-(j_1-j_2)^2]/(J +1)\}^{1/2}$$

6. Coupling of three commuting angular momenta.
   A general expression has been derived by applying (B.35) twice for the reduced matrix element of the irreducible tensor operator built up from the coupling of three irreducible tensor operators acting on different spaces:

$$\langle j_1 j_2 j_{12} j_3 J\|\{\{T_{k_1}\otimes T_{k_2}\}_{k_{12}}\otimes T_{k_3}\}_k\|j'_1 j'_2 j'_{12} j'_3 J'\rangle =$$
$$= \{(2J+1)(2J'+1)(2k+1)(2j_{12}+1)(2j'_{12}+1)(2k_{12}+1)\}^{1/2} \times$$

$$\times \begin{Bmatrix} j_{12} & j'_{12} & k_{12} \\ j_3 & j'_3 & k_3 \\ J & J' & k \end{Bmatrix} \begin{Bmatrix} j_1 & j'_1 & k_1 \\ j_2 & j'_2 & k_2 \\ J_{12} & J'_{12} & k_{12} \end{Bmatrix} \times$$

$$\langle j_1\|T_{k_1}\|j'_1\rangle\langle j_2\|T_{k_2}\|j'_2\rangle\langle j_3\|T_{k_3}\|j'_3\rangle$$

Subject Index

ferromagnet 164
ferromagnetic coupling 2, 4–8, 10, 13, 18, 30, 79, 98, 181, 186
fine structure 179
fluctuation–dissipation theorem 141
forbidden transition 175, 177
Fourier transform 141, 143, 144
fourth order operators 114, 118
fractal 132

gadolinium(III) 29
g anisotropy 150
gaussian distribution 140
— lineshape 137, 139, 143, 145, 157, 158
— modulation 141, 142, 147
Goodenough 5, 6, 9, 12, 15, 30, 186
g shift 163, 165

half field transitions 160
Heisenberg 3, 38, 44, 106
— representation 145
heterodinuclear species 60
homodinuclear species 60
homogeneous broadening 137
Hubbard 44
hydrogen molecule 1
hyperfine 48, 49, 55, 68, 72–74, 76, 78, 79, 101, 109–111, 119, 135, 150, 153, 177, 179, 180, 187

interaction representation 145
interchain coupling 158, 159
intercluster exchange 114
intermediate exchange 77–84, 87
intermolecular exchange 112
iridium(IV) 124, 129, 133
iron(II) 106, 131, 132
iron(III), high spin 109, 111, 131, 132
—, low spin 39

Jahn–Teller 184

Kahn 13
Kanamori 5, 6, 9, 12, 15, 27, 30, 31, 186
Kramers 78, 79, 81, 124, 184
Kubo–Tomita theory 139, 152, 159

lanthanide ions 33
lattice wave 128
ligand field 20, 39, 129
linear response 141
linewidth alternation 191
longitudinal wave 128, 129
lorentzian lineshape 144, 151, 155, 157, 158
low dimensional lineshape 156–158

magic angle 158
magnetic dimensionality 135
— dipoles 22
— orbital 5, 13, 17, 18, 26, 35, 186
— susceptibility 2, 20, 106, 112, 113, 116, 132, 146, 175
manganese(II) 8, 23, 25, 29, 60, 61, 63, 64, 65, 150, 158, 163, 164, 167
manganese(III) 42
manganese(IV) 42
memory function 139
mixed valence 42, 44, 45
modulation amplitude 141, 142
molecular orbital 2, 3, 9–11, 13, 16, 25, 27
molybdenum(III) 168, 170
monomeric impurity 74
Moryia 24, 32, 33
Mossbauer 109
^{55}Mn 61

^{14}N 187
nearest-neighbor interaction 148, 153
nickel(II) 33, 42, 45, 49, 64, 79, 82–84, 182–185
nitronyl nitroxide 164
nitroxide 25, 187, 189
NMR 27, 121, 139, 145
nonorthogonalized orbitals 16, 18
non-secular 147, 150, 151, 154, 187

one-dimensional 135, 139, 148, 155, 158, 160
one-phonon process 123
operator equivalent 101, 102
Orbach relaxation 121–124, 128, 130, 132, 134
orbitally degenerate 38, 44
organic biradical 25, 187, 190, 191
— radical 28, 29
orthogonalized orbitals 11, 44
overlap density 14
oxovanadium(IV) 7, 30, 66, 67, 74, 171, 172, 186

phonon 33, 121, 123, 124, 132
phonon bottle-neck 123, 132, 134
Pilbrow 68
point dipolar approximation 25, 27, 181
polarization 5, 15
porphyrin 39
potential energy surface 43
projection operator 35

quadrupole, electric 33
—, nuclear 48
quartet 125, 133, 137, 139
quintet 98, 133, 180, 181

radical ion 49, 191
Raman relaxation 121–124, 132, 134
recoupling coefficient 88
recurrence relationship 102
Redfield theory 190
reduced matrix element 52, 53, 56, 90, 102
relaxation 121
— function 142, 143, 145, 156
rhenium 42

second moment 148, 149, 154
— quantization 34, 36, 37, 38
secular 147, 148, 150, 151, 154, 158, 166
septet 175
short range order effects 162
silver(II) 66, 74
singlet 1, 2, 4, 5, 10–12, 18, 20, 60, 117, 125, 127,
  128, 130, 137, 175, 177, 180, 181, 189
Slater 36
slow modulation 142, 143, 144
SO(3) group 52
spin degeneracy 98
— diffusion 155
— hamiltonian 20, 34–36, 48, 51, 66, 79, 81, 86,
  89, 90, 100, 102, 114–116, 118
— labeling 66
— lattice relaxation 131, 132, 186
— orbit coupling 27–29, 38, 39, 125
— probe 187
Spirulina maxima 131
statistical ensemble 140
Stevens 34
strain 128
strong exchange limit 49–66, 86–91, 100, 152,
  157, 190
superexchange 1, 2, 4
superhyperfine 48
supertransferred hyperfine 53
symbol 3–j 52, 66
— 6–j 88, 92
— 9–j 88
— 12–j 88
— 15–j 88
— 18–j 88

T_1 121, 123–125, 129–131, 133
T_2 121

TEMPOL 25
tensor, antisymmetric 21
— operator 52, 90, 99
— operator irreducible 52, 53, 66, 89
—, symmetric 21
tetrad 98, 115, 133
three-dimensional 135, 139, 148, 155, 157, 163
titanium(III) 40, 42, 67, 172, 173
TMMC 158, 160, 163, 166
transfer integral 4
transition intensity 48, 76
— metal ion 28, 49
— probability 72
transverse wave 128, 129
triad, general 91, 107, 108
—, linear 91
—, regular 91, 93
—, symmetric 91, 93, 102, 107, 108, 110
triplet 1, 2, 4, 5, 10–12, 14, 15, 17, 18, 20, 25, 54,
  60, 98, 114, 116–118, 125, 127–130, 133, 137,
  175, 177, 180, 181, 189
two-dimensional 135, 139, 148, 155, 163
two-phonon process 123, 124

unitary transformation 87

^{51}V 186
vacuum state 36
Valence Bond 2, 10
vanadium(II) 172
van Vleck 3, 106, 139
vector coupling 88, 126
vibronic effects 33, 43, 46
vitamin B_{12} 67

weak exchange limit 66–77, 137, 152, 157, 191
Wigner–Eckart theorem 52, 66, 89

Xα 16

Zeeman 42, 48, 51, 69, 79, 86, 101, 125–127, 129,
  135, 136, 145–147, 149, 151, 185, 187
zero field splitting 25, 30, 48, 49, 51, 54, 57, 59, 62,
  64, 67, 79, 101, 116, 119, 122, 154, 168, 172,
  173, 177, 179, 181, 183, 184, 191
zinc(II) 60, 61, 83, 84, 109, 130, 184, 185

J. G. Verkade, Iowa State University, Ames, IA, USA

A Pictorial Approach to Molecular Bonding

1986. XIII, 282 pp. 231 figs. Hardcover DM 125,–
ISBN 3-540-96271-9

Contents: The Orbital Picture for Bound Electrones. – Atomic Orbitals. – Diatomic Molecules. – Linear Triatomic Molecules. – Triangular and Related Molecules. – Bent Triatomic Molecules. – Polygonal Molecules. – Octahedral and Related Molecules. – Tetrahedral and Related Molecules. – Bipyramidal and Related Molecules. – Prismatic Molecules. – Appendices 1–5. – Index.

J. Simon, Paris; J.-J. André, Strasbourg, France

Molecular Semiconductors

Photoelectrical Properties and Solar Cells

Editors: J. M. Lehn, C. W. Rees

1985. XIII, 288 pp. 166 figs. 41 tabs. Hardcover DM 198,–
ISBN 3-540-13754-8

Contents: List of Symbols. – Basic Notions of Solid State Physics. – Photoelectric Phenomena in Molecular Semiconductors. – Metallophthalocynines. – Polyacetylene. – The Main Other Molecular Semiconductors. – Conclusion. – References. – Subject Index.

D. L. Andrews, University of East Anglia, Norwich, UK

Lasers in Chemistry

1986. XII, 176 pp. 115 figs. Softcover DM 68,–
ISBN 3-540-16161-9*

Contents: Principles of Laser Operation. – Laser Sources. – Laser Instrumentation in Chemistry. – Chemical Spectroscopy with Lasers. – Laser-Induced Chemistry. – Appendix 1: Listing of Output Wavelengths from Commercial Lasers. – Appendix 2: Directory of Acronyms and Abbreviations. – Appendix 3: Selected Bibliography. – Subject Index.

Springer-Verlag Berlin
Heidelberg New York
London Paris Tokyo
Hong Kong

Distribution rights for the socialist countries: Akademie-Verlag Berlin

Springer

R. L. Carlin, University of Illinois, Chicago, IL, USA

Magnetochemistry

1986. XI, 328 pp. 244 figs. 21 tabs. Hardcover DM 118,–
ISBN 3-540-15816-2

Contents: Diamagnetism and Paramagnetism. – Paramagnetism: Zero-Field Splittings. – Thermodynamics. – Paramagnetism and Crystalline Fields: The Iron Series Ions. – Introduction to Magnetic Exchange: Dimers and Clusters. – Long-Range Order. – Ferromagnetism and Antiferromagnetism. – Lower Dimensional Magnetism. – The Heavy Transition Metals. – The Rare Earths or Lanthanides Ions. – Selected Examples. – Some Experimental Techniques. – Formula Index. – Subject Index.

I. Gutman, Kragujevac, Yugoslavia; O. E. Polansky, Mülheim a. d. Ruhr, FRG

Mathematical Concepts in Organic Chemistry

1986. X, 212 pp. 28 figs. Hardcover DM 139,–
ISBN 3-540-16235-6*

Contents: Introduction. – Chemistry and Topology: Topological Aspects in Chemistry. Molecular Topology. – Chemistry and Graph Theory: Chemical Graphs. Fundamentals of Graph Theory. Graph Theory and Molecular Orbitals. Special Molecular Graphs. – Chemistry and Group Theory: Fundamentals of Group Theory. Symmetry Groups. Automorphism Groups. Some Interrelations between Symmetry and Automorphism Groups. – Special Topics: Topological Indices. Thermodynamic Stability of Conjugated Molecules. Topological Effect on Molecular Orbitals. – Appendices 1–6. – Literature. – Subject Index.

Distribution rights for the socialist countries: Akademie-Verlag Berlin

D. Britz, University of Aarhus, Denmark

Digital Simulation in Electrochemistry

2nd rev. and ext. ed. 1988. X, 229 pp. 35 figs. 3 tabs. Softcover DM 68,– ISBN 3-540-18979-3

(The 1st edition was published as Vol. 23 of the series "Lecture Notes in Chemistry")

From the contents: Basic Equations. – Diffusional Transport – Digitally. – Calculation of Boundary Values. – Advanced Methods. – Accuracy, Efficiency and Choice. – Coupled Homogeneous Chemical Reactions. – Miscellaneous Topics. – Programming and Example Programs.

Springer-Verlag Berlin Heidelberg New York London Paris Tokyo Hong Kong

Springer